THE UNCONSCIOUS QUANTUM

Metaphysics in Modern Physics and Cosmology

VICTOR J. STENGER

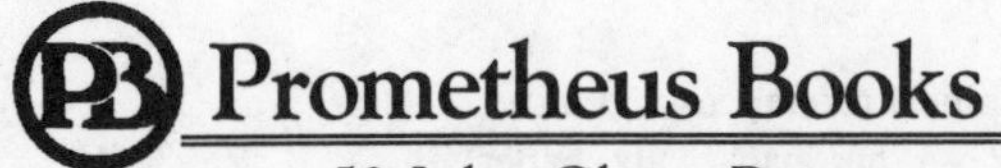

59 John Glenn Drive
Amherst, NewYork 14228-2197

Published 1995 by Prometheus Books

Inquiries should be addressed to Prometheus Books, 59 John Glenn Drive, Amherst, New York 14228–2197, 716–691–0133. FAX: 716–691–0137.

99 98 97 96 95 5 4 3 2 1

Library of Congress Cataloging-in-Publication Data

Stenger, Victor J., 1935–
The unconscious quantum : metaphysics in modern physics and cosmology / Victor J. Stenger.
p. cm.
Includes bibliographical references and index.
ISBN 1–57392–022–3 (hardback. : alk. paper)
1. Physics—Philosophy. 2. Metaphysics. 3. Quantum theory. 4. Cosmology.
I. Title.
QC6.S8126 1995
113—dc20 95–32599
CIP

Printed in the United States of America on acid-free paper

Contents

Preface 7

Acknowledgments 15

1. Cosmic Mind 17
2. Copenhagen, Complementarity, and Chance 44
3. Paradox 66
4. Hidden Variables 100
5. Nonlocality, Holism, and the Arrow of Time 127
6. One or Many? 165
7. Restoring Reality 192
8. Cosmythology 213
9. The Edge of Chaos 250
10. Shining Through? 268

Bibliography 295

Index 309

Preface

> Can it be that so many men, of various times and nations, outstanding minds among them, have devoted so much effort, and indeed fervor, to metaphysics, when this consists of nothing more than words strung together without sense?
>
> —Rudolph Carnap[1]

When an untitled manuscript by Aristotle was found following his treatise *Physics*, first-century editors assigned it the title *Metaphysics.* Since then, this term has been associated with those studies that are believed to transcend the material world of physics, cutting to the core of a supposed ultimate reality that goes deeper than appearances.[2]

Most people take for granted that such a reality exists, that some kind of perfect world lies beyond the realm of imperfect matter. In religious terms, this world is heaven, or more generally the domain of the spirit. In Western philosophy, metaphysical reality is patterned after Plato's world of perfect "forms" from which our senses draw a distorted image of the true reality. Most mathematicians and scientists view their theorems and laws in Platonic terms, although few will admit it.

If a realm exists beyond matter, how can we know about it? By definition, it is inaccessible to our senses, at least in any direct way. In religion, the world beyond is accessed by revelation or some other mystical channel that bypasses the senses. In ostensibly more rational discourses on metaphysics, the Platonic world is thought to be accessed by reasoning from fundamental insights that are themselves not directly amenable to the empirical testing methods of conventional science.

Since physics is concerned with the material world as it is presented to our senses and the scientific instruments that enhance the power of those senses, metaphysics is concerned with notions that transcend the senses and may not be constrained by physical law. On the assumption that mind is somehow separate from matter, metaphysics can also refer to studies of the relationship between mind and matter.

Finally, to the lay public, metaphysics is associated with psychical and occult ideas, as evidenced by the volumes that can be found on the shelves marked "Metaphysics" in popular bookstores.

By the twentieth century, science had progressed to the point where it was the dominant thought system, at least in Western, secular circles. Among philosophers, in the meantime, metaphysics had gradually fallen out of favor. Its coffin seemed to be permanently sealed by the rapid developments in physics that occurred early in the century. Einstein's theory of relativity, and the quantum theory that followed on its heels, demonstrated that our most fundamental notions of space, time, and matter were intimately tied to the way in which we chose to make observations and do experiments.

Einstein showed that the observed tick rate on a clock depends on the motion of the clock relative to the observer, implying that time intervals are simply defined by counting the clock ticks between events. Distance and mass were also found to be relative, to depend like time on the observer's frame of reference. In similar fashion, quantum mechanics demonstrated that observations at the microscopic scale could not be made without interfering, sometimes catastrophically, with the object being observed. For example, the position of an electron in an atom cannot be measured without destroying the atom. Philosophers asked: Does it even make sense to assume the existence of quantities that cannot be measured?

From the time of Galileo and Newton, science has been firmly grounded on observation and measurement. The empirical method was effectively promoted by eminent philosophers such as John Locke and David Hume. However, the new physics of the twentieth century encouraged some philosophers to go much farther. Members of a school of philosophy called *logical positivism,* centered in Vienna, came to the conclusion that the study of anything except the measurable was necessarily empty of content.

For the logical positivists, Einstein's demonstration of the relativity of time suggested the notion that perhaps all metaphysical discourses on the meaning of time were themselves meaningless. A similar conclusion was drawn for our concepts of space, matter, and energy. Physical properties now appeared to be nothing more than arbitrary human inventions.

The logical positivists did not claim that the concepts of time, space, and energy were useless. Quite the contrary. Defined operationally by well-prescribed, repeatable measurement procedures, these quantities formed the framework upon which physics and the rest of science was constructed. Since these methods had more than proved their utility, it seemed that the instrumental definitions must be the correct ones and previous speculations about space, time, matter, and energy were based on muddled thinking that should now be tossed on the rubbish heap of outmoded ideas.

The logical positivists insisted that all our concepts rest ultimately on observation. To the extent that our speculations about ultimate reality rested on observation, they were not metaphysics but physics. To the extent that metaphysics did not rest on observation, it was meaningless. This left no room at all for metaphysics; it became a fruitless enterprise.

But humans have never hesitated to pursue fruitless enterprises, and metaphysics proved no exception. The logical positivists did not convince everyone—including themselves—that metaphysics was a waste of time, even if time was just a count of ticks on a clock. Furthermore, after developing the theory of general relativity in 1916, Einstein had discarded his original positivist position (if, indeed, he ever had one).

General relativity seemed to imply that space and time are not simply arbitrary human inventions but represent some aspect of a real, if invisible, backbone of the universe. Einstein showed that gravity could be reinterpreted in terms of an underlying, non-Euclidean curvature of spacetime. That curvature was not directly observable, but existed in abstract form within the elegant tensor equations of general relativity. Thus spacetime curvature seemed to possess a reality akin to Plato's forms, an underlying, invisible, perfect order that lay beyond the grubby data collected by scientific instruments.

Philosophers also found much to criticize in the more extreme conclusions of the logical positivists and the school soon fell out of favor, opening up metaphysics once more as a legitimate subject for philosophical discourse. More recently, there has been an ironic twist of the tale, which leads to the theme of this book.

As relativity evolved to a form compatible with the Platonic model of reality, the quantum revolution broke new metaphysical ground. Quantum mechanics was found to produce apparent paradoxes when interpreted in terms of familiar, classical concepts. Quantum phenomena seemed to have the property that they cannot be fully described at one point in space without knowing about conditions that exist at the same instant at other points. A decision made at one place, such as orienting a particle detector in a certain direction, appears to affect what will be observed in another detector at another place. This can happen within a time interval too small for a signal to go between the two detectors without travelling faster than the speed of light, the limiting speed of objects according to Einstein.

So despite Einstein's retreat from positivism with general relativity, quantum mechanics provided continued support for a positivist, or instrumentalist, view that was taken to imply that physical properties do not have an objective reality independent of their observation. Since consciousness comes into the act of human observation at some point, many modern authors then suggested that quantum mechanics provides for the connection between the human mind and the universe that has been claimed for millennia by mystics East and West.

Quantum mechanics, as a mathematical theory, has proved to be of immense practical utility. But it has come to be recognized very slowly that this success was not to be regarded as any demonstration of the validity of the attempts that have been made to put words behind the mathematics, to provide what is called an *ontological interpretation.* Several interpretations are equally capable of yielding the same empirical results. Since none provides its own unique predictions, this can only mean that all the interpretations of quantum mechanics are equivalent, at least until someone shows us how to improve on one, or falsify the others.

If an interpretation were to be found that allows quantum effects to be described in familiar terms, it would be a preferable choice. This has happened before. In the nineteenth century, atoms were shown to underlie the thermodynamic behavior that had been established for macroscopic systems. The atomic model of matter was adopted as the fundamental underpinning of macroscopic phenomena.

Thus it seems sensible to seek a comparable deterministic model to underlie quantum mechanics. Presumably this would be the simplest model, requiring the fewest new assumptions, and so would be accepted by the principle of parsimony. But theoretical physics does not proceed in a vacuum; it requires experimental input, and in particular empirical anomalies, to guide the way.

After the better part of a century of experimentation with the most sensitive instruments ever built by humankind, no evidence for a subquantum world akin to the world of atoms has yet been found. Indeed, a subquantum world with the familiar properties of classical physics may even be incompatible with the data. As has been shown theoretically, and verified experimentally, any subquantum model will have to be capable of producing instantaneous effects over great distances, in apparent violation of Einstein's principle that nothing can move faster than light.

Although the metaphysical elements of the various interpretations of quantum mechanics imply quite different conclusions about the nature of physical reality, they have been muddled together in a new genre of popular literature I call *quantum metaphysics.* This book takes a detailed look at both quantum mechanics and quantum metaphysics in an attempt to find a parsimonious explanation for their apparent paradoxes and cosmic implications.

We will also review some of the other discoveries in modern physics and cosmology claimed to have metaphysical consequences. We will consider the *big bang* and the origin of the universe, to see if our very existence required a miraculous violation of the laws of physics. We will explore the so-called *anthropic coincidences* that seem to imply a special place for humankind in the universe.

Recent developments in *chaos theory* that have been stirred in with the quantum mixture to give yet another claimed basis for a great holistic "paradigm shift" away from reductionist physics will also be examined. We will find that no such shift is in fact implied by chaos, but that the new sciences of complexity of which chaos is a part offer the means to understand the process of *emergence* by which highly interacting material systems develop the properties we associate with life and mind.

Finally, we will discuss some of the arguments made by mathematician Roger Penrose that certain principles of mathematics and computer science lead us to return once again to a Platonic view of the universe. To some, "Penrose mysticism" suggests that perhaps the human mind can reach beyond sensory experience to the realm of true reality, to read the "mind of God." However, we will see that Penrose insists that he is proposing new physics, not mysticism. This discussion will enable us again to examine the nature of human thinking and the meaning and role of consciousness in the cosmos.

Much of the literature of modern metaphysics, with important exceptions such as the books of Penrose, is written in a "gee whiz" fashion for a popular audience. That audience is duly impressed by the mysteries of quantum mechanics and eager to believe the implications drawn that human consciousness holds the key to reality and their personal immortality. My task is much more difficult, if not impossible, since I am telling people things that many do not want to hear: that according to our best knowledge, the world of matter is all that exists.

Still, I feel it must be done. As science and critical thinking become increasingly watered down in our educational system, and opposing forces exploit the consequent public gullibility, the duty of every scientist is to speak out in protest. The antiscientists who pursue a political agenda, and the pseudoscientists who pursue the dollar, need to be fought at every turn. Scientists cannot continue to ignore these issues. As I write this, major basic research projects are being cancelled or cut back, granting agencies are being told to do more applied work, and unscientific alternative medicine is creeping into health-care systems. In some cases, such as "quantum healing," the public is being led to believe that science and mysticism are converging on a new "paradigm" in which matter and spirit, body and mind, are one.

I have attempted to write this book for the broadest possible audience without watering down my arguments to the gee-whiz level. I recognize that this audience is not likely to include everyone. Those who pick up this book will probably already be familiar, at least in general terms, with the popular and semipopular literature on the new physics and metaphysics. Judging by the number of books and articles on the subject, I can safely assume that this group is still a large one.

The reader of this book does not have to be a physicist or philosopher, just someone who has an interest in the issues and is capable of following a chain of reasoning. While I present all the arguments from scratch, the going may be difficult for those hearing them for the first time. Terms that may be unfamiliar are indicated in boldface type where they are first definitively discussed, so the reader can refer back to that section as needed.

I have included some mathematical discussions for completeness. Clearly boxed off and using a special typeface, these are designed for those with some background in physics who are then able to explore the subject in more precise detail. The difficulty of the material in these special sections varies from rather simple, freshman-level physics to an occasional graduate-level discussion. However, the main text does not depend on the mathematical sections, and the latter can be skipped without losing the thread of thought.

I believe this book fits into, but hardly begins to fill, a niche that is currently almost empty. I know of no other work that addresses the metaphysical claims of the new physics from a relatively detailed critical perspective. While some fine popular books written by prominent physicists can be found that are critical of various aspects of the new metaphysics, these discussions usually amount to little more than flippant dismissal. None of these books probes very deeply into the questions raised by the much larger collection of works by credulous authors. Many of the latter have been written by physicists and other scientists and so require a scientist's response. I hope other scientists and philosophers will join me in providing the reading public with a more balanced view of the significance of developments in twentieth-century physics and cosmology.

Notes

1. Carnap 1931.
2. For a history of metaphysics as applied specifically to physics, see Trusted 1991.

Acknowledgments

After finishing the first crude draft of this manuscript in early 1994, I sent out a request over the Internet for volunteers to act as "devil's advocates" and provide me with no-holds-barred, critical reviews of each chapter. I got more than I bargained for, but the comments, corrections, and suggestions received were invaluable. The result you see before you is the product of many revisions prompted by comments from my "devils " and many others, though none bear responsibility for any remaining errors.

I want to especially thank two people who carefully read every chapter, most in several versions, and commented extensively. These super-devils were Gerald Huber, a mathematics student at the University of Regensburg in Germany, and Taner Edis, then a graduate physics student at Johns Hopkins University. Both impressed me with the maturity of their thinking and their deep knowledge of a wide range of related issues. And each was tireless in his criticisms, almost all constructive. Their influence will be found throughout this book. At this writing I have still not met either face-to-face, such are the wonders of the information superhighway.

I was fortunate also to receive comments from two physicists who have developed their own interpretations of quantum mechanics: Leslie Ballentine and John Cramer. I hope I have represented them accurately.

I am grateful for the critiques and other useful input obtained from many of my physics and astronomy colleagues in Hawaii and around the world: Hugh Bradner, Peter Crooker, Richard Crowe, Rocky Kolb, Sandip Pakvasa, Leo Resvanis, David Schramm, Paul Sommers, Peter Smith, and

Xerxes Tata. Paul was particularly generous in sending me a number of relevant papers by himself and others, and debating several issues in detail with me by way of electronic mail.

Michael Lilliquist made many excellent suggestions on the subject of consciousness, as well as other useful contributions. Harvey Brown, Greg Mulhauser, and Keith Parsons provided important philosophical insights. William Jefferys straightened me out on several matters of statistical theory and Jeeva Anandan patiently explained several problems at the interface between physics and philosophy.

Finally I would like to list the other Internet devils who provided miscellaneous comments that I have implemented herein: Chip Denman, Bill Denker, Mike Easterbrook, Edward Gracely, Elaine Gunzerath, Michael Hagen, Thomas Kagle, Michael Martin, Mitch Porter, Mark Rupright, Bill Salvatore, Donald Simanek, and Richard Young. Again, I have never met any of them and hope I will some day.

It should not be assumed that my devils agree with all or any of the conclusions in this book. I deliberately sought opinions from people I knew held opposing views, including York Dobyns, Jean-Pierre Pharabod, and Jack Sarfatti among others. That is not to say they agree with each other!

I am very grateful to the Rutherford Appleton Laboratory of the Central Laboratory of the Research Councils of the United Kingdom, and the Nuclear Physics Laboratory of Oxford University, for their hospitality during the last phase of this project.

Finally I must acknowledge the continuing influence that my wife Phylliss has on my work. As an outstanding university teacher of English composition and literature, she has taught me much about expository and argumentative writing and I have tried to obey her caution against *ad hominen* attacks. I remain one of her many loving and grateful students.

1

Cosmic Mind

> All the choir of heaven and furniture of earth, in a word all those bodies which compose the mighty frame of the world, have not any substance without the mind. . . . So long as they are not actually perceived by me, or do not exist in my mind, or that of any other created spirit, they must either have no existence at all, or else subsist in the mind of some Eternal Spirit.
>
> —Bishop Berkeley

Science and Ego

As scientific instruments have probed farther into the reaches of space and time, and deeper into sensory realms beyond the puny range of human experience, humanity has gradually receded from their view. Where our unaided eyes perceive humans as the center of existence, telescopes and microscopes reveal no special role for their inventors in the grand scheme of things. So vast is the universe we see with our instruments, and so small is humankind, we are forced to conclude that the earth could explode tomorrow and the rest of the universe would hardly take note.

The insignificance of humanity is almost impossible for most humans to accept. It was bad enough in the sixteenth century, when Copernicus suggested that the earth may not be the center of the universe. It became worse in the nineteenth century, when Darwin proposed that we are an accidental mammalian species and not some unique creation of God. And this painful message was only reinforced in the twentieth century, when astronomers declared that the sun is but one of ten billion trillion stars in a universe at least a hundred billion trillion kilometers in extent, and geologists showed that recorded history is but a blink of time—a microsecond in the second of earth's existence.

Whether we like it or not, the most economical conclusion to be drawn from the complete library of scientific data is that we are material beings composed of atoms and molecules, ordered by the largely chance processes of self-organization and evolution to become capable of the complex behavior associated with the notions of life and mind. The data provide us with no reason to postulate undetectable vital or spiritual, transcendent forces. Matter is sufficient to explain everything discovered thus far by the most powerful scientific instruments.

But what about you and me? Simple, everyday observation tells us that we are individually mortal and that our bodies must someday lose their abilities to move, act, and think as we dissolve back into the earth from which we arose. Still, we find it very difficult to accept inside what the data outside say about our individual selfhood. The message of our senses and instruments conflicts too profoundly with what our inner voices insist.

Humans, for evolutionary reasons, or no reason at all, possess egos that listen largely to their own counsel, most often ignoring other, conflicting messages. These egos are so massive that they are the foci toward which all other bodies gravitate. The ego can hardly conceive of a universe in which it is not an active participant. Ask yourself: Can you imagine a universe without you? As much as I try to be objective, to accept the judgment of reason, I still find it very difficult to develop that image.

From the time of its first murmurs, science's message of humanity's insignificance has been resisted by powerful forces within church and state. Religion is always ready to affirm the inner message and provide the comforting promise of immortality. And the state has always found religion useful in keeping the populace in line, to provide divine justification for its actions.

And so, while science may have triumphed in some intellectual circles, and while few deny science's remarkable power and utility, most modern humans simply ignore the unwelcome implications of scientific discovery. The alternative, soothing message of the feel-good religions of today, from modern evangelical Christianity to the cults of the New Age, is far more appealing: *You are the image of God, if not God himself. You are one with the entirety of existence. Your physical death means nothing! You will live on beyond death, as an inseparable component of the essence of existence.*

Still, some other sense, a spark of reason, hints that this may be a hopeless delusion. It seems that the objective outer message of our senses cannot but conflict with the subjective inner message of ego. They cannot both be correct. How can we decide between the two? Can the two views be made compatible?

Ego has shown no signs of changing for thousands of years, while science is characterized by progress, flexibility, and the continual discarding of old ideas to make room for new discoveries. Scientists readily admit that their conclusions are tentative. Wouldn't it be wonderful if science could only finally confirm what our inner voices have been telling us all along—that we really are immortal personalities with a meaningful if not leading role in the cosmos?

A host of recent authors have proclaimed that this revolution in scientific thought has in fact occurred, that the new physics of the twentieth century has discovered that human consciousness, not matter, is the fundamental substance of the universe. This notion has struck a responsive chord. But is that chord being played on the fine strings of a heavenly harp, or on the last bits of straw grasped at by an ego incapable of accepting reality?

Convergence?

For more than a decade now, gurus of the New Age and preachers of the New Christianity have been telling us that developments in twentieth-century physics and astronomy—quantum mechanics, big-bang cosmology, the so-called anthropic coincidences, and the new sciences of chaos and complexity—are leading toward a convergence of the differing views of the universe provided by the outer voices of science and the inner voices of ego. They proclaim that the discoveries of modern physics imply a central role for human consciousness, and for a universe created with humans in mind. In their view, human beings are not tiny, negligible points in space and time but an integrated part of a greater, cosmic whole, elements of an infinite field that spreads throughout all of space and time.

In some New Age writings, our bodies are said to exist in symbiotic relation to Gaia, goddess earth, and through Gaia to the rest of the universe. Our minds are supposedly tuned in to a greater *cosmic mind* that reaches inside to the smallest particle, outside to the farthest galaxy, back to an infinite past, and ahead to an eternal future.

In New Christian thought, our spirits tune into the cosmic mind of Jesus. The phrase "mind of God" has become fashionable in books and magazine articles that attempt to link modern science to religion, with science interpreted as the process of discovering the laws that God laid down in creating the universe. A huge literature has been generated as modern Christian writers and the secular media attempt to reconcile science and religion.

In reality, most of the arguments are not new. They encompass elements as old as history, and probably prehistory. They hark back to the idealistic philosophy of ancient India, to Plato and Pythagoras, and to the deism of the Enlightenment. But today's cosmic mind has been repackaged by an appeal to twentieth-century science for its authority.

The new wrinkle on venerable Eastern and Platonic/Christian mysticism exploits certain interpretations of quantum mechanics, the revolutionary theory of physics that was developed early in this century. Traditional religious myths, East and West, call on scripture or the utterances of charismatic leaders as their authorities. By contrast, the new mythology is supposedly grounded on up-to-date scientific knowledge. Since the seventeenth century, a materialistic, reductionist view of the universe had formed the foundation of the scientific revolution. Now this is to be cast aside by a new spiritual, holistic science.

The Development of Quantum Mechanics

Quantum mechanics was developed early in the twentieth century to explain certain anomalous phenomena associated with light and atoms. By the 1930s, its mathematical structure had evolved almost to the point where it exists today as the major theoretical tool of physics and chemistry. Calculations using the mathematical formalism of quantum mechanics have been tested against countless laboratory measurements for almost a century, without a single failure.

Quantum mechanics is often associated with "uncertainty." Nevertheless, it is capable of calculations to a high degree of precision. For example, the magnetic moment of an electron, which measures the strength of the electron's magnetic field, is calculated in quantum electrodynamics, an extension of quantum mechanics, to be 1.00115965246. Its measured value at this writing is 1.001159652193 ± 0.0000000010. Thus, the calculation is correct to at least one part in ten billion. We have neither measured nor calculated the earth's magnetic field with anything approaching this accuracy.

Among its many applications, quantum mechanical calculations have made possible lasers, transistors, computer chips, superconductors, plastics, thousands of new chemicals, and nuclear power. Today's high-speed computers are products of quantum mechanics. Quantum mechanics lies at the heart of physics, chemistry, biology, and life itself. It may provide the key to understanding the origin of the universe, showing how everything can have come from nothing.

While the methods of quantum mechanics have proved their utility, no consensus exists on what quantum mechanics "really means." Some argue that the question itself is meaningless, that the mathematics speaks for itself.

Descriptions of quantum mechanics are conventionally cast in terms of the *Copenhagen interpretation.* This interpretation was primarily the offspring of Niels Bohr and Werner Heisenberg, who, along with Erwin Schrödinger (who did not support Copenhagen), were the revered main inventors of quantum mechanics. Today an evolved Copenhagen interpretation remains the consensus view among most physicists, who see no reason to change a theory that has worked well over a great period of time and has never been demonstrated to be incorrect by either experimental facts or mathematical proof.

As we will see, however, the Copenhagen interpretation contains more than the minimum number of assumptions that is needed to provide a foundation for quantum mechanics as it is actually practiced by scientists. Copenhagen includes the added assertion that quantum mechanics is *complete*; Bohr and his colleagues of the Copenhagen school claimed that no theoretical structure can be found that is capable of making predictions about observable phenomena that does not fit within the framework of quantum mechanics. This was not meant to imply that quantum mechanics can now explain everything; just that any new theories must not contain elements that violate the basic precepts of quantum mechanics.

This assertion is disputed by the proponents of so-called *hidden variables* theories. They seek a deeper theory that lies beyond conventional quantum mechanics. We will be investigating these issues in great detail in this book.

On the Fringes

While the mathematical formulation and methods for the practical application of quantum mechanics have remained largely unchanged and unchallenged for six decades, the deeper philosophical significance of quantum mechanics has continued to be debated. On the fringes of this debate we find numerous popular articles and books that promote a stupendous notion: Our egos could be right after all. Humans and human consciousness may indeed constitute the fundamental essence of reality. If you were to judge by the space occupied by this genre on the shelves of popular book stores, you would conclude that it has become mainstream science.

On the contrary, the pragmatic, mainstream physicist's attitude toward the new quantum metaphysics has generally been to ignore it, figuring it will simply die away like any other popular fad. Most physicists prefer to leave deliberations on the "deeper significance" of quantum mechanics to the philosophers who make their livings discoursing on the meanings of words, and never seem to settle anything anyway. Physicists like to think of themselves as people of action, not words.

Unfortunately, arguments over words have a much greater impact on human life than most physicists prefer were the case. Words are not benign. Words generate action. Words sell products, inspire devotion, incite riots, and start wars.

Words also help physicists get the large sums of money needed to build their action toys. As a practicing researcher in high-energy particle physics and astrophysics for over thirty years, I spend much of my time writing proposals, progress reports, technical notes, and scientific papers. I attend several international conferences each year where I listen to speakers, present my own work, and exchange ideas in hallway and dinner-table conversations—all utilizing the medium of words. Often these discourses are philosophical in nature, addressing the meaning of the research being conducted and its value to science and society.

The jargon of quantum mechanics has inspired some people to extract mystical messages that were never intended. In particular, deep meaning has been found in the unfortunate way physicists often describe the process of measurement. Sometimes they make it sound as though the conscious act of observation, by itself, creates the quantity that is being measured.

You will frequently read the statement that physical objects do not possess a certain property until that property is measured: An electron in an atom has no position until that position is determined by measurement; a photon has no polarization until it passes though the polarizing sheet that is used to measure polarization.

The source of this strange assertion is the practical fact that physical notions, such as position and polarization, are *operationally* defined in terms of the apparatus that makes the measurement of the associated quantity. These measurements are performed according to a well-prescribed procedure that can then be repeated independently by someone else. This is what gives science its claim on objectivity.

Thus distance (the quantity of space) is what we measure with a meter stick. Time is what we measure with a clock. Polarization is what we measure with a polarimeter. All these operational quantities were defined by human beings. Is there any reason to assume that any has an intrinsic reality that exists in the absence of its measurement? As we will see, there is ample reason to assume at least some aspect of reality when the results obtained are predictable and repeatable.

The idea that properties are brought into being by the act of their measurement clashes with our intuitive notion that the universe possesses an objective reality independent of the observer. Surely, as Einstein insisted, the moon is still there when no one is looking.

But many authors have construed quantum mechanics, with its strict use of operational terms, to imply a central role for the human mind in affecting the very nature of reality itself. Let me give a sampling of some of the expressions of this viewpoint.

Physician Robert Lanza has written that, according to the current quantum mechanical view of reality, "We are all the ephemeral forms of a consciousness greater than ourselves." The mind of each human being on earth is instantaneously connected to every other—past, present and future—as "a part of every mind existing in space and time." In Lanza's view, quantum mechanics tells us that all human minds are united in one mind and "the entities of the universe—electrons, photons, galaxies, and the like—are floating in a field of mind that cannot be limited within a restricted space or period."[1]

Physicist Fritjof Capra has long been an influential proponent of mystical interpretations of quantum mechanics. He first expressed his ideas in 1975 in *The Tao of Physics*, which drew strained parallels between modern physics and Eastern mysticism.[2] Quantum mechanics, in Capra's view, "reveals the basic oneness of the universe" in a manner that harmonizes with the Hindu notion of *Brahmin* the "unifying thread in the cosmic web, the ultimate ground of being: 'He on whom the sky, the earth, and the atmosphere are woven (Mondaka Upanishad, 2.2.5).' "

Capra's film *Mindwalk*, which was shown in major theaters in 1992 and is available in video stores, gives considerable insight into his hopes for the potential social and philosophical impact of this new perspective. Written by Capra and directed by his brother Bernt, *Mindwalk* is based on *The Tao of Physics* and a later book, *The Turning Point*.[3] In the film, an American politician, played by the fine actor Sam Waterston, comes to France after losing his bid to be president. There, he and his friend, an expatriate poet played by John Heard, wander into the spectacular fortress of Mont St. Michel in the English Channel. Soon they meet a disillusioned physicist, played by Liv Ullman, and for the rest of the film the two men roam around the fortress, slack-jawed with astonishment at the profound ideas Ullman pours forth: The world is in trouble from overpopulation and pollution. Americans eat too much red meat. Wow! The presidential candidate had not heard about this before.

The problem, according to Ullman, is a crisis in perspective. Humanity still follows the mechanistic reductionism of Descartes and Newton, viewing the world as being like the old clock in the fortress tower. However, a new, holistic physics called *systems theory*, in which the universe is seen as one interconnected whole, has now overthrown evil reductionism. If humanity will only adopt this revolutionary perspective and realize that we are all one with each other, the earth, and the cosmos, then the planet will be saved from self-destruction. What a magnificent thought, the politician gushes. Why don't you come back to America with me, professor, and join my staff? Let's put these new ideas to work for humanity.

Finally, outside the fortress, on the spit of land that joins it to the mainland, Ullman is asked to explain life. She says, "Life is self-organization." Poet Heard is so overwhelmed by this deep concept that he flops down in the sand, repeating the line over and over: "Life is self-organization, life is self-organization."

Unfortunately, this is the only hint of the most far-reaching idea that appears in Capra's *The Turning Point*. There he suggested that all material systems, from humans to animals, plants, the earth, and the cosmos itself, are part of one gigantic mind. Holistic physics provided him with a model for the vague notion of cosmic consciousness: We are all one with the cosmos, speaking to each others' minds with extrasensory perception (ESP), able to break down the barriers of space and time and the laws of physics. We can achieve anything, perform miracles, if we just think we can.

Capra's ideas have taken hold within the New Age movement in America. In her 1980 New Age bible, *The Aquarian Conspiracy*, Marilyn Ferguson says that new scientific knowledge has revised "the very data base on which we have built our assumptions, institutions, our lives." Promising far more than "the old reductionist view," the new scientific perspective "reveals a rich, creative, dynamic, interconnected reality."[4]

Capra has not been alone in claiming parallels between the new physics and Eastern mysticism. In *The Dancing Wu Li Masters,* Gary Zukav says physicists "are dancing with Kali, the Divine Mother of Hindu mythology." Zukav sees the new physics as suggesting that "there really may be no such thing as 'separate parts' in our world."[5]

In a chapter called "The Dancing Moo-Shoo Masters" from his recent book *The God Particle,* physicist Leon Lederman has spoofed the notion that physics has any connection with the philosophies of the ancient Orient. He calls Capra's and Zukav's conclusions "bizarre."[6]

The idea of a cosmic field of mind merging physics with Hindu mysticism has also been promoted by Maharishi Mahesh Yogi and his Transcendental Meditation movement. Trained at one point as a physicist, the yogi also claims modern physics as his authority. In newspaper ads placed around the country in the 1980s, the Maharishi very specifically associated his version of cosmic consciousness with the Grand Unified Theory (GUT) field of particle physics that was in fashion at that time.

Unfortunately, reality intervened. Theoretical particle physicists, applying the simplest version of GUT, made a very firm, testable prediction that the proton was unstable with a very long but measurable lifetime. After a series of accurate, multimillion-dollar experiments, proton decay was not found at the expected level.[7] As a result of this and other precision tests, Grand Unified Theories have fallen out of fashion and the Maharishi's association of the GUT field with the cosmic mind has been discarded.[8]

At this writing, GUT has been replaced as the Maharishi's cosmic field by the currently more trendy *superstrings.* If superstring theory is found wanting, as I suspect it will be, I am sure the yogi will find some other physics fashion to exploit. He can always claim, like another Yogi named Berra, that he never said half the things he said.

One of the Maharishi's disciples, Dr. Deepak Chopra, is perhaps the most successful of a growing group of authors who have appropriated the quantum as the foundation for alternative, nonmedical methods of healing based on the belief that mind can overcome the limitations set by the laws of physics and biology. Chopra's 1989 book was entitled *Quantum Healing: Exploring the Frontiers of Mind/Body Medicine.*[9] His latest bestseller is called *Ageless Body, Timeless Mind: The Quantum Alternative to Growing Old.*[10] Placing the word "quantum" in the title of a book may not guarantee it for the best-seller list, but it's worth a try.

In spring 1994, Chopra visited Honolulu to give all-day seminars on "Quantum Healing." At the time, an English department colleague of mine assured me that Chopra has "helped a lot of people" with his holistic methods.[11]

Of course, promising a halt to aging is a dangerous thing. Let's see what Chopra looks like in ten years. He already looks older in the photograph on the dust jacket of the latest book compared to the earlier one. Hopefully Chopra will not suffer the fate of Dr. Stuart Berger, author of *Forever Young,* who died at age forty weighing 365 pounds after falling off his diet of steamed broccoli.[12]

In a similar vein, Johns Hopkins University psychiatrist Patricia Newton uses the quantum as the basis for what she says is an Afrocentric approach to healing. In a talk presented before a medical conference in 1993, Newton said that traditional healers "are able to tap that other realm of negative entropy—that superquantum velocity and frequency of electromagnetic energy and bring them as conduits down to our level. It's not magic. It's not mumbo jumbo. You will see the dawn of the twenty-first century, the new medical quantum physics really distributing these energies and what they are doing."[13] Shirley MacLaine could not have put it better.

I do not deny a certain limited value in the traditional healing methods from many cultures. Surely, over the ages, useful treatments for a host of aches and pains were discovered by trial and error. It appears that many of these methods trigger the well-established placebo effect and perhaps other mechanisms by which the human body heals itself. No doubt Western medicine can improve its methods for treating the "whole person." I simply wonder what it all has to with the quantum.

In *The Tao of Physics,* Capra also made a strong association between the unbroken wholeness he saw in Eastern philosophy and a similar-sounding theory of physics that also was once quite the vogue, but has now dropped from sight. Few of today's graduate students in physics would even recognize the name of this faded concept: *bootstrap theory.*

Dating from the 1960s, when Capra worked as a theoretical physicist in Berkeley, bootstrap theory speculated that all the properties of physical systems could be derived from a set of equations whose input assumptions were little more than some general rules of mathematical smoothness ("analyticity") and self-consistency.

While this was a nice thought, and it once gave Capra a vague basis for his speculations, bootstrap theory simply did not work. It failed to describe the data while the conventionally reductionist theories of quarks and leptons, now referred to as the *Standard Model*, eventually did. For that purely pragmatic reason, not for any lack of popular or aesthetic appeal, bootstrap theory no longer appears in physics textbooks.

Being a failure, bootstrap theory does not provide a very convincing model for Capra's holistic universe. By its vividly contrasting success, the quark-lepton model provides every reason to continue to look to reductionist ideas to provide the framework for understanding the physical world. However, I caution the reader against making the connection between reductionism and Newtonian determinism that is found in so much New Age literature. A nondeterministic but still reductionistic universe is perfectly possible.

ESP and Quantum Mechanics

Many authors, including Capra and the others mentioned above, have argued that so-called psychic, or **psi** phenomena provide an empirical basis for a connection between the human mind and the cosmos. They refer to the numerous reports of experiences that people label as psychic: premonitions, out-of-body and near-death experiences, miraculous cures, stigmata, poltergeists, "mystical" experiences, past-life regression, remote viewing, and others. These are taken, in sum, as a strong indication that the mind is something beyond matter, that it has the ability to overcome the laws that rule the behavior of normal material objects.[14]

Einstein once said that he would not believe in extra sensory perception (ESP) unless it was observed to fall off with distance. This view was based on the well-established physical principle of energy conservation. If a mind is radiating some form of "psychic energy" in all directions, then that energy should spread out over an area that increases with the square of the distance from the source.[15]

Since the 1930s, unsuccessful attempts have been made by parapsychologists to measure a distance effect for ESP.[16] In most sciences, the failure of an experiment to confirm a theoretical prediction is taken as a strike against the theory. However, those whose personal beliefs are unshakable by facts will always find a way to rationalize such failures.

One way to explain the absence of an ESP distance effect is to argue that the psi signal is some type of encoded message akin to a radio broadcast. Such messages can be transmitted without degradation over large distances, though they still have a finite range. This is not implausible in itself. However, Einstein's point was that the observation of a distance effect would have been a strong point in favor of ESP and perhaps converted him into a believer. This did not happen.

With the failure of distance experiments to produce an effect, some psi believers began to develop the idea that ESP is a nonphysical phenomenon, unbound by limitations of space, time, or energy. Instead of interpreting the lack of a distance effect as a failure of the ESP hypothesis, they took it as positive evidence that ESP is not a phenomenon akin to electromagnetic radiation. If ESP violates conventional principles of physics, then perhaps it goes beyond conventional physics toward a broader, all-encompassing theory of mind and the universe. Perhaps, but the absence of evidence for ESP can prove little one way or the other.

In 1974, American physicist Jack Sarfatti was working in London with distinguished quantum theorist David Bohm. Before his death in

1992, Bohm was the central figure in quantum mysticism. His name will appear often on these pages, in both this later role and his earlier one as a major contributor to the development of quantum physics.

Bohm, Sarfatti, and prominent author Arthur Koestler were among those present on June 21, 1974, when the famous Israeli psychic, spoon-bender Uri Geller, gave a demonstration of his powers in London. Geller succeeded in bending a metallic disc and triggering a strong burst from a Geiger counter held in his hand.[17]

The next day, the performance with the Geiger counter was repeated before Koestler and author Arthur C. Clarke, among others. According to a press release put out by Sarfatti that was widely distributed, Koestler was "visibly shaken" and reported a strong sensation simultaneous with the burst. The previously skeptical Clark was also impressed and challenged magicians to "put up or shut up" in duplicating Geller's feat. At the time, Sarfatti said that Geller had demonstrated "genuine psycho-energetic ability" under "relatively well-controlled and repeatable experimental conditions."

Before he died, Koestler endowed a chair in parapsychology at Edinburgh University. Clarke, who has been an influential science popularizer and science fiction writer for decades, has hosted a series of mildly skeptical television programs on "mysterious phenomena."

As for Geller's London demonstrations, plausible explanations can be found that do not rely on the invocation of supernatural forces. Martin Gardner has pointed out that Geller could have simply hidden a small amount of harmless radioactive substance, such as a radium watch dial, on his body to cause the Geiger counter to read a higher level of radiation.[18] Geller's performances are accompanied by much writhing and twisting that offers him ample opportunity to put a magician's skills to use.

Sarfatti tells me he no longer believes that Geller has the power to affect physical objects with his mind. Apparently this happened after magician James Randi duplicated Geller's tricks for Sarfatti.[19]

For thousands of years people have told stories and related personal anecdotes that have convinced them that the mind has special powers that reach beyond the world of matter. Despite this, science has yet to accept the reality of psi as a fact. Beyond anecdotal tales and magician's tricks, which have little scientific value except as data for studies of anecdotal tales and magician's tricks, psychic phenomena have a history of scientific and semiscientific investigations dating back to the mid-nineteenth century. I have previously written about these studies, and the claims made that they support the existence of psychic phenomena, in my book *Physics and Psychics: The Search for a World Beyond the Senses.*[20]

My conclusion agrees with that of a 1987 inquiry by the National Research Council of the U.S. National Academy of Sciences: After almost a century and a half of study, "the best scientific evidence does not justify the conclusion that ESP—that is, gathering information about objects or thoughts without the intervention of known sensory mechanisms—exists."[21]

Unsurprisingly, the parapsychological community emphatically disagrees with this conclusion.[22] They continue to insist that the sum total of observations over the years is a strong indication that "something must be there" beyond the reach of conventional, materialist science. The subject refuses to die, as each discredited claim is replaced by new ones from a different variation of psi experiment.

In one way, parapsychology does mimic conventional science: most attention is focused on the latest fashions. One current parapsychological fashion is the *ganzfeld experiment,* in which a subject in a sensory-deprived state attempts to read the mind of another. Recently, strong positive, replicable results have been claimed.[23] However, leading experts still find these experiments flawed and no single experiment is by itself convincing.[24] Work is continuing, especially in Edinburgh, where a major effort is underway to see if previous results can be replicated. It remains to be seen whether these and other ganzfeld experiments will yield results any more reliable than those of their predecessors in psi science, or simply follow precedent and fade away as the next psi fashion moves into their place.

Significant effects in recent years have also been claimed in experiments that study whether humans (and, in some cases, animals and even cockroaches) can affect the output of random event generators (**REG** experiments, or sometimes **RNG**, for random number generator). These touch especially on the subject of this book because quantum fluctuations are sometimes used to produce the random events that form the data base. Thus any significant deviation from expectations would be direct evidence for a quantum-mind connection, provided all experimental artifacts could be ruled out.

Although hundreds of REG experiments have been reported,[25] the largest data samples have been collected by Helmut Schmidt[26] and by the group headed by Robert Jahn at the Princeton Engineering Anomalies Research Center (PEAR).[27] Both projects claim significant deviations from expectation at a level than cannot be explained by statistical fluctuations or experimental artifacts.

Still, the two sets of experiments do not agree quantitatively, and so cannot claim to independently replicate each other. In fact, you could even argue that since they quantitatively disagree, they thereby disconfirm each other. Schmidt reports that of the order of one percent of his hits are above expectations, while the PEAR result is approximately one-tenth of a percent high. In either case, the effect claimed is small and becomes noticeable only after a huge number of trials. Also, it is not clear whether the PEAR studies even replicate themselves, because the size of the effect from early trials differs from that of later trials.

I discussed the status of the REG experiments through 1990 in *Physics and Psychics*.[28] At that time, critics had found a number of deficiencies in the experimental protocols and noted that most of the PEAR effect was essentially due to a single operator, who just happened to be the first subject as well as a primary member of the research team.[29]

The PEAR group remains very active and claims to have answered its critics. However, its members continue to report results in a cumulative fashion and it is not clear from their papers that these are not affected by biases that may have been introduced in the early, developmental phases of the experiment.

History is full of reports of exciting new results obtained in the preliminary stages of scientific experiments, which disappear as the experiments are improved. Two examples that come immediately to mind are the ESP work by Joseph Banks Rhine at Duke University in the 1930s and the recent reports on cold fusion.

As noted, a severe criticism of the PEAR protocols is that experimenters also act as operators and their results are included anonymously in the cumulative data sample. While the experimenter-operators are subjected to the same controls as the others, this still strikes me as an unwise procedure that leaves them open to the suspicion, however unfair, that they have somehow "cooked up" the effect. Indeed, as mentioned, the results are less significant when the experimenter data are removed, though they are still claimed to be significant.

This is not to say that any cheating has occurred, but given the history of ESP research, this must remain an economical explanation until it is ruled out to the highest degree. Normal scientific protocols in which the experimenters are kept from having any influence on the specific outcome of an experiment are strongly called for in this case. The researchers can still serve as subjects to test out the equipment and experimental procedures, but their data runs should be excluded from the samples used to test for an effect.

Even if the PEAR experimental protocols are assumed to be adequate, the significance of the result remains arguable. Using standard ("classical") statistical tests, there have been reports of probabilities of the order of 10^{-4} for the result being simply due to statistical error.[30] As best as I can tell from reading the papers, this is intended to mean that only one experiment in ten thousand similar ones would have given the same deviation, or a greater one, as the result of normal statistical fluctuations.

This type of measure of statistical significance, called the **significance level**, is widely used, including within my own field of particle physics. However, statistics experts argue that it is not always appropriate, depending as it does on hypothetical, nonexistent experiments. Some recommend that other techniques, including but not limited to those referred to as **Bayesian**, should be used to determine the level of significance of an experimental result.[31] In a Bayesian analysis, different *a priori* hypotheses are tested against the data.

A Bayesian analysis of the PEAR data has been done by PEAR researcher York Dobyns. The result was a range of significance, depending on assumptions, that included "no effect" as a strong possibility.[32] Dobyns used this result to argue that the classical (significance-level) method should be taken as more reliable in this situation than the Bayesian method, because it is insensitive to assumptions.

However, astronomer William Jefferys has responded that the classical method involves hidden and less well-formulated assumptions, and that Bayesian methods at least make their assumptions clear.[33] The Bayesian analysis of PEAR data suggests that the classical result is too optimistic by a factor of at least ten and perhaps hundreds. A significance level of 10^{-4} merits attention, although any effect must certainly be independently confirmed when one is claiming an important new result. Any lower significance level, say 10^{-3}, should not create a stir. In the hundreds of experiments done yearly, statistical fluctuations will produce many artifactual effects at the 0.1 percent significance level.

I conclude that, as with the ganzfeld experiments, we are forced by scientific method to adopt a skeptical, wait-and-see attitude toward the random event generator experiments. Under normal standards, no one has a right to claim evidence for a quantum-mind connection based on these results, though this has been done.[34] Even weak claims will be blown out of proportion in the public media. Experiments of such momentous implication must be independently replicated at the same quantitative level, with believable statistics and far tighter experimental procedures, before they can be used to support the mystical belief in a cosmic mind.

Most parapsychologists believe the evidence for psi is strong enough to conclude that the phenomena are real. I think they are dead wrong. In my mind, all these years of searching with no convincing evidence should be taken as a clear indication that psi does not exist. Parapsychologists and I disagree on this, but conscientious psi researchers cannot deny that a broad scientific consensus has yet to be assembled in support of their position.

Still, the average person is likely to wonder how so many observations of mysterious phenomena reported in thousands of books, articles, and newspaper stories over many years could be wrong. Movies and television continue to exploit the public's thirst for such tales, with programmers paying token lip-service, at best, to the very real doubts that exist in every case where evidence for psychic forces is claimed.

Many of the stories used have already been proved to be hoaxes or downright fabrications. But they are rarely reported as such in the popular media, in what can only be described as scandalous behavior on the part of the authors and producers of these fables.

Undoubtedly, some narratives are honest reports of unusual happenings and are simply misinterpreted as requiring the intervention of magical forces beyond the familiar world of matter. People tend to look for mysterious explanations when the improbable occurs; they are more interesting, more comforting than the mundane. But with billions of people in the world, improbable events occur somewhere on a daily basis. When the critical, skeptical methods of conventional science are applied to the observations labeled as psychic, and when those data are sufficiently clear to form a judgment, more economical explanations not involving extraordinary new hypotheses have so far always been found.

The average person is not scientifically trained and generally unaware of a primary rule of scientific discovery: *Extraordinary claims require extraordinary evidence.* To demonstrate an extraordinary claim, like the miracle of ESP, extraordinary evidence must be obtained. Only after every alternate, mundane explanation has been ruled out with the highest degree of certainty can one begin to entertain hypotheses that introduce new elements that go beyond current science. So far, the evidence for psi phenomena has been ordinary at best.

Nonetheless, people continue to try to make something of nothing. In recent years, some proponents of psi phenomena have interpreted quantum mechanics as providing a basis for instantaneous ("nonlocal") psychic communication across the universe. As physicist Amit Goswami has put it:

> The farther away the point, the less intense is the signal reaching it. In contrast, nonlocal communication exhibits no such attenuation. Since the evidence indicates that there is no distance attenuation of distant view-

> ing, distant viewing must be nonlocal.[35] Thus it is logical to conclude that psychic phenomena, such as distant viewing and out-of-body experiences, are examples of the nonlocal operation of consciousness.
>
> Any attempt to dismiss a phenomenon that is not understood merely by explaining it as hallucination becomes irrelevant when a coherent scientific theory can be applied. Quantum mechanics undergirds such a theory by providing crucial support for the case of nonlocality of consciousness; it provides an empirical challenge to the dogma of locality as a universal limiting principle.[36]

In this book, I take up Goswami's challenge.

As we see from his quotation, quantum mechanics offers believers in ESP a hypothetical basis for their continued insistence that something must exist beyond the world of conventional physics. That something is usually associated with human consciousness, which is assumed to possess qualities that cannot be explained from purely material, physical considerations.

Arthur Koestler once remarked that "the apparent absurdities of quantum physics . . . make the apparent absurdities of parapsychology a little less preposterous and more digestible."[37] Again, quantum mechanics provides the metaphor. A "quantum mechanics of consciousness" has been proposed in which consciousness is represented by the quantum mechanical wave function.[38]

Recently, quantum physicist Henry Stapp has written a paper, published in the prestigious journal *Physical Review,* suggesting that a new version of quantum mechanics can account for the REG results through an interaction between consciousness and the quantum wave function.[39] I will come back to this, and many other claims, later in this book.

The quantum-consciousness connection and its association with mystical notions of wholeness provide a metaphor that believers in the existence of psychic powers use to lay a veneer of scientific respectability over ideas that require a drastic revision in our existing models of reality.[40] However, as we will see, that veneer is so thin as to be invisible. Quantum physics is supported by solid experimental evidence, but psi phenomena are not and the admitted absurdities of parapsychology remain absurd.

Aether and Spirit

The cosmic mind, viewed from the paranormal perspective, is some sort of invisible field that pervades the universe. Human minds are supposedly linked to this field, able to excite it and receive excitations from it. This is far from a new idea. In fact, a very similar notion developed in the nineteenth century, for much the same purpose as New Age ideas are promulgated today.

As science gradually became established, people sought ways that it might be reconciled with their traditional beliefs, or even used to buttress those beliefs. In the nineteenth century, some scientists associated spiritual or psychic forces with the **aether**, a substance that was thought to fill all space and provide the medium for the transmission of light from distant stars. Going beyond physics, these scientists suggested that the aether provided the mechanism by which humans connected to an imagined world beyond matter—the world of the spirit.

The belief in a universal, cosmic fluid pervading space has even older roots. To the ancient Greeks, *quintessence* was the rarified air breathed by the gods on Olympus. Aristotle used this term for the celestial element, the stuff of the heavens, and said it was subject to different tendencies than the stuff of earth. Quintessence was not bound by the same laws as ordinary matter.

When Newton was prompted to explain the nature of gravity, he replied that gravity might be transmitted by the invisible aether.[41] He further suggested that the aether also may be responsible for electricity, magnetism, light, radiant heat, and the motion of living things that he, like his contemporaries, thought was the consequence of some source beyond inanimate matter.

Today, with knowledge not available to Newton, we can account for life as a purely material phenomenon with no need to invoke any special life-force. Despite this, and the complete absence of scientific support for the existence of immaterial, vital forces, we still hear of ch'i, ki, prana, and psychic energy, usually in association with alternative healing. Again the ego is doing the thinking, assuming that something special must account for the wonder of its own existence.

Newton had envisioned matter and light as particulate in nature, though they appear continuous to the human eye. Gravity, however, seemed to be something else, acting invisibly—holistically—over the entire universe. (It should be noted, however, that the gravitational force falls off inversely with the square of distance, unlike the imagined psychic fields.)

In the mid-nineteenth century, the mathematical concept of the field was developed to describe the apparent continuity of matter, light, and gravity. A field has a value at each point in space, in contrast to the properties of a particle, which are localized to an infinitesimal region of space.

Pressure and density in a fluid are two examples of how the field concept is successfully applied in practice. Although matter is discontinuous at the atomic and molecular level, these "matter fields" provide for an accurate description of the behavior of solids, liquids, and gases because, on the everyday scale, matter appears continuous to a very good approximation.

As the phenomena of electricity and magnetism became better understood, they also were described in terms of fields. Then, in 1867, James Clerk Maxwell had one of those rare insights that punctuate the history of science. He discovered that the equations uniting electricity with magnetism called for the propagation of electromagnetic waves in a vacuum. Furthermore, these waves moved at the speed of light.

Waves were already very familiar phenomena in physics. In (apparently) continuous media such as air, pressure and density propagate as sound waves when the media are excited. For Maxwell's electromagnetic waves, the question arose: What's doing the waving? The analogy was drawn that all of space out to the most distant stars was filled with an elastic medium—the aether—whose excitation produced the phenomenon of light.

Electromagnetic waves beyond the narrow spectrum of visible light were predicted, soon observed, and put to use in "wireless telegraphy." One of the early workers in this area was the English physicist Oliver Lodge. While making major contributions to physics and engineering, Lodge joined William Crookes, Alfred Russel Wallace (codiscoverer of evolution), and other notable nineteenth-century scientists in extending their horizons to search for phenomena that transcend the world of matter.

If wireless telegraphy was possible, why not wireless telepathy? If electrical circuits could generate and detect ethereal waves, why not the human brain? Coincidentally, certain people who claimed to possess the ability to communicate with other minds, living and dead, had just appeared on the scene. They were called spiritualist mediums a century ago; today their spiritualist descendants are known as psychics or channellers.

Unfortunately, most scientists lack the specific skills needed to distinguish fact from illusion in the world of magic. The universe does not lie; people lie. And so Lodge and other nineteenth-century psychical researchers unwittingly allowed themselves to be fooled by the tricks of professional fortune tellers and sleight-of-hand artists posing as spiritualists. They permitted their wishes and dreams to govern their senses and reason. Lodge, desperately wanting to believe in life after death, had written passionately about imagined communications with his son Raymond, killed in Flanders in 1915. Sadly, he accepted the wildest claims of mediums and skilled stage magicians.[42]

Spiritualism offered scientists like Lodge a way to reconcile science with a belief in immortality. The resurrection of the complete body had always been the primary tenet of Christianity. If only Jesus' soul had gone to heaven, why would his body have been missing from the tomb? The Catholic church has insisted that the Virgin Mary's body also ascended, atom by atom, to heaven. As for the rest of humanity, our bodies have to await the Judgment Day for our complete resurrection.

By the nineteenth century, however, it had become clear that it was absurd to think of all the atoms of a human body reassembling on Judgment Day. Our atoms are being replaced moment by moment anyway. So the idea of a "spiritual body," separate and distinct from matter, was developed.[43] Lodge proposed that the aether was the substance of spirit. As he put it:

> The body of matter which we see and handle is in no case the whole body; it must have an etheric counterpart to hold it together, and it is this etheric counterpart which in the case of living bodies is, I suspect, truly animated. In my view, life and mind are never directly associated with matter; and they are only indirectly enabled to act upon it through their more direct connection with an etheric vehicle which constitutes their real instrument, an ether body which does not interact with them and does not operate on matter. . . . An etheric body we possess now, independent of accidents that may happen to its sensory aggregate of associated matter, and that etheric body we shall continue to possess, long after the material portion is discarded. The only difficulty of realizing this is because nothing etheric affects our present senses.[44]

Few of the faithful today realize that the notion of a separate "spirit" and "body" was a fairly recent development in Christian thinking, though it goes back ages in India and Greece. That is not to say that the idea of a spirit or soul is new to Christianity, but simply that the sharp distinction between body and spirit, or body and mind, now commonplace in Christian thinking, was a modern innovation that cannot be found in the scriptures or early teachings of the Church.

Relativity and Quantum Mechanics

Near the turn of the century, Albert Michelson and Edward Morley sought to find experimental evidence for the electromagnetic, or "luminiferous," aether and succeeded in showing instead that it did not appear to exist. Shortly thereafter, in 1905, Einstein developed his theory of relativity, which demonstrated that the concept of an aether was mathematically and logically inconsistent with Maxwell's equations of electromagnetism. Einstein concluded that electromagnetic waves, including light, could not be the vibrations of an aether. Still, Oliver Lodge remained firm in his belief that a universal cosmic fluid existed that could be excited by the human mind. To Lodge, the aether was a necessity, the cosmic glue without which "there can hardly be a material universe at all."[45]

Lodge was similarly unhappy with what he was hearing quantum physicists, like Planck and Bohr, say about the fundamentally discrete, quantized, nature of all phenomena. He deplored "the modern tendency . . . to emphasize the discontinuous or atomic character of everything."[46] But progress passed him by, as evidence accumulated that matter is composed of discrete atoms, that electricity is the flow of electrons, and that light is a current of particles called photons.

By the time Lodge died in 1940, both the luminiferous aether and material continuity were already long in their graves. Today the electromagnetic aether is no longer a candidate for the stuff of spirit. The aether simply does not exist. In its place, even more ephemeral aether fields have been imagined as sources for spiritual quintessence—the field of the quantum wave function, the "quantum potential," or perhaps, as Danah Zohar suggests, the vacuum itself.[47]

Like Lodge, Ernst Mach, and many other capable physicists of the early century, Einstein was uncomfortable with quantum mechanics, calling it "spooky." In 1935, he and two collaborators, Boris Podolsky and Nathan Rosen, wrote a paper arguing that quantum mechanics was incomplete because it does not provide for a description of what they called "physical reality."[48]

Einstein and his collaborators pointed out that, following conventional quantum mechanics, an experiment performed at one point in space seems to immediately determine the outcome of another experiment performed at a different point, even when the separation between these points is such as to require a signal moving faster than light to carry information from one to the other in the elapsed time interval. In fact, a signal must move at infinite speed to connect two simultaneous events separated by any distance, even one as small as an atomic diameter. This distance could also be billions of light years, if all events past and future are to be connected.

Yet quantum mechanics seems to allow for just such an instantaneous correlation between separated events. This has provided a scientific basis, at least in some minds, for the notion that the universe is one simultaneously connected whole. Einstein referred to this quantum connectivity as a "spooky action at a distance," noting that it was incompatible with his claim that no signals can move faster than light.

Like so many of the strange effects of quantum mechanics, this apparent paradox, which we will be examining in great detail, is a consequence of the wave-particle duality in which physical systems seem to behave either like waves or particles, depending on which type of property is being measured. Again, the distinction is between the discrete, localized properties of a particle and the continuous, distributed properties of a wave field.

It is not commonly appreciated that instantaneous correlations between separated events were already present in prerelativistic, prequantum physics. Before Einstein, no theoretical limit existed on the speeds of bodies. Furthermore, classical waves, even those moving at finite speed stimulated by tossing a pebble in a lake, can produce correlations between separated phenomena. You can imagine such a wave carrying information in the modulation of its amplitude or frequency, just as with sound and radio waves.

As a radio wave propagates outward, all the information carried by the waveform spreads through space. At any given time, two separated receivers on the wave front obtain that identical information; they simultaneously hear the same program. The two receivers can be said to be correlated, but that relationship is not a causal one in which an action at the place of one receiver generates a result at the place of the other receiver. Observers at the receiver positions cannot instantaneously signal each other unless that signal can move at infinite speed.

So, independent of quantum mechanics, observations at separated points in space can still be correlated. This correlation, however, does not imply superluminal signalling nor any other miracle; no physical law is violated. Two points in space can receive the same information when that information originates from the same point.

Quantum mechanics, on the other hand, has suggested to some that measurements made at one point in space can instantaneously affect the outcome of measurements at another point. This notion, which was expressed in the Goswami quotation above, is termed *nonlocality.* It implies some sort of superluminal signalling, in violation of Einstein's assertion that nothing can go faster than light. As we will see in the following chapters, the consequences of nonlocal communication are so profound as to turn most of our concepts of space and time on their heads. Indeed, the realization by Einstein that motions at infinite speed made it impossible to assign points in space and time a unique reality led him to assert that a maximum speed, the speed of light, exists.

In 1964 John S. Bell, stimulated by the ideas of David Bohm, showed how it was possible to experimentally test the spooky way quantum mechanics seemed to allow for superluminal action at a distance.[49] Bohm, following a largely forgotten suggestion of Louis de Broglie a quarter century earlier, had proposed an alternative interpretation of quantum mechanics in which yet-undetected entities were responsible for the wavelike behavior of particles.[50] Following convention, I will call these entities *hidden variables,* though the term is not particularly enlightening.

Bell showed the way to experimentally decide between the most important class of hidden variables, those that are both "local" and "real" as are the variables of classical physics, and the conventional interpretation of quantum mechanics. *Local variables* do not violate Einstein's relativity and involve no superluminal signalling. *Real variables,* in this context, are like the familiar variables of classical physics, being simultaneously measurable and behaving in predictable ways.

Now, after a series of precise experiments, the issue has been decided: hidden variables that are both local and real are ruled out.[51] Real, nonlocal hidden variables, such as those introduced by de Broglie and Bohm, remain possible alternatives to the conventional interpretation of quantum mechanics.

But nonlocality implies superluminal connections at some level, and at least an apparent violation of relativity. Since experiment has not yet shown any such violation, a more economical interpretation of the results on experimental tests of Bell's theorem is simply that no hidden variables exist. Popular literature, however, would lead you to think that nonlocality is a demonstrated fact of nature. As I will explain in great detail in these pages, nonlocality exists only in theory. No superluminal motion or communication has ever been observed.

Experiment, not theory, will decide whether nonlocality is indeed a fact of nature. So far, it is not known to be a fact. Those quantum interpretations that incorporate nonlocality claim, with a certain illogic, that the superluminal transfer of information is still impossible. However, I fail to see how nonlocality can imply anything meaningful other than communication, or other motion, faster than the speed of light.

The New Holism

With experiment ruling out local hidden variables, a new holism has begun to develop. For example, Bohm's nonlocal quantum potential, which we will describe later, seems to imply an interconnectedness between separated phenomena that does not exist in reductionist physics. In the new holism, a revised quantum mechanics provides the mechanism by which signals can move faster than light, making possible instantaneous connections across the universe.[52]

However, the nonlocality of hidden variables, or other variations on nonlocal, causal mechanisms underlying quantum mechanics, is a nonlocality within that specific interpretation and not necessarily within quantum mechanics itself as a theory that describes the results of observations.

If the apparent empirical violation of Bell's theorem is to be construed as evidence for nonlocality in nature, which is by no means demonstrated, then that nonlocality is contained in hidden variables or other structures that play no role in quantum mechanics as it is currently practiced. Any theory of hidden variables is thus a new theory, a *subquantum* theory that must lie deeper than quantum theory.

This has not discouraged many authors from finding other mystical messages within the conventional Copenhagen interpretation of quantum mechanics. They conclude that we can never adequately describe, in scientific terms, the "irreducible whole." This obscure concept has been related to the "being-in-itself" of that master of obscurity, philosopher Martin Heidegger.

For example, in their book *The Conscious Universe*, astrophysicist Menas Kafatos and philosopher Robert Nadeau associate being-in-itself with the quantum wave function:

> If the universe were, for example, completely described by the wave function. . . . One could then conclude that Being, in its physical analogue at least, had been "revealed" in the wave function. We could then assume that any sense we have of profound unity with the cosmos or any sense of mystical oneness with the cosmos, has a direct analogue in physical reality. In other words, this experience of unity with the cosmos could be presumed to correlate with the action of the deterministic wave function which determines not only the locations of quanta on our brain but also the direction in which they are moving.[53]

The vision of the new holists is not so appealing as it may first appear. The field of cosmic mind, whether aether, wave function, or quantum potential, is completely deterministic. In whatever manifestation, holistic physics possesses the very Newtonian, mechanistic character that is so decried by New Age authors.

In the view of quantum holism, although we humans are proscribed from ever being able to predict the exact outcome of events, those events are predetermined nevertheless. In a holistic universe, everything is intimately and instantaneously connected to every event past and future, here on earth and far out in space, with no room for chance or choice.

I ask myself: Do I really want to be one with the universe, so intimately intertwined with all of existence that my individual existence is meaningless? I find I much prefer the notion that I am a temporary bit of organized matter. At least I am my own bit of matter. Every thought and action that results from the remarkable interactions of my personal bag of atoms belongs to me alone. And so these thoughts and actions carry far greater value than if they belonged to some cosmic mind that I cannot even dimly perceive.

The mystical holist trades the real, pulsating life of the outer world for what he perceives as an inner world of peace. But that peace is the peace of a prison. Science has always provided the means for breaking us free from the prisons of ignorance and superstition. I hope to convince the reader that science has not suddenly reversed its course and become yet another set of shackles for humanity to carry. On the contrary, science continues to provide the key that unlocks all of our chains so that our bodies and minds are free to roam the universe.

Notes

1. Lanza 1992, pp. 24–26. For my response, see Stenger 1993.
2. Capra 1975.
3. Capra 1982.
4. Ferguson 1980, p. 145.
5. Zukav 1979, p. 314.
6. Lederman 1993.
7. GUT predicted that the average proton lifetime was of the order of 10^{32} years. The current experimental limit is greater than 10^{33} years.
8. Technically, only one particular GUT was falsified, so all possible GUTs are not ruled out. But when the simplest model failed, theorists started looking elsewhere.
9. Chopra 1989.
10. Chopra 1993.
11. For a critical review of Chopra's ideas, see Butler 1992, pp. 110–18.
12. *Newsweek,* March 23, 1994, p. 81.
13. Patricia Newton, talk before the 98th Annual Meeting of the National Medical Association, San Antonio, Texas, 1993. Quotation provided by Bernard Ortiz de Montellano (private communication).

14. Palmer 1986.

15. A focused beam will fall off less rapidly, but still will be expected to decrease in intensity as one moves away from the source. The more focused, the lower the decrease, but also the less likely that the beam will intercept a receiver. For a discussion of Einstein's view on ESP, see Gardner 1981, pp. 151–57.

16. Rhine 1954. For a more recent attempt, see Dunne 1992.

17. See *Science News* 106, July 20, 1974, p. 8, and *New Scientist,* October 17, 1974, pp. 170–85. See also Gardner 1981, p. 94, for his recounting of the events. In a private communication with me, Sarfatti has confirmed the accuracy of these reports.

18. Gardner 1981, p. 94.

19. See Randi 1973, 1985 and Gardner 1981, note 7, p. 104.

20. Stenger 1990. Uri Geller filed three lawsuits against me in 1992 over this book. All were settled in my favor.

21. Druckman 1987.

22. See, for example, Palmer 1989, p. 10.

23. Bem 1994.

24. Blackmore 1994, Hyman 1994.

25. Druckman 1987, p. 185.

26. Schmidt 1969, 1992, 1993.

27. Jahn 1986, 1987, 1991, 1992; Dunne 1992.

28. Stenger 1990, pp. 180–84. Other critiques can be found in Hansel 1989, Druckman 1987, pp. 184–90, and Alcock 1990.

29. Alcock 1990, pp. 6, 107.

30. Dunne 1992.

31. For a nice introduction to Bayesian methods of inference, and its connection to Occam's razor, see Jefferys 1992a.

32. Dobyns 1992.

33. Jefferys 1992b.

34. Jahn 1986, Stapp 1994.

35. Distant viewing, or remote viewing, is a formerly fashionable version of ESP. Like all other previous ESP fashions, it has been thoroughly debunked.

36. Goswami 1993, p. 136.

37. See, for example, Oteri 1975. The Koestler quotation can be found on p. 268. See also Puharich 1979.

38. Jahn 1981, 1986; Schmidt 1969, 1993.

39. Stapp 1994.

40. For a review of the early history of quantum theory and ESP, see Gardner 1981. This article originally appeared in the *New York Review of Books,* May 17, 1979. The reprint also contains letters reacting to the review and Gardner's response to them. See also Stenger 1990, pp. 246–50 for my review of Evan Harris Walker's quantum theory of psychokinesis given in Puharich 1979.

41. For a history of the idea of the aether, see Cushing 1989, pp. 272–311.

42. For further discussion and references, see Stenger 1990, chapter 7.

43. Lamont 1990.

44. Lodge 1929, p. 14.

45. Lodge 1920.

46. Lodge 1914, p. 21.
47. Zohar 1990, p. 225.
48. Einstein 1935.
49. Bell 1964.
50. Bohm 1952.
51. Aspect 1982.
52. See, for example, Talbot 1991.
53. Kafatos 1990, p. 124.

2

Copenhagen, Complementarity, and Chance

There is no quantum world. There is only an abstract physical description. It is wrong to think that the task of physics is to find out how nature is. Physics concerns what we can say about nature.

—Niels Bohr[1]

Lighting the Path

In the late nineteenth century, James Clerk Maxwell wrote down four equations that could be used to calculate the electric and magnetic fields produced by any assemblage of electric charges and currents. Using a fifth equation, devised by Hendrik Lorentz, the electric and magnetic forces on charges and currents could be calculated and then, with Newton's laws, the motions of these charges and currents could be predicted.

Eventually it was recognized that this array of equations provided a complete theoretical framework for the computation, at least in principle, of all phenomena associated with electricity and magnetism. Since Newton's theory of gravity was already a highly successful part of established physics, the physical forces directly involved in all of human experience were thus encompassed by what is now called classical physics. Everything that was then known about physical phenomena at their most fundamental level was thereby "explained."

Even today, after the discovery of nuclear forces, Maxwell's theoretical framework remains intact as the basis for understanding electricity and magnetism. Quantum effects must be incorporated at the atomic and subatomic level, and for certain very special macroscopic phenomena. Otherwise, Maxwell's equations have been applied successfully for over a century, from the development of electronic communications and computers to exploring the neural network of the human brain. Although new developments like quantum mechanics have revised our ideas about the fundamental nature of electromagnetism, it would be a mistake to conclude that the techniques of classical electromagnetism developed in the last century have no further use.

Maxwell's equations, however, did more than just provide an accurate description of previously observed phenomena and stimulate new technology. They also unified electricity and magnetism and predicted the existence of electromagnetic waves. These waves moved at exactly the speed of light, leading to the conclusion that light is a form of electromagnetic wave.

Further, the prediction was made that a previously unimagined, invisible component of the universe existed beyond the range of human sensory experience. When radio waves were observed in the laboratory, this prediction was triumphantly confirmed. The discovery of radio waves provides a profound example of how the mathematical description of certain restricted classes of observations can be used to deduce the existence of new, unexpected phenomena. It also demonstrates how seeking fundamental knowledge about the physical universe can lead to unexpected benefit to humankind. Today's world of computers and instant telecommunications is the direct result of the application of Maxwell's equations, as they evolved with the development of quantum mechanics.

But what was the nature of Maxwell's electromagnetic waves? By analogy with sound waves, which occur as vibrations of air and other material media, electromagnetic waves were assumed to be vibrations of some medium. However, since light reaches us from distant stars, electromagnetic waves cannot be "airwaves," the term once used to describe radio transmissions. Sounds, not radio or light, are airwaves. An invisible, frictionless aether, visualized as pervading all of space, was imagined as the medium whose vibration produced electromagnetic waves. As discussed in chapter 1, the notion of an aether goes back to Aristotle and formed an important part of Newton's thinking. So it was natural to ascribe the new phenomenon to "aether waves."

Nevertheless, this seemingly reasonable picture had a serious inconsistency. Maxwell's equations implied that light travelled at a fixed, "absolute" speed in any given medium.[2] This violated the longstanding principle of

Galilean relativity, which says that all speeds are relative. This principle was established by Galileo to explain why we do not notice the earth's motion around the sun. According to Galilean relativity, if light waves are the vibrations of the aether, then movement of the light source or detector through the aether should change the apparent velocity of the aether waves of light.[3]

In a series of experiments starting in 1881, Albert Michelson and Edward Morley attempted to detect the motion of the earth through the aether by measuring differences in the speed of light as the earth circles in its orbit around the sun. They failed to find any differences, a result consistent with a fixed speed of light and Maxwell's equations but inconsistent with Galilean relativity.

In 1905, Albert Einstein discovered that in order to maintain a fixed speed of light, he had to alter our preconceptions about some of the most basic properties of the universe. He showed that the time intervals measured by two clocks will be different when those clocks are moving at different speeds with respect to the observer. Following on earlier ideas of Lorentz and George Fitzgerald, Einstein also concluded that the length of an object will be observed to be different when the object is viewed at rest and when it is viewed as moving.

By similar reasoning, Einstein proved that a measurement of the mass and energy of a body will also depend on its relative motion. And so, familiar quantities that were previously assumed to be fixed properties of material bodies were shown to depend on the motion of the observer. On the other hand, the speed of light was fixed.

Furthermore, Einstein demanded that the speed of light be an absolute limit to the speed of material bodies (but see the discussion of tachyons later in this book). As a body approaches the speed of light, the energy required to further accelerate it approaches infinity.[4]

The theory of relativity showed that mass and energy could be converted into one another, as expressed by the equation $E = mc^2$. At rest, a particle of **rest mass** m_o possesses a **rest energy** m_oc^2. And so, the previous picture of a physical universe composed of mass and energy was replaced by a simpler, one-component universe of mass/energy. According to relativity, one basic type of material stuff exists, whether you wish to call it mass or energy.[5]

Toward the Quantum

These developments in electromagnetism and relativity laid the foundation for the quantum revolution that soon followed. Maxwell's equations provided such a powerful and successful theoretical structure that any observed deviation from their predictions could be reasonably labeled an anomaly. Such anomalies were found, and these pointed toward a new theory that went beyond classical physics.

When we come across anomalies that cannot be explained by conventional ideas no matter how hard we try, we are forced out of our previous line of thinking. This can often lead to major changes in the theoretical framework within which we cast our ideas. Around the turn of the century, empirical anomalies that were inexplicable in Maxwellian electromagnetic theory were observed in the radiation from heated solids and gases and in the photoelectric phenomenon in which light produces an electric current.

The photoelectric effect provided a particularly dramatic example of the failure of classical wave theory. Light hitting a metallic plate kicks electrons out of the plate to produce a measurable electric current. The current is found to be largely independent of the intensity of the light, occurring only when the frequency of the light exceeded a certain minimum. It is as if twenty-foot ocean waves are unable to dislodge a pebble from a beach when the wave crests arrive once a minute, but two-foot crests arriving every ten seconds immediately carry the pebble away.

The quantum theory was launched in the first year of the twentieth century with Max Planck's explanation of the anomalous radiative behavior of light. In order to account for the spectrum of light radiated from a hot body, Planck introduced the idea that light occurs in discrete bursts, called **quanta**, whose energies are proportional to the frequencies of the light light.

This idea was utilized by Einstein in 1905, the same year he published his theory of relativity, to explain the photoelectric effect. Einstein proposed that light was composed of particles, later dubbed **photons**, and that the energy E of a photon was related to the frequency of the light ν by Planck's formula $E = h\nu$, where h is Planck's constant.

In Einstein's model, the photoelectric effect results when an incoming photon of sufficiently high energy strikes a single electron, knocking it out of the metal. Higher-frequency light, such as ultraviolet light, contains higher-energy photons and so may be able to produce a current where visible light cannot. The particle-collision picture also explained why no time delay is observed in the photoelectric effect, as might be expected if the electron had to wait to absorb enough energy by some continuous process. The electron is struck once, and comes right out.

In 1911, Ernest Rutherford proposed a model of the atom patterned after the solar system, with electrons orbiting around a central nucleus the way planets orbit the sun. However, Rutherford's model violated the laws of classical physics. Unlike planets, electrons have electric charge and should radiate energy as they circle a nucleus.

Two years after Rutherford's proposal, Niels Bohr proposed a new physics principle, asserting that only certain atomic orbits are allowed, namely, those for which the angular momentum L of the electron is an integer multiple of Planck's constant h divided by 2π. This quantity was originally called the **quantum of action**. In the conventional notation, the quantum of action is written as $\hbar = h/2\pi$ and appears frequently in quantum mechanical equations. Using the principle of quantized action, Bohr was able to calculate the discrete frequencies of light observed to be emitted by the hydrogen atom.

The emitted energy $E = h\nu$ (by Planck's formula) corresponds to the energy difference between the two levels. Bohr did not believe in photons at first, offering a rather more complicated explanation. However, he soon came to realize that photons, emitted or absorbed during the transitions between energy levels in an atom, provided a practical description of radiation processes.

As mentioned on chapter 1, Sir Oliver Lodge and other physicists of his time were highly skeptical of the new quantum ideas. Lodge particularly objected to the replacement of his beloved "continuity" with the discrete, localized picture of matter and energy. Quanta were truly weird, and their inventors could not provide a simple visual model that could help people understand the new physics in familiar terms. Despite this, the new quantum physics could be used to calculate previously unexplained measurements and predict others. It had to have something to do with reality.

Quantum theory could not be ignored because it described so many empirical facts that could otherwise *not* be accounted for. Planck's theory fit the spectrum of radiation from so-called "black bodies" at different temperatures and could be used to determine the constant h. Einstein's photon theory, using the same value of Planck's constant h, fit the photoelectric data. Bohr calculated the frequencies of light emitted by hydrogen, again with the identical value of h. By contrast, Maxwell's equations and other existing principles of classical physics could not even explain what was going on qualitatively.

Undoubtedly, light behaves like waves—bending around corners and spreading out through small apertures. As a simple illustration, hold your thumb two feet in front of your eye and line up its edge with a bright light. You will not see the expected sharp outline of your thumb, but a fuzzy outline with some light appearing as if coming from your thumb. The light has bent around the edge of your thumb before reaching your eye. This effect is called **diffraction** and is a common wave phenomenon.

But light also behaves like particles, exciting single, localized cells in very sensitive light sensors. Countless experiments have since repeatedly confirmed the discrete, particle-like nature of electromagnetic radiation. On the other hand, countless experiments have also continued to confirm the wave nature of light as derived from Maxwell's equations.

Wave-Particle Unity

In 1924, Louis de Broglie suggested that electrons and other objects conventionally regarded as particles should also exhibit wave properties. De Broglie noted that the wavelength (λ) associated with a photon of momentum p is given by $\lambda = h/p$. He then made the bold assumption that the same formula applies for all particles. That is, if a particle has a momentum p, it will behave, under appropriate experimental conditions, as a wave with **de Broglie wavelength** $\lambda = h/p$.

> The momentum of a body is given by $p = mv$, where m is the inertial (relativistic) mass of the body and v is its velocity. For a photon, $v = c$ (the speed of light). By Einstein's equation, its energy $E = mc^2$.[6] So, photons have a momentum related to their energy given by $p = E/c$. According to Planck's equation, photon energy is proportional to light frequency: $E = h\nu$. If we substitute this in the previous equation we get $p = h\nu/c$. Now, from wave theory, $\lambda = c/\nu$ is the wavelength that corresponds to a frequency ν. Thus it follows that $\lambda = h/p$.

The wavelengths of familiar bodies in our everyday experience are too small to produce noticeable diffraction effects. However, this is not the case for electrons, such as those in the accelerated beam that produce the image on a television or computer screen. De Broglie's hypothesis was quickly confirmed by experiments with electrons. Wavelike behavior has since been observed for other particles, nuclei, and atoms as well. And so, all bodies, not just photons, look like waves when detectors are designed to measure wave properties, and look like particles when detectors are designed to measure particle properties. Being neither particles nor waves, they are sometimes called **wavicles**.

This so-called **wave-particle duality** has led many people to think that quantum physics has demonstrated the validity of mind-brain duality, or soul-body duality. For example, the philosopher J. B. S. Haldane remarked in 1934, "If mind is to be regarded as expressive of the wholeness of the body, or even the brain, it should probably be thought of as a resonance phenomenon, in fact part of the wave-like aspect of things."[7] More recently, physicist Brian Josephson and his collaborators have argued that the "field of mind" is a life process that underlies all natural phenomena.[8] But that connection is remote, to say the least. Are we supposed to think of the brain/body as a particle, and the mind/soul as a wave?

However, we can see how wave-particle duality leads people to infer that quantum mechanics requires a drastic revision in our thinking about nature. If the choice of whether an entity is a particle or a wave depends on the apparatus being used, then how can these concepts represent anything that is objectively real, that is, possesses a reality independent of the observer?

A simple answer is that neither waves nor particles are "real." If observations do not support the notion of pure particlelike objects, or pure wavelike objects, but rather waviclelike objects, then it is the waviclelike objects that are objectively real and the others that are not.

Imagine the entity being emitted from a source, and then well after it has left the source we decide whether we will look at it with a particle meter or a wave meter. It would seem that the nature of the object, particle or wave, is being determined by a conscious act. But, this conclusion is based on the assumption that the particle and wave concepts represent two distinct forms of objective reality. If, instead, they represent a single objective reality, then no difficulty arises. The particle and wave meters simply measure two different properties of the object, like a ruler measuring an object's length and a thermometer measuring its temperature. However, as we will see, an important complication arises when two measurements cannot be simultaneously performed because they interfere with one another.

I agree with a number of commentators who have noted that the use of the term wave-particle duality in quantum mechanics is a misnomer. More accurate terms might be wave-particle *unity* or wave-particle *synthesis*. The discovery that waves and particles were different aspects of a single reality was a unifying principle, not a divisive one. Instead of two separate classes of physical entities, waves and particles, we find that only one class exists, which, under appropriate conditions, seems to exhibit one property or the other.

Many of the great principles of science have the feature of bringing under one umbrella concepts that previously were held to be different: Newton unified terrestrial and extraterrestrial laws of motion and gravity in physics. Darwin unified biology. John Dalton unified chemistry. Faraday and Maxwell unified electricity, magnetism, and optics. Einstein unified space and time, and matter and energy. De Broglie unified waves and particles. More recently, electromagnetism has been unified with the weak nuclear force.

One often hears the lament that as we dig deeper the world becomes more mysterious, more difficult to understand. But this observation does not fairly represent the history of science from the time of Thales, which has been largely a progression of mystery being replaced by understanding. The world appears mysterious when we try to frame our understanding in terms of preconceived ideas and prejudices. To achieve understanding, we must cast aside our preconceptions to make room for the next set of more powerful, often simpler, generalizing concepts.

De Broglie's electron waves provided a crude, intuitive explanation for Bohr's atom, with its mysterious, discrete electron orbits. The electron orbits can be pictured as analogous to the standing waves set up when plucking a guitar string. Allowed orbits in Bohr's model correspond precisely to those electron waves that fit along the orbit's circumference. The energy levels of an atom may be thought of as the fundamental and harmonic notes at which an atom resonates.

> Imagine an electron orbit of radius r. A standing wave will be set up around the circumference when $n\lambda = 2\pi r$ or when the angular momentum $L = rp = n\hbar$, where $\hbar = h/2\pi$ and n is a positive integer. Recall $p = h/\lambda$.

Wave Packets

In the early years of quantum theory, de Broglie and Erwin Schrödinger championed the idea that a "real" field Ψ can be associated with a particle. This was somewhat analogous to the way the electric and magnetic fields are, presumably, "real" fields associated with photons. Exactly what is meant by "real" is a subject of considerable philosophical debate that we will be returning to again and again in this book. For the present purpose, let us just see how the idea works in practice and not worry about labels. In particular, how can a localized particle be represented by a wave?

The combining ("superposition") of waves of different amplitudes,

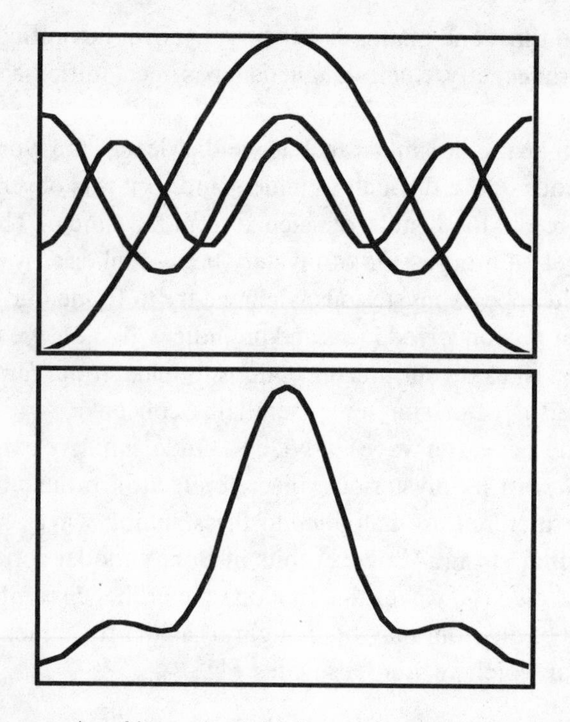

Fig. 2.1. An example of how sinusoidal waves of different frequencies can interfere to produce a localized disturbance. In the top figure, three sinusoidal waves of frequencies f, 2f, and 3f are shown. Adding these together gives the more localized wave shown in the bottom figure. This is an example of a wave packet. As we add higher frequencies, we will get an increasingly sharper peak.

phases, and frequencies can produce a very localized disturbance called a **wave packet** (see figure 2.1). For example, short-duration electrical pulses are composed of many different high-frequency components. Fast electronic circuits must be carefully designed so that these frequencies are all transmitted from point to point, to maintain the pulse shape.

De Broglie and Schrödinger imagined that the localized disturbance we call a particle is in fact a wave packet. However, this idea, reasonable as it seems, was discarded for a number of technical reasons. I will mention just one: the problem of dispersion, which is familiar in optics and acoustics.

Optical or acoustical pulses, propagating through some medium, can be represented by wave packets. Such pulses spread out as they propagate because of the different speeds of the various components that make up the packet. For example, nearby thunder will produce a sharp clap; from far away one hears only a long rumble. Wave packets, with the important exception of photons in a vacuum, would also exhibit dispersion. Thus, a localized, microscopic particle moving freely through space will rapidly spread out and become unlocalized.

However, this loss of localization is not observed in experiments that detect the motion of individual particles. For example, the track of an electron in a cloud chamber maintains its narrow width as the electron moves through the chamber fluid. Since individual particles did not appear to exhibit the dispersion associated with wave packets, early critics concluded that the wave-particle duality could not be resolved with a purely wave-based description in which individual particles are represented by highly localized wave packets. However, this criticism was perhaps unfair since a particle moving through a cloud chamber is continually interacting with the chamber fluid and is not analogous to a particle moving in a vacuum.

The Uncertainty Principle

In 1927, Werner Heisenberg began to think about what constituted the act of measurement itself. First, he realized that in order to observe something, we must interact with it. For us to see an object, light must bounce off it into our eyes. But a light beam is a stream of photons, and photons carry momentum that is partially or wholly transferred to objects they collide with. A body, when hit by another body, recoils under the impact. For most large objects, a negligible recoil occurs from the impact of a light beam. However, for subatomic particles such as electrons, the transferred momentum can produce an appreciable change in motion.

Now you might imagine using photons of very low momentum to produce a negligible recoil. However, low momentum is associated with large wavelength, from de Broglie's equation $\lambda = h/p$, and limitations are then imposed by diffraction. For example, we can't see an atom with visible light because the wavelength of visible light is too large. We must use a shorter wavelength, but the implied higher momentum leaves us in a "can't-win" situation: it is impossible, in practice and in principle, to measure the position of a particle to arbitrarily high precision without increasingly disturbing the particle's motion. This is the **Heisenberg uncertainty principle**: The product of the uncertainty Δp of a body's momentum and uncertainty Δx of a body's position must be greater than or equal to a constant: $\Delta p \Delta x \geq \hbar/2$, where $\hbar$ is the quantum of action, Planck's constant h divided by 2π. In this book, where high precision is not needed, I will use the more familiar, approximate form $\Delta p \Delta x \geq h$.

The Heisenberg uncertainty principle had a profound effect on physics, seemingly overturning its most basic precept, the predictability of the motion of a body when the forces on that body are known. In classical, Newtonian mechanics, the motion of a body is predicted by solving its equation of motion subject to the initial values of the body's position and velocity. If the world is composed only of material bodies obeying Newtonian mechanics, we are forced to conclude that everything that happens in that world is predetermined by the laws of mechanics.

In the usual metaphor, the Newtonian universe is like a vast machine, a clockwork mechanism, with no room for chance. Material bodies behave the way they must behave under the strict enforcement of the laws of physics. Everything that happens was set down to happen in exactly that way at the Creation. This mechanistic, deterministic model of the universe is called the *Newtonian world machine.*

Heisenberg's uncertainty principle, however, implies that the position and velocity of a body cannot be measured simultaneously with arbitrary precision. Since each quantity must be known at some time in order to predict a body's motion after that time, that motion cannot be exactly calculated.

The universe being composed of moving and interacting particles, quantum mechanics would seem to say that it cannot be a Newtonian machine. But this depends on whether the uncertainty principle implies that the positions and momenta of particles are inherently uncertain, or just that we cannot measure them with certainty. As we shall see, this question forms the major disagreement between conventional, nondeterministic quantum mechanics and the deterministic subquantum theories of so-called hidden variables.

Formulation

By the early 1930s, a precise mathematical structure for the theory of quantum mechanics was developed essentially to the point at which it is used today. The process began in 1925–26 when Heisenberg invented a rather abstract formulation that, with the help of Max Born and Pascual Jordan, was shortly reformulated using matrix algebra. This **matrix mechanics** made the theory only a little less abstract, but provided the tools needed to enable physicists to make useful calculations.

Almost simultaneously, Schrödinger invented **wave mechanics**, which relied on the more familiar methods of partial differential equations that were widely used in classical mechanics and electrodynamics. Maxwell's equations, for example, can be expressed as partial differential equations, as can the wave equation that governs the propagation of electromagnetic or acoustical waves.

Arguing by analogy, Schrödinger introduced a field called the **wave function** to describe the wave properties of particles. This wave function evolves with time according to the **Schrödinger equation**.

The Schrödinger equation is the basic dynamical equation in quantum mechanics. If H is the Hamiltonian operator of the system (energy operator), then the wave function $\Psi(t)$ evolves with time according to

$$i\hbar\partial\Psi/\partial t = H\Psi \qquad (2.1)$$

This is usually called the time-dependent Schrödinger equation in the textbooks, to distinguish it from the time-independent Schrödinger equation used to solve standard problems such as the hydrogen atom. The latter is a special case of the former, however, and only valid nonrelativistically in the usual form you see it written. Equation 2.1 is more fundamental and holds true relativistically. The Schrödinger equation, as used in this book, will refer to equation 2.1.

Matrix mechanics and wave mechanics were shown by Schrödinger to be equivalent in their common applications, but his wave mechanics became more widely applied because of the greater familiarity of its mathematical techniques and its more intuitive nature. Wave mechanics is easier to teach, with the result that it is the only quantum mechanics that most scientists other then physicists ever learn, if indeed they learn any at all. Matrix mechanics, on the other hand, proved to be more powerful, especially when dealing with particle spin and other phenomena that lack classical analogues. Schrödinger's method did not encompass spin.

Although Bohr had already calculated the energy levels of the hydrogen atom with his original ad hoc theory of quantized action, Schrödinger's equation and Heisenberg's matrices determined these from deeper principles. They also enabled further progress in understanding the structure of atoms and contained a much greater range of predictive power.

Very importantly, the new quantum mechanics provided the basic framework within which the periodic table of the chemical elements, the foundation of chemistry, could be explained. The quantum numbers so familiar to chemistry students arose out of the mathematical solution of Schrödinger's equation for the atom. Bohr's mysterious quantization rules were shown to follow from mathematical boundary conditions on the wave function.

In 1932, John von Neumann placed quantum mechanics on a firm logical and mathematical foundation, writing down five axioms and then showing how the calculational rules that had been previously developed in a more intuitive way followed by deduction. The Heisenberg uncertainty principle also could be derived from the basic axioms.

Although the methods developed by von Neumann and Paul Dirac provided the most general and powerful formulation of quantum mechanics, the philosophical issues we are concerned with here can be discussed in terms of the more familiar Schrödinger language. The wave function Ψ used to characterize the wave properties of a particle arises as a solution of the Schrödinger equation. In quantum mechanics, Ψ is said to represent the "state" of a quantum system, but exactly what that means is debated to the present day.

Both Heisenberg's and Schrödinger's original formulations of quantum mechanics were nonrelativistic. That is, they dealt only with particle speeds much less than light. This specifically excluded photons, but also only approximately represented electrons in atoms, which can move at an appreciable fraction of the speed of light, where Einstein's relativity must be applied.

In 1928, Dirac showed how to extend the scheme to relativistic electrons. In the process, he gave an explanation for the phenomenon of electron spin that was absent in Schrödinger's nonrelativistic equation. Dirac found it necessary to include electrons with negative energy that were interpreted as antielectrons, particles identical to electrons except with opposite electric charge. In 1932, the previously unimagined antimatter component of the universe was confirmed by the observation of antielectrons (**positrons**) in cosmic rays. Eventually, antiprotons, antineutrons and many other antiparticles would be discovered. As was the case with electromagnetic waves, another component of the universe was uncovered first in mathematical equations and only later confirmed by experiment.

The Statistical Postulate

In 1926, Max Born proposed what was to become a primary postulate of quantum mechanics in the von Neumann scheme. According to this postulate, the wave function is used to compute the probability P for a particle to be found in a particular state. This probability was to be proportional to $|\Psi|^2$, the square of the magnitude of the wave function Ψ (which is a complex number in general).

Born originally thought in terms of specific energy or angular momentum states, but his postulate was extended by Wolfgang Pauli to include the probability for finding a particle at a particular position.

> Pauli proposed that the probability P for finding a particle in an infinitesimal volume element ΔV located in a specific region of space is equal to the square of the magnitude of the wave function Ψ computed at that point multiplied by ΔV: $P = |\Psi|^2 \Delta V$. Since we can measure volume in any units we wish, such as cubic femtometers (1 fm = 10^{-15} meter, about the size of a proton), no loss of generality occurs if we assume a unit volume, $\Delta V = 1$, and simply write $P = |\Psi|^2$ and understand it to mean probability per unit volume, that is, **probability density**.

The association of the wave function with probability was not as bizarre as it may seem. In classical electromagnetic theory, if **E** is the electric field associated with a beam of light, then $|\mathbf{E}|^2$ is a measure of the intensity of a beam of light. Similarly, the intensity of a beam of particles hitting a screen is proportional to $|\Psi|^2$ with the probability hypothesis.

The role of statistics in quantum mechanics was supported by Einstein's own calculation of the probabilities for atomic transitions. Still, the uncertain nature of its predictions was one of its aspects that Einstein found unsatisfying about quantum mechanics. Einstein is well known for having said, "God does not play dice." However, as we will see, Einstein did not object so much to the statistical form of the existing formulation of quantum mechanics as to the notion that statistics was the final word. He found it hard to accept that no underlying causal laws determined the behavior of individual quantum particles at the most fundamental level.

Actually, statistics enters quantum mechanics only in an indirect way. The Schrödinger equation predicts the exact value of the wave function Ψ at future times given its value at some initial time. Probability enters with the Born postulate (or the *projection postulate* in the von Neumann formalism) when the time comes to make a prediction on the expected value of some measurement.

Thus quantum mechanics is often said to be "deterministic" in that its basic equation, the Schrödinger equation, precisely determines the time evolution of the wave function. However, it is *indeterministic* in the sense that knowledge of the wave function is not always sufficient to predict the outcome of a measurement. This is only possible when, as happens sometimes but not always, the probability for a given outcome is zero or unity.

Since we are doing science here, and science is concerned with the results of measurements, we should restrict the use of the term "deterministic" to those theories, like classical mechanics, that predict the outcomes of measurements with certainty.[9] Since in conventional quantum mechanics the wave function is not a measurable quantity, conventional quantum mechanics is indeterministic. Alternate theories in which the wave function becomes measurable, at least in principle, are deterministic.

By the probability postulate, the wave function allows for the prediction of the average motion of a system of particles. Quantum fluctuations can cause individual particles to deviate from the average. These fluctuations become fractionally smaller as the number of particles in the system increases. By the time we reach the macroscopic scale the fluctuations are usually negligible. In that limit, quantum mechanics leads to the same equations of motion as conventional Newtonian mechanics.

Although some macroscopic systems exhibit quantum effects, for most practical purposes macroscopic phenomena retain, even with quantum mechanics, the same predictable behavior as in classical mechanics. If quantum phenomena should turn out to be as bizarre as dreamed by the most fervent quantum mystic, they may still have no profound effects on the bodies and brains of macroscopic human beings. It is important to keep this in mind. Strange behavior at the atomic and subatomic scale does not necessarily translate into strange behavior at the everyday scale.

No controversy exists over the fact that conventional quantum mechanics cannot, in all cases, predict the precise behavior of individual quantum systems. The question that continues to be debated is whether this is the final story, or whether some deterministic theory going beyond quantum mechanics and operating on the level of individual quantum systems may someday be found.

Complementarity

In the philosophical outlook of Bohr and Heisenberg, concepts such as position can only be defined in terms of the procedure that one carries out to measure position. Otherwise, the notion of position is meaningless. As Heisenberg liked to remind Einstein, this point had been implemented in his own relativity, where space, time, and energy were found to depend on the reference frame of the apparatus that was used to measure these quantities. This seemed to imply that at least these concepts had only operational meaning and did not necessarily correspond to any more fundamental "reality." According to Heisenberg, Einstein remarked when confronted with his own early views, "Perhaps I did use such a philosophy earlier, and even wrote it, but it is nonsense all the same."[10]

Heisenberg had joined Bohr in Copenhagen in 1926. At first, the two disagreed vehemently, not over the uncertainty principle per se, but over its meaning. Bohr had been thinking along related lines on the meaning of the wave-particle duality. He argued that these were alternate, incompatible descriptions of nature. By contrast, Heisenberg essentially proposed that the two descriptions should be used together, as they are classically. In Heisenberg's picture, quantities such as position and momentum or energy and time must be assigned uncertain values.

Heisenberg recalled "breaking out into tears" under the pressure from Bohr, but eventually, with the help of Pauli, they reached agreement. Pauli convinced the combatants that their dispute was only over the precedence of concepts—whether the measurement or the definition of a concept comes first. Heisenberg added a postscript to his paper and he and Bohr were reconciled. However, they did not continue to agree on every aspect of quantum mechanics.

Bohr shortly thereafter elucidated his principle of **complementarity**, in which apparently conflicting classical concepts such as waves and particles are regarded as equally valid, complementary but incompatible descriptions of the same reality. In most books on quantum mechanics, complementarity is treated as an outgrowth of Heisenberg's uncertainty principle. This was not precisely the case. Nevertheless, we can safely assume that the debate with Heisenberg stimulated Bohr into attempting to formulate what was, until then, a vague idea rattling around in his head.

Still, Bohr never placed the principle of complementarity on a firm, logical foundation. Philosopher Max Jammer has given the following definition, which I have paraphrased slightly (but insignificantly):[11] A theory is complementary if it contains at least two descriptions of its substance-matter, neither of which taken alone accounts exhaustively for all phenomena within the theory's range of applicability. They are mutually exclusive in the sense that their combination into a single description would lead to logical contradiction.

Usually one reads about the complementarity of measured quantities such as momentum and position, or energy and time. For example, the momentum of a particle can be measured exactly, as long as you make no attempt to measure its position. In this case, the particle is imagined (using the old wave packet picture) as a sinusoidal wave of a single wavelength, spread throughout space. Alternatively, you can measure the position of a particle with infinite precision, if you avoid asking about its momentum. Then the particle's momentum is completely uncertain; that is, all wavelengths are equally represented in an infinitesimally narrow wave packet.[12]

The Copenhagen Interpretation

And so Bohr, Heisenberg, Born, and their colleagues developed what has for many years been called the Copenhagen interpretation of quantum mechanics. Some disagreement still exists on what it really does say, and the writings of its founders do not give a consistent picture. Bohr, however, is regarded as the interpretation's main progenitor and his writings should be consulted first.

The following statement by John G. Cramer, reproduced here complete with his references to the original work (the specific reference numbers are mine), is a good summary that I believe is consistent with what has come to be associated with the Copenhagen school in various reviews produced over the years.[13]

THE COPENHAGEN INTERPRETATION

(C1) The uncertainty principle of Heisenberg:[14] this includes wave-particle duality, the role of canonically conjugate variables, and the impossibility of simultaneously measuring pairs of such variables to arbitrary accuracy.

(C2) The statistical interpretation of Born:[15] this includes the meaning of the state vector given by the probability law ($P = \Psi\Psi^*$) and the predictivity of the formalism only for the average behavior of a group of similar events,

(C3) The complementarity concept of Bohr:[16] this includes the "wholeness" of the microscopic system and macroscopic measurement apparatus, the complementary nature of the wave-particle duality, and the character of the uncertainty principle as an intrinsic property of nature rather than a peculiarity of the measurement process.

(C4) Identification of the state vector with "knowledge" of the system by Heisenberg:[17] this includes the identification itself and the use of this concept to explain the collapse of the state vector and to eliminate simple nonlocality problems.

(C5) The positivism of Heisenberg: this includes declining to discuss "meaning" or "reality" and focussing interpretive discussions on observables.

I have already discussed most of these elements of the Copenhagen interpretation. I have also used most of the terms in the above summary, but now let me make them a bit more precise and define some of the terms not mentioned previously.

In C1 the term "canonically conjugate variables" refers to the sets of observables, such as position and momentum or energy and time, that are related in classical mechanics by equations of motion and cannot be measured simultaneously in quantum mechanics. These can be read as "complementary variables" or "incompatible variables." In the quantum formalism, they are represented by mathematical objects (operators or matrices) A and B that do *not* obey the commutative law, AB = BA. That is, they are not familiar real numbers but another type of mathematical object (differential operators and matrices each fit the bill). The results of their measurements, however, remain real numbers.

The **state vector** in C2 can be read as "wave function" for most of our purposes; we will occasionally use this more general way of referring to quantum states.

The "wholeness" of the microscopic system and macroscopic measurement apparatus mentioned in C3 refers to the assumption that the quantum description of the system being observed depends on all the details of the observation process. It also makes a distinction between the first being microscopic (or better, "quantum") and the second being macroscopic (or better, "classical"). This distinction between observer and observed is a major disadvantage of the Copenhagen view, as we will see.

The identification of the state vector, or wave function, with our "knowledge" of the system being observed is specified in C4, which also introduces the term **wave function collapse** (sometimes called **wave function reduction**), the misinterpretation of which in many naive discussions of quantum mechanics has lead to much of the metaphysical confusion we will be talking about in this book. Let me take a moment to explain it further.

The history of the notion of wave function collapse is obscure, appearing in early discussions and eventually being worked into the standard formalism by von Neumann in what is called the *projection postulate*. Wave function collapse, as it appears in current discussions of quantum mechanics, is illustrated in figure 2.2.

A source of electrons is directed toward a screen that has an opening that allows the electrons to move toward a pair of detectors A and B. The wave function that describes the electrons that pass through the aperture is a modulated sinusoidal wave (a wave packet) that is approximately as

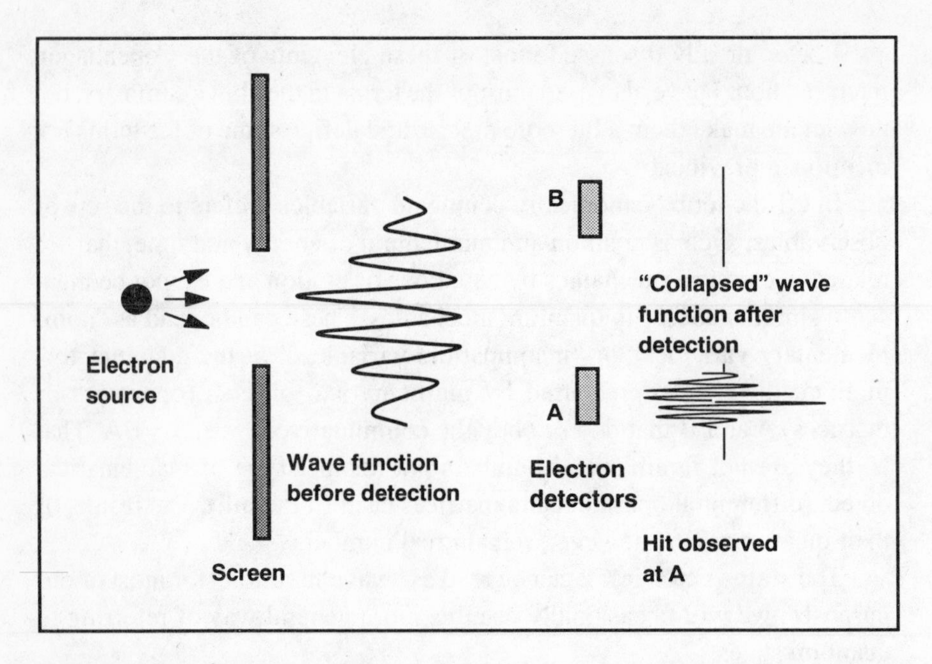

Fig 2.2. Wave function collapse. The electrons emerging from the aperture in the screen are described by a modulated sinusoidal wave function approximately as wide as the aperture. When an electron is detected at A, conventional quantum jargon says that the wave function collapses to the wave packet shown that is localized at A. The wave function at B instantaneously collapses to zero.

wide as the aperture, meaning that the probability of an electron appearing far from the aperture is small.

The detectors are capable of registering individual electrons. Assume the beam of electrons has very low intensity, so that only one passes though the aperture to the detectors in a specified time interval. Suppose detector A obtains a signal and detector B does not during this time interval. In the standard language of quantum mechanics, the wave function is said to "collapse" to the form shown behind the detectors. Again that wave function is a modulated sinusoidal wave packet, but its width is now reduced to the size of the detector and centered on detector A.

Conventional quantum mechanics provides no mechanism for wave function collapse; it does not follow from the Schrödinger equation. In the conventional view, collapse is brought about by the act of measurement. As we will find, this is the primary source of the connection that is made between quantum and consciousness.

In the next chapter I will relate how Einstein objected to the idea of wave function collapse, saying that an instantaneous, "spooky action at a distance" is being assumed to take place. However, the Copenhagen interpretation, if it is accurately represented above, avoids this problem by making it clear that the wave function is simply a mathematical object and not a physical entity.

Mathematical quantities can do whatever the equations tell them to do. As abstractions rather than concrete objects, they are not bound by any laws of physics. Still, much has been read into wave function collapse, in particular the notion that it happens as the result of a conscious act of the human mind. We will be discussing this issue in great detail.

Finishing off the list, the "positivism" specified in C5 is associated with the philosophies of David Hume and Auguste Comte, and traceable back to Epicurus. Heisenberg's positivism was probably strongly influenced by physicist Ernst Mach.

Implications of Copenhagen

Physicist Henry Pierce Stapp of the Lawrence Berkeley Laboratory has written extensively on quantum mechanics and its philosophical and metaphysical implications. Stapp's recent writings, which we will sample later, introduce many highly controversial ideas, including the association of consciousness with wave function collapse. However, Stapp's 1972 article on the Copenhagen interpretation was reviewed by both Heisenberg and Bohr's close companion Léon Rosenfeld. They generally agreed with the article's conclusions; their comments are included as appendices to Stapp's article.[18]

Stapp claims that Bohr and Heisenberg rejected the traditional presumption that nature could be understood in terms of elementary spacetime entities. In their revolutionary view, a complete description was to be given in terms of probability functions, such as the wave function.

Stapp agrees with the summary above that the wave function, in Copenhagen, is not a real field in the sense that the electric field is real in classical physics. As Stapp puts it, "In the Copenhagen interpretation the notion of an absolute wave function representing the world itself is unequivocally rejected. . . . The probabilities involved are the probabilities of specified responses in the measuring devices under specified conditions."[19]

While many writers, such as Cramer quoted above, associate the Copenhagen interpretation with philosophical positivism, Stapp prefers to label Bohr's and Heisenberg's views as *pragmatic.* Stapp uses William James's pragmatic definition of truth, "Truth is what works," to argue that quantum mechanics represents reality, as long as its ideas bring order to our physical experiences. He argues, "A scientific theory should be judged on how well it serves to extend the range of our experience and reduce it to order. It need not provide a mental or mathematical image of the world itself."[20]

A good part of quantum metaphysics rests on the belief that, during the act of observation, some incorporeal entity called the human mind "causes" the properties being observed to come into existence. This idea at least partially arises from the strong connection that is made between a property and the procedure for its measurement that is part of the Copenhagen interpretation. If we rely on Stapp's review, as seconded by Heisenberg and Rosenfeld, the founders of Copenhagen had no such preposterous notion in mind.

However, Heisenberg and Rosenfeld were looking back forty years from a modern perspective that may have clouded their memories. Jammer discusses how Copenhagen supporter Pascual Jordan, according to a 1935 essay by Edgar Zilsel, very emphatically declared that observations not only *disturb* what is to be measured, they *produce* it! Jordan is quoted as saying, "We ourselves produce the results of measurement."[21]

Jammer explains that the view in which properties such as position and momentum "are not attributes possessed by the particle in the classical sense but are the result of interactions with the measuring device or instrument of observation—became in the early 1930s the characteristic feature of the complementarity interpretation."

According to Jammer, one of the first to recognize the profound implication of this reasoning was Paul Jenson, who pointed out that it leads to the "far-reaching conclusion that, not only in microphysics but quite generally, any physical state of affairs would be the outcome only of observation; furthermore, an objective set of facts, independent of the sense organs and brains of men or manlike creatures would not exist." [22]

It is rather ironic that in the years since writing of the pragmatic nature of orthodox quantum mechanics, Stapp has become one of the chief spokesmen of the view that a strong connection exists between quantum and consciousness.

Notes

1. As quoted by Aage Peterson in French 1985.

2. Here we will be talking only about the speed of light in a vacuum. The speed of light in a medium is normally less than that in a vacuum, which accounts for the phenomenon of refraction.

3. According to Maxwell's equations, the speed of a beam of light from a flashlight in a moving car will be the same whether that speed is measured on the ground or in the

car. These two points of view, that of an observer on the ground and that of the driver of the car, are called **frames of reference.** When a physical quantity has the same measured value independent of the frame of reference in which the measurement is made, it is said to be **invariant**. Examples of such quantities, called **scalars**, include the electric charge and rest mass of a particle, and the atomic weight of a chemical element. By contrast, the speed of a bullet fired from a moving car is relative, that is, noninvariant; it depends on whether you measure the speed with respect to the car or with respect to the ground.

4. Technically, Einstein's relativity allows for tachyons, particles that always move faster than the speed of light. An additional hypothesis of causality must be introduced to rule out these objects, which have not, so far, been observed. This will be discussed in a later chapter.

5. The physics convention today is to call the basic stuff *energy* and reserve the term "mass" for the mass equivalent of the rest energy—the *rest mass.* However, in this book I will use mass to mean what it has traditionally meant in prerelativistic physics: the quantity of inertia of a body that resists changes in its motion. When important, I will make the distinction between rest mass and inertial mass explicit.

6. Photons have zero rest mass, but still possess inertial mass within the framework of relativity. See previous note.

7. Haldane 1934.

8. Conrad, Michael, D. Home, and Brian Josephson 1988. "Beyond Quantum Theory: A Realist Psycho-Biological Interpretation of Physical Reality" in Van der Merwe 1988, pp. 285–93.

9. Of course, measurement errors are inevitable. However, in a deterministic theory, no first principle prevents them from being arbitrarily small.

10. Heisenberg 1958.

11 Jammer 1974, p. 104.

12. Here the words "infinite" and "infinitesimal" should not be interpreted with mathematical literality, but as physicists do, namely as "very large" or "very small" compared to the other relevant quantities involved.

13. Cramer 1986.

14. Heisenberg 1927, translated in Wheeler 1983.

15. Born 1926.

16. Bohr 1928.

17. According to Cramer, the knowledge element of Copenhagen was not explicit in the early papers of Bohr and Heisenberg but came out during general discussions at the 1927 Solvay conference as a response to Einstein's objections. In Heisenberg's later writings he makes this position clear. See, for example, Heisenberg 1958a.

18. Stapp 1972.

19. Stapp 1972, p. 1102.

20. Stapp 1972, p. 1104.

21. Jammer 1974, p. 161 and references therein.

22. Jammer 1974, p. 162.

3
Paradox

> If one is of the opinion that a theory of the structure of quantum mechanics is something final for physics he has either to renounce the space-time localization of the real or to replace the idea of a real state of affairs by the notion of probabilities for the result of all conceivable measurements.
>
> —Albert Einstein[1]

The Great Debate

Starting in 1927, Einstein and Bohr engaged in a historic running debate, conducted within a framework of great mutual respect, on the nature of quantum mechanics. Bohr's personal chronicle of the contest can be found in a volume of essays entitled *Albert Einstein: Philosopher-Scientist,* published in 1949.[2] Even after Einstein's death in 1955, the debate continued, with Bohr arguing with Einstein in his mind, asking himself what Einstein would have said on a particularly complex question.

The result of the conflict between these two great scientists was a far deeper understanding of the issues than would likely have happened had either deferred to the other's immense authority. That is not to say that the discussion was free of the occasional sarcastic remark that enlivens scientific discourse. I have already quoted, as have a thousand authors before me, Einstein's famous statement that, "God does not play dice." In one version that particularly applies to the theme of this book, Einstein is reported to have said, "It seems hard to look at God's cards. But I cannot for a moment believe that He plays dice and makes use of 'telepathic' means (as the current quantum theory alleges He does)."[3]

In another place, Einstein says, "The Heisenberg-Bohr tranquilizing philosophy—or religion?—is so delicately contrived that, for the time being, it provides a gentle pillow for the true believer from which he cannot be easily aroused. So let him lie there."[4]

Einstein had, in fact, done much to create the quantum mechanics he so doggedly challenged. His photon theory of light was the first example of wave-particle unity. But when Philipp Franck told Einstein that the Bohr-Heisenberg picture was "invented by you in 1905" Einstein replied, "A good joke should not be repeated too often."[5]

The fifth Solvay Institute conference in Brussels, held in October 1927, must have been the experience of a lifetime for those ordinary scientists fortunate enough to have been there. In attendance were most of the founders of quantum mechanics: Planck, Einstein, Bohr, de Broglie, Heisenberg, Schrödinger, Pauli, Born, and Dirac. Also present were other recognizable figures from physics history: Bragg, Brillouin, Lorentz, Compton, Debye, Ehrenfest, and Kramers among others.

At Solvay, de Broglie made a proposal that the wave properties of particles can be understood by viewing the wave function as a kind of **pilot wave** that guides the particle along its path. This idea received little attention except for strong criticism by Pauli. Einstein reacted favorably, but with scant community support de Broglie did not pursue pilot waves further. As a result, they lay dormant until resurrected by David Bohm, in a different form, in 1952. We will return to de Broglie's pilot waves when we discuss hidden variables theories of quantum mechanics.

The 1927 Solvay conference made history, not so much because of the scheduled talks but rather as the launching pad for the Bohr-Einstein debate. As Jammer explains it, the conference placed the issue of the meaning of "the new quantum theory" at the forefront of foundational research, where it has remained ever since. Bohr maintained that the existing quantum mechanical description of microphysical phenomena exhausted all possibilities of accounting for observable phenomena, while Einstein suggested that it was incomplete as it stood and could be carried further.[6]

Einstein did not give a prepared presentation at Solvay and said little during the formal sessions. But in the general debate at the end of the meeting, and in informal discussions around the dining table at the hotel, he expressed his concern that quantum mechanics seemed to have abandoned the causal description of events in space and time.

Each morning Einstein came down to breakfast with a new *gedanken-experiment* (thought experiment) that he said contradicted quantum mechanics. According to Otto Stern, Heisenberg and Pauli would usually brush it off. But Bohr "reflected on it with care and in the evening, at dinner, we were all together and he cleared up the matter in detail."[7]

During the final session at Solvay, reacting to comments by Born, Einstein stood up to question the notion of collapse of the wave function that was being formulated at that time and that dogs quantum mechanics to this day. Recall the discussion in chapter 2; the collapse of the wave function was illustrated in figure 2.2.

Einstein's argument was basically this: Suppose a beam of electrons is incident on a screen that contains a tiny slit, as shown in figure 3.1. Beyond the slit is a parallel plane of detectors that are capable of measuring individual particles. The electrons emerging from the slit will produce the familiar diffraction intensity pattern composed of a central maximum with alternating bands of decreasing overall intensity on either side.

In the classical wave theory of light, diffraction results from the constructive and destructive interference of electromagnetic wavelets emanating from different portions of the slit. In quantum theory, the wave function of a beam of electrons is visualized as interfering in an analogous manner. The intensity of the resulting wave function diffraction pattern at a particular point on the screen then gives the probability for a particle hitting that point.

Now suppose a detector at A in figure 3.1 registers the signal of a single electron. Einstein observed that another detector, at a different position B, equally as likely could have registered an electron if the incident wave, prior to detection, had comparable probability of hitting either A or B. In the common language still used in quantum mechanics today, the electron's wave function is said to collapse the moment the electron is detected at A. Since no electron is observed at B, the wave function at that place must have collapsed to zero.

Einstein argued that for the wave function to collapse at B the moment a measurement was made at A, a signal must have passed from A to B at infinite speed. He suggested that the quantum description is necessarily incomplete and must be supplemented by some mechanism that produces the localization of the wave packet prior to its detection, without violating Einstein's principle that nothing can move faster than light. Neither the Schrödinger nor Heisenberg forms of quantum mechanics provide a mechanism for wave function collapse, nor has such a mechanism been successfully implemented since (though many serious attempts have been made).

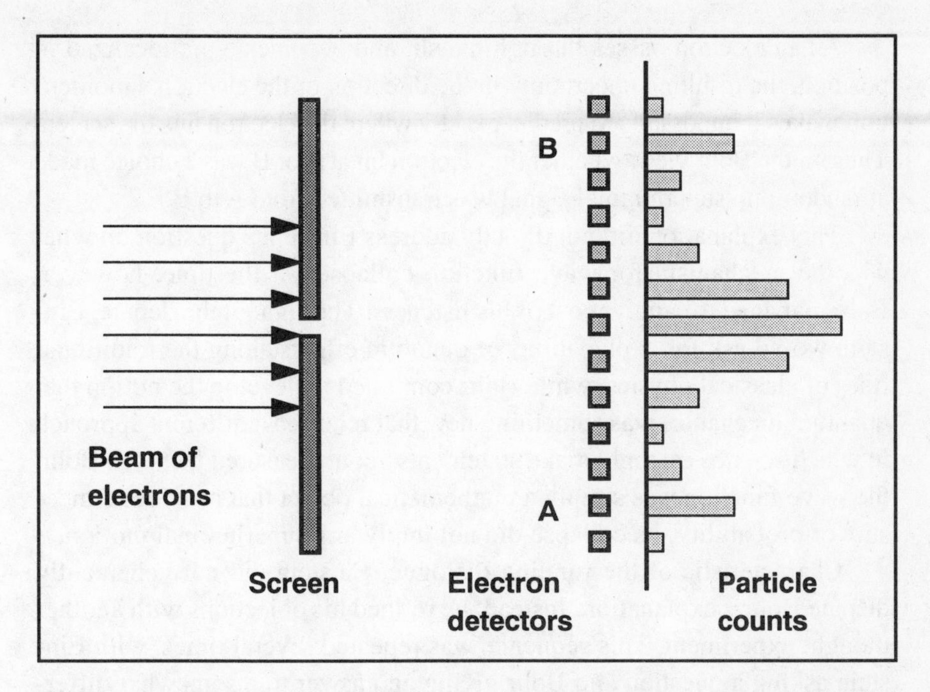

Fig. 3.1. Single slit diffraction of electrons. A beam of electrons impinges on a screen containing a slit. The number of counts in the detectors, plotted to the right, show maxima and minima resembling the intensity bands for the diffraction of light. Detectors at A and B have equal probabilities for seeing an electron. The wave function at B instantaneously collapses to zero when the electron is detected at A. Einstein called this a "spooky action at a distance."

Bohr's answers to this challenge, and the others that quickly followed, were based on detailed application of the uncertainty principle to the actual measurements that must be performed to carry out the *gedankenexperiment* in practice. Bohr explained that the localization of the particle, caused by its transmission though a slit, results in an increase in the uncertainty of its momentum according to the uncertainty principle.

If the slit has a width b, then the particle, upon passing through the slit, must, by the uncertainty principle, have an uncertainty in the component of momentum perpendicular to the original direction of motion $\Delta p \approx h/b$. The particle will then be deflected away from its original path by an angle $\theta: \approx \Delta p/p \approx (h/b)/p = \lambda/b$, using the de Broglie relationship $\lambda = h/p$ in the last step. This angle gives the width of the diffraction peak observed on the screen. The equation $\lambda/b = \theta$ for the diffraction peak for light has been known for centuries, and the uncertainty principle leads to the same result for all particles—not just photons.

As an electron passes through the slit and becomes more localized in position, the resulting uncertainty in the direction of the electron's momentum makes it impossible to predict exactly where the electron hits the screen. Thus, in the Bohr view, whether the electron hit at A or B was a choice made at random; no superluminal signal was transmitted from A to B.

This explanation did not directly address Einstein's question on what was the mechanism for wave function collapse. At the time, however, Bohr managed to satisfy most of his listeners. Throughout the debate, Einstein would ask for explanations of quantum effects along the traditional lines of classical physics, while Bohr continued to develop the notion that quantum mechanics was something new that required a different approach in which science can only make statements about measured facts. To Bohr, the wave function was simply a mathematical object that provides a measure of probability; its collapse did not imply any superluminal motion.

Characteristic of the running dialogue, Einstein did not vehemently dispute Bohr's explanation. Instead, he refined his objections with another thought experiment. This sequence was repeated several times, with Einstein asking a question and Bohr giving an answer to a somewhat different question that still seemed to satisfy almost everyone else, Einstein then replying with yet a new thought experiment.

Einstein asked what would happen if the particles that pass through the slit have carefully controlled momenta. Wouldn't that determine the particle's trajectory while at the same time producing a diffraction pattern, thus simultaneously demonstrating both particle and wave properties in violation of Bohr's complementarity principle? Bohr again pointed out that the uncertainty in the particle's momentum would lead to an uncertainty in position, as given by the uncertainty principle. This uncertainty corresponded exactly to the separation of the light and dark bands in the diffraction pattern.

Thirty years later I would have the pleasure of hearing Richard Feynman lecture on quantum mechanics. Feynman employed the double slit experiment to demonstrate the sharp departure of quantum mechanics from classical physics and we will return to it often in this book.[8]

The double slit experiment had been used by Thomas Young in the early nineteenth century to convincingly demonstrate the wave nature of light. This experiment is actually easier to analyze than the single slit experiment, provided the slits are made much smaller than the wavelength of the light passing through. In this case, a pattern of equally spaced, alternate light and dark bands results on the screen from the constructive and destructive interference of the waves from the two slits.

Feynman examined how quantum mechanics describes this experiment in terms of photons or, equally well, for other particles such as electrons and how the results contrast with what one would observe in a similar experiment conducted with bullets (see figure 3.2). As I recall Feynman explaining it to his attentive audience, the detection of a single electron at a point on the screen does not tell us which slit it actually passed through. In fact, we must treat the electron as if it had passed through both slits. The observed pattern can be described in terms of the interference of waves from each slit, just as in classical wave theory.

Now, Feynman said, place a detector behind one slit and operate it in coincidence with the detectors along the screen. In that case you can sense whether a electron has passed through the slit. When you do this, believe it or not, the interference pattern disappears.

This example illustrates the argument Bohr made—before, during, and after the Einstein debates—that the type of measurement being performed determines the nature of what is being measured. When you do the experiment in such a way that a wave property is measured, you get a wave. When you do the experiment in such a way that a particle property is measured, you get a particle.

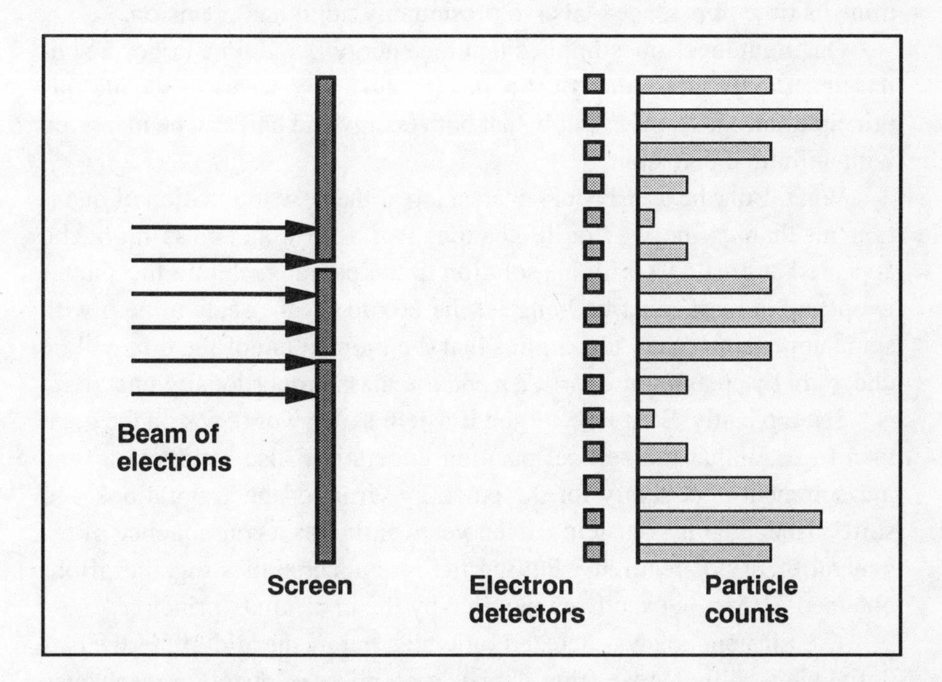

Fig. 3.2. Double slit interference of electrons. A beam of electrons impinges on a screen containing two slits. The number of counts in the detectors, plotted to the right, show maxima and minima resembling the intensity bands for the double slit interference of light.

This became the place where others made the quantum-consciousness connection, with the human mind seeming to decide whether an object is a wave or a particle. Bohr did not agree. He insisted that consciousness has nothing to do with quantum mechanics. Rather, he claimed, an object contains two complementary aspects that will show up objectively when either is measured. Bohr never suggested, to my knowledge, that the reality of a property was brought into existence by the act of its measurement.

Moving back to the great debate, since Bohr's responses to Einstein were centered on the uncertainty principle, Einstein decided to attack that head on. At the sixth Solvay conference in October 1930, Einstein presented a thought experiment that he believed contradicted the uncertainty principle.

Einstein visualized a box containing photons (as any otherwise-empty box does) with a hole that could be opened or closed by a shutter under the control of a clock (see figure 3.3). The box would be weighed at a specified time, the shutter would then open just long enough to let one photon out, and then the box weighed again. In principle, the difference in measured masses m would give the energy of the photon, by $E = mc^2$, to unlimited precision (or so Einstein thought) while the clock would determine its time of passage—also to presumably unlimited precision.

Quantum mechanics implies that the energy E and time t obey a similar uncertainty principle, $\Delta E \Delta t > h$. Einstein's box seems to violate this principle since it seems to imply that both energy and time can be measured with unlimited precision.

When Bohr heard Einstein's latest jab at the very foundation of quantum mechanics, he was quite shaken. But after a sleepless night, he appeared at breakfast with his solution to the paradox: Before the shutter is opened, a mass must be hung on the box to set the scale to zero with some uncertainty Δx. This implies that the momentum of the box will be uncertain by an amount $\Delta p = h/\Delta x$ and the mass proportionally uncertain.

Triumphantly, Bohr next turned Einstein's own work back on the great man to show that the vertical position uncertainty also results in a time measurement uncertainty for the clock by virtue of the gravitational red shift, discovered by Einstein fifteen years earlier as a consequence of his general theory of relativity. Putting the two uncertainties together, Bohr obtained $\Delta E \Delta t > h$, exactly as required by the uncertainty principle.[9]

The Einstein box was debated for while, but by the mid-1930s the battlefield had shifted away from the basic premises of quantum mechanics such as the uncertainty principle, which seemed to stand up to the tests of real experiments as well as those of the thought variety, to more philosophical issues about the meanings and interpretations of the words used

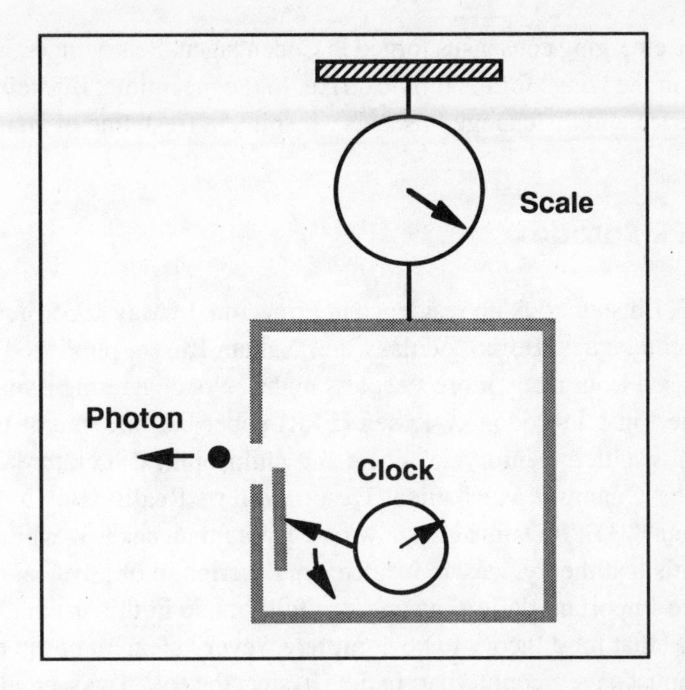

Fig. 3.3. Einstein's photon-in-a-box thought experiment. A box containing photons has an opening with a shutter controlled by a clock within. The shutter is opened just long enough to let one photon escape. The box is weighed before and after determining the energy of the photon. Einstein proposed that this violated the energy-time uncertainty principle since both are capable of being measured with unlimited precision. Bohr was able to show that it did not.

to describe quantum mechanics. Einstein's basic contention, however, remained consistent throughout: Quantum mechanics gives an incomplete description of reality.

While Bohr argued that every word in our language refers only to our perceptions based on sensory data, Einstein insisted that we must nevertheless try to reach deeper toward an objective reality behind appearances. Einstein's view of objective reality is best summarized in the story told by his biographer Abraham Pais: "We often discussed his notions on objective reality. I recall that during one walk Einstein suddenly stopped, turned to me and asked whether I really believed that the moon exists only when I look at it."[10] No doubt, Einstein did not believe such nonsense.

Pais has said about this period in Einstein's life: "After 1933 it was his almost solitary position that quantum mechanics is logically consistent but that it is an incomplete manifestation of an underlying theory in which an objectively real description is possible."[11] Even de Broglie went along with the emerging consensus forged in Copenhagen. Schrödinger, however, waited in the wings for his turn to strike. In the meantime, Einstein moved to America and the change in scenery inspired a fresh line of attack.

The EPR Paradox

In 1933, Einstein took up residence in Princeton. In May 1935, he and two young colleagues, Boris Podolsky and Nathan Rosen, published a paper that caused Bohr many more sleepless nights. Not containing a single citation, the Einstein-Podolsky-Rosen (EPR) paper became one of the most cited in twentieth-century scientific and philosophical literature. Entitled "Can the Quantum Mechanical Description of Reality Be Considered Complete?"[12] EPR claimed to show that quantum mechanics, while correct as a statistical theory, was an insufficient description of physical reality.

Two important definitions are carefully made in this paper. The first specifies that for a theory to be **complete**, "every element of the physical reality must have a counterpart in the physical theory." The second defines **physical reality**.

Rarely do papers in the *Physical Review,* or other scientific publications, deal directly with terms like reality. Usually, they leave that to the philosophical journals. However, EPR introduced reality into the discussion and so the authors were forced to define it, at least for the purposes of their paper.[13]

Einstein and his co-authors defined physical reality as follows: "If without in any way disturbing a system we can predict with certainty (i.e., with a probability equal to unity) the value of a physical quantity, then there exists an element of physical reality corresponding to this physical quantity."

In his various statements, Einstein seems to have used the terms physical reality and **objective reality** equivalently. Objective reality generally represents the collective, reproducible, and predictable experience of more than one human being, aided, in the case of scientific data, by impersonal instrumentation. Since scientific progress is very much dependent on a consensus being reached among independent, often skeptical and critical, observers, physical reality is necessarily objective.

Many books, some of them bestsellers, talk about subjective, "alternate" realities such as the experiences during dreams, meditation, or drug-induced hallucinations. Those claims are dubious at best and beyond our concern because they are, by admission, nonobjective.

Still, the EPR definition of reality is a very restrictive one. It demands predictivity with unit probability—that is, certainty. When I lecture someone will often say to me: "Nothing is certain." I usually respond, "Are you sure of that?" But these questioners do have a point, even if they cannot prove it without creating a logical paradox. If we are to associate reality with predictivity, that predictivity need not be accurate 100 percent of the time. Being able to predict at a rate slightly better than random chance, or the house odds, which are somewhat less favorable, is sufficient to make a good living as a professional gambler. This is real enough for most people and in this book I will use a less restrictive definition of reality that does not require certainty, just reasonable predictability.

In fact, EPR did not insist that the authors' was an inclusive definition of reality. They merely noted that if they could predict something with certainty, then the ingredients that went into the prediction must have something to do with what most reasonable people would agree is real.

The EPR paper uses an example in which two particles, A and B, are emitted from the same source (see figure 3.4). For simplicity (although this is not necessary), assume that the source is at rest. If after the particles have moved away from one another a measurement is made of the position of particle A, then the position of B at that time is known with certainty. B can then be said to be in a state that corresponds to a particle at that position, even though its position has not been directly measured. Since a subsequent measurement of the position of B will yield the predicted result with certainty (within measuring errors), then that position must possess an "element of physical reality," according to the EPR definition—even though it has not been directly measured.

Now, if instead of position we measure the wavelength, or equivalently, the momentum of particle A, then the wavelength or momentum of particle B at that time is likewise known with certainty, since the total momentum is zero if the source is at rest. In this case, B can be said to be in the state that corresponds to a particle of that specific momentum. Again, no measurements have been performed on particle B. Nevertheless, its momentum is predictable with certainty. Thus the momentum of B also possesses an element of physical reality.

Now, in quantum mechanics, the wave function that represents the state of an ensemble of particles of definite position contains no information about the particles' momenta. If you try to predict the momenta from that wave function, you will find that all values of momentum are equally likely. In other words, if a particle "really" has a certain momentum, the wave function can't tell you what it is and so does not carry complete information about the state of the particle.

Similarly, the wave function representing a state of an ensemble of par-

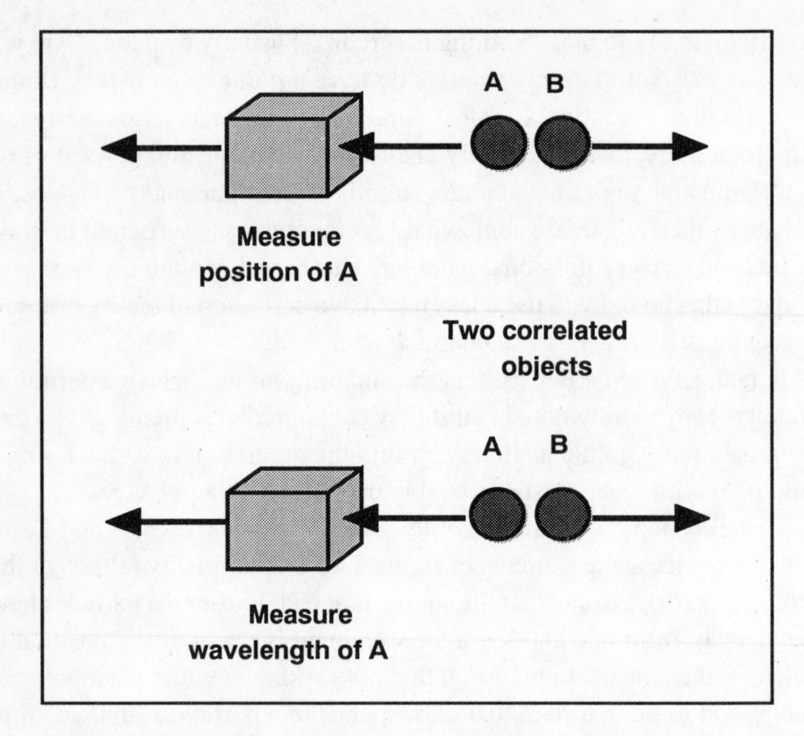

Fig. 3.4. The original EPR proposed experiment. Particles A and B emerge in opposite directions from a source with a known total momentum, which can be taken to be zero. If the position of A is measured, the position of B is then known. If, instead, the wavelength (momentum) of A is measured, the wavelength (momentum) of B is known. Two incompatible quantities have thus been determined without direct measurement. Since the result of their measurements is now predictable with unit probability, the position and wavelength of B must possess an element of "objective reality." Since quantum mechanics does not include incompatible variables in its description, it must be incomplete.

ticles of definite momentum gives a completely indeterminate position. In this case, all values of position are found to be equally likely. If you are to maintain the classical notion that a particle always possesses a "real" position, even when not being observed, then the wave function is again incomplete.

Einstein and his co-authors offered a choice: Either incompatible observables like position and momentum cannot have simultaneous reality, or the quantum mechanical picture of reality is not complete.

Let me dissect their argument. Suppose that the observables have simultaneous reality. Then, if the theory is to be complete, the description of particle states in terms of wave functions, or whatever else you might introduce into the formalism, must be such that both quantities are simultaneously predictable. But the uncertainty principle and the wave functions of conventional quantum mechanics do not allow this. So, in this case, the theory cannot be complete. An underlying subquantum theory still awaits discovery.

Alternatively, conventional quantum mechanics is complete and incompatible observables cannot be simultaneously real. This is the view often attributed to the Bohr and Copenhagen positions, although that is not quite the case, as we will see in a moment. As for Einstein, in any case, his many statements make it clear that he took the first viewpoint, in which position and momentum are simultaneously real and quantum mechanics is simply incomplete.

An important point must be emphasized that is easy to miss in the usual discussions of the EPR paradox. EPR did not propose an actual experiment that would yield predictions of simultaneous values of position and momentum of particle B, in direct contradiction of the uncertainty principle. In the authors' scenario, if the position of A is measured, then the position of B is determined. If instead the momentum of A is measured, then the momentum of B is determined. The issue EPR raise is that both position and momentum must still be regarded as simultaneous elements of reality, even when they are not measured.

The term **counterfactual definiteness** is now used to label the assumption that incompatible observables, though not simultaneously measurable, still have real values prior to their measurement, because they could in principle be measured. To assume otherwise, it seems, is solipsism, with the world being the way we choose it to be. As we will see, counterfactual definiteness is one of the key assumptions that differentiates the various interpretations of quantum mechanics.

EPR did not reject the uncertainty principle. Apparently, by 1935, Einstein had become convinced of its validity. Thus he tacitly accepted defeat in his earlier debates with Bohr, at least in those attempts to refute the uncertainty principle. And it must be emphasized again that Einstein never questioned the validity of the quantum formalism and its ability to make calculations and predictions to compare with experiment. Despite the common impression that Einstein opposed quantum mechanics, he did not object to the practice of quantum mechanics.

The questions raised by the EPR paradox are over interpretation rather than practice. Basically, EPR are saying that a complete theory must contain a precise mathematical description of all the elements of physical reality, not just those that happen to be measured in a given experiment. The Copenhagen response is that quantum mechanics is a complete theory since it accurately describes whatever measurements you chose to perform and all measurements you can perform. The argument, as usual, is over words rather than experimental facts. Philosophically, Einstein was a realist while Bohr was a positivist or pragmatist, depending on how you define these philosophical schools.

Bohr Responds

Bohr dropped everything he was doing to work on a suitable reply to the EPR paper. He published his answer in October 1935, using the same title used by EPR.[14]

Basically, Bohr reiterated his idea of complementarity, arguing that the EPR definition of physical reality "contains an essential ambiguity" when applied to quantum phenomena. The procedure of measurement defines the properties of physical quantities, so Bohr argued that EPR were unjustified in assigning the term "physical reality" to properties without considering their measurement. He added that one cannot analyze a quantum system in terms of independent, individual parts, as EPR did. The system must be analyzed as a whole, with all aspects of the experimental arrangement taken into consideration.

In Bohr's perspective, quantum mechanics is perfectly capable of predicting the outcome of actual measurements that may later be made on object B in the EPR experiment. Remember that EPR proposed two experiments, not one. In one experiment the position of A is measured. In that case, the position of B is predicted. In another experiment, the momentum of A is measured and the momentum of B predicted. The predictions for either experiment will be verified when the actual measurements on B are performed in that experiment.

True, the momentum of B in the first experiment and the position of B in the second are unpredictable. But these are again two different experiments, and this unpredictability also agrees with what is observed. Do these experiments over and over again, measuring one observable for A and its complement for B, and you will get a random distribution of the B observable. Quantum mechanics describes exactly what we see in the real world. In some cases it predicts unpredictability, and this is what we find.

In Bohr's view, quantum mechanics is what is now called a **contextual** theory. The outcomes of measurements depend on the entire experimental setup. Quantum predictions are made for specific experimental arrangements, not single, isolated detectors within arbitrary experimental arrangements.

Bohr managed to convince most physicists that EPR required no revision of Copenhagen quantum mechanics. Even Einstein's loving biographer, Abraham Pais, was convinced by Bohr's rebuttal, saying, "Most physicists (myself included) agree with this opinion."[15]

I suspect, however, that the continuing consensus of support for Bohr's view is more the consequence of the unbroken line of success of the applications of the formal methods of quantum mechanics than any force of logical argument, or a conscious attempt by physicists to sign up with one or another philosophical school. Quantum mechanics works, so why quibble about what seems to be a verbal disagreement leading to no measurable effects?

Einstein, Podolsky, and Rosen never abandoned the belief that in their 1935 paper they had proved that quantum mechanics, as it was formulated, was an incomplete description of physical reality.[16] Certainly Einstein saw no need to discard locality, which was his personal invention. And he had no wish to forsake quantum mechanics, which he had also helped invent. He only said it was not the final story.

Intrinsic Properties

Discussions on quantum mechanics tend to focus on the simultaneous unmeasurability of complementary quantities such as position and momentum. As a consequence, the layperson may be led to the false conclusion that the quantum world is filled with fuzzy entities that disappear in one place and appear simultaneously at another. This is far from the case.

Many properties of matter are fixed and, for practical purposes, permanent. They can be determined without their respective measurements interfering with one another. These include rest mass, electric charge, magnetic moment, and spin.[17] The rest mass of a particle measures its inertial and gravitational properties in the reference frame in which the particle is at rest. The electric charge measures the particle's electric strength. Most particles, even those that are electrically neutral, are tiny magnets and their magnetic moments measure their magnetic strengths. The spin of a particle measures its intrinsic angular momentum.

These measurable properties are used to define what constitutes a particle. For example, an electron is defined as a physical entity having a certain rest mass, electric charge, magnetic moment, and spin. These quantities serve to identify the particle in the measuring apparatus; any measurement yielding a different value for any one of these quantities, outside measurement errors, would signal the presence of a different particle.

In contrast to position and momentum, measurements of the intrinsic properties of particles can be done simultaneously without disrupting one another. In fact, that is why we regard these properties as intrinsic. They certainly possess the element of physical reality called for in the EPR definition. Most popular presentations of quantum mechanics, in their rush to excite the reader with the wonders of uncertainty and nonlocality, tend to ignore the fact that material bodies possess many unambiguous features that are not the slightest bit ephemeral.

Spin

Among the nonephemeral, intrinsic properties of microscopic bodies, perhaps the most interesting is spin. The spins of atoms, nuclei, and elementary particles do not seem at all like the spins of macroscopic bodies. In fact, they exhibit some of the most profound phenomena found in the quantum regime. They also afford simple and practical tests of the EPR paradox.

Any rotating body, such as the earth or a bicycle wheel, has angular momentum about its rotation axis. The principle of conservation of angular momentum keeps the earth rotating and a moving bicycle from falling over. In both classical and quantum physics, the total intrinsic angular momentum of a body is called its spin.

The magnetism of a microscopic body is closely related to its spin. Crudely speaking, we can think of the orbiting charges inside a body as forming a tiny electromagnet. For example, the electrons circulating in an atom produce an electric current very much like the current in a loop of wire. This current produces a magnetic field similar to that of a bar magnet, with north and south poles aligned with the particle's spin vector.

A particle's magnetic field will interact with an external field just as a compass needle interacts with the earth's field; the north-south axis tends to line up with that of the external field, with opposite poles attracting one another in the familiar fashion. Hence the spin of a particle will also tend to line up with any applied external field.

In 1921, O. Stern and W. Gerlach sent a beam of silver atoms through a nonuniform magnetic field. According to classical electromagnetic theory, the beam should have spread out into a more diffuse beam as a result of the magnetic interaction between particles and field. Stern and Gerlach instead observed two distinct beams, which were interpreted in terms of the spin of the orbiting outer electron in the atom.

This result was explained as the consequence of **quantization of angular momentum**. The basic quantum of angular momentum is $\hbar =$

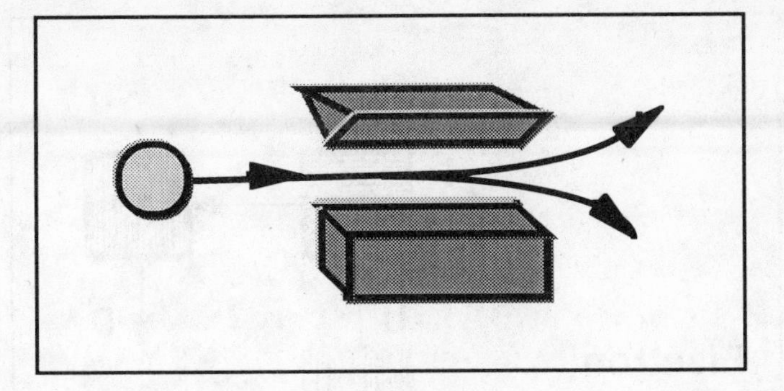

Fig. 3.5. A beam of electrons will be split in two by an inhomogeneous magnetic field.

$h/2\pi$, identical to Bohr's quantum of action described earlier, and the components of spin can differ only in units of the quantum of action.

Suppose we send a beam of electrons through a Stern-Gerlach apparatus, as sketched in figure 3.5.[18] The two beams that emerge in the experiment lead us to conclude that the electron has a spin equal to half a quantum. When that spin is projected along a given axis, it can have a value +½ quantum when the spin is along that axis, and -½ when it is opposite. The difference is one quantum.

Not all particles, nuclei, and atoms give two beams in a Stern-Gerlach experiment. Sometimes we see three, four, or more. The interpretation is that each body has a spin S, in quantum units. S can be an integer or half-integer, where 2S+1 beams emerge from the magnet.[19]

Let us now send an electron beam into a Stern-Gerlach magnet oriented so that the two emerging beams are "up" and "down" in a vertical plane (see figure 3.6). All the electrons in the up beam can be said to be in a state U, and those in the down beam said to be in a state D.

Now we take the up beam alone, and send it into a second Stern-Gerlach magnet that is rotated by 90° about the beam axis with respect to the first magnet. We will observe two emerging beams that are "left" and "right" in a horizontal plane. By the same logic followed above, we can label the state of the left beam L and the right beam R.

Since only up electrons from the first magnet are being used, it would seem reasonable to add the U state label to the specification of the states of the left and right beams and call the first UL and the second UR. That is, the electrons in the left beam out of the second magnet are "up-left," and those in the right beam "up-right."

Now let's take the up-right electrons and feed them through a third

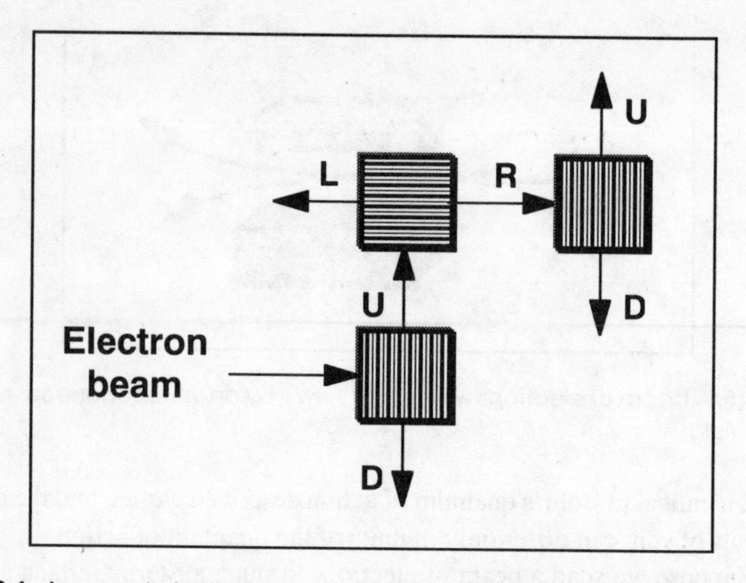

Fig. 3.6. An unpolarized electron beam passes through a Stern-Gerlach magnet that separates it into a beam U of electrons spinning up and a beam D of electrons spinning down. Beam U is sent through another magnet oriented perpendicular to the first and beams L with spin left and beam R with spin right emerge. Beam R is then sent into a third magnet oriented the same as the first and both U and D beams emerge. This illustrates that the components of angular momentum are incompatible observables.

magnet oriented exactly as the first. Common sense dictates that only one beam will emerge—in the up direction. But common sense fails, as it often does in quantum mechanics. Both up and down beams emerge!

The quantum explanation is that the components of angular momentum in perpendicular directions are incompatible observables like position and momentum. Measurements of the spin components along two different axes cannot be made simultaneously.

I am not sure this result is all that profound. Although the proof can be found in the most widely used graduate classical mechanics textbook, most physicists have forgotten that the incompatibility of the components of angular momentum occurs even in classical mechanics. If one regards the components of angular momentum to be the momenta canonically conjugate to the angles of rotation about each axis, then it can be shown that only one can be conserved at a given time.[20] I have never seen any discussion of this fact in quantum literature, which uniformly attributes the incompatibility of components of angular momentum to be a quantum phenomenon.

Classically or quantum mechanically, the spin component states of an electron can be specified with respect to only one direction. In the example above, the two spin states are either U/D or L/R. When we take our up beam and run it through the second magnet, its "upness" is destroyed. The right and left beams that emerge each contain both up and down electrons.

The Bohm-EPR Experiment

David Bohm was perhaps the premier figure of legitimate scientific stature in the recent history of quantum metaphysics. In his last years, he speculated about a radically different, holistic view of nature called the *implicate order* that he proposed to replace the traditional reductionist view.[21] This will be discussed in a later chapter. However, mysticism came late in Bohm's life. Long before, he had established himself as an unusually brilliant and original physicist. He was also an interesting character.

Bohm was a Pennsylvanian who received his Ph.D. from the University of California at Berkeley in 1943. As a graduate student he had already distinguished himself by showing an exceptional ability to solve very complex physics problems.[22] Shortly thereafter, while an assistant professor at Princeton, Bohm wrote the very fine book called *Quantum Theory*.[23] During my graduate student days, I purchased a copy that remains nearby on my bookshelf today. While much of the book follows conventional Copenhagen quantum mechanics, and served as a text, Bohm also went to great pains to explain fundamental concepts in a way lacking in most of today's textbooks, which focus mainly on how to apply mathematical formalism to solve standard problems.

Most significantly, Bohm discussed the Einstein-Podolsky-Rosen paradox and proposed a practical way it could be tested using electron spin. From his analysis of this experiment, Bohm concluded that, "No theory of mechanically determined hidden variables can lead to *all* the results of the quantum theory."[24]

During the dark days of anticommunist hysteria in the United States, Bohm fell victim to the purges of the House Un-American Activities Committee and Princeton refused to renew his contract. Bohm moved to Brazil, where he developed his own ideas on hidden variables and wrote a pair of papers published in the *Physical Review.*[25] Eventually Bohm settled in England, where he was a professor at Birkbeck College of the University of London at his death in October 1992.[26]

The Bohm-EPR experiment, illustrated in figure 3.7, starts with two electrons spinning opposite to each other with zero total angular momen-

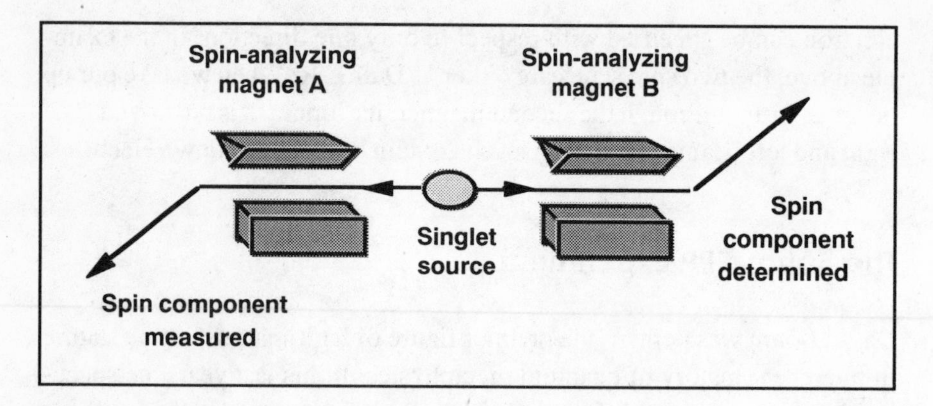

Fig. 3.7. The EPR Paradox experiment proposed by David Bohm. Electrons from an initial singlet (total spin zero) state are analyzed by magnets that are oriented to deflect the particles one way when they are spinning along a particular axis, and the opposite way when they are spinning opposite. Once the spin component at one end is measured, the component at the other end is determined.

tum. In the conventional language used to describe this situation, we say that the two electrons exist in a composite state called the **singlet**. The electrons then move off in opposite directions. At the ends of two beam lines A and B of equal length, Stern-Gerlach magnets are used to determine the spin components of each electron in the pair emitted from the source.

Let us use the same language we used above in describing the behavior of electrons passing through Stern-Gerlach magnets. If the two magnets in the Bohm experiment are oriented so they select vertical spin components, then angular momentum conservation demands that if spin up is measured at A, then spin down will be measured at B.

If instead both magnets are oriented to select horizontal components, when a left electron is measured at A, a right electron will be measured at B. Regardless of the magnet orientations, the angular momentum must sum to zero. When the orientations of both are the same, the components at either end must cancel.

Suppose that the magnets at both ends are oriented to measure the vertical component of spin, but that measurements are made only at the end of beam line A. Then we can predict with certainty that whenever A measures up, B will be opposite. If A is up, B will be down. If A is down, B will be up. If we rotate the magnets by 90°, we say measure A is left. Then we predict with certainty that B is right.

So we have another example of the problem raised by Einstein, Podolsky, and Rosen. Using the EPR definitions of completeness and physical reality, the vertical and horizontal spin components of the electron travelling beam line B, presumably undisturbed by any measurements at that end, must in the EPR view have "simultaneous reality," or quantum mechanics is incomplete.

Bohm noted that two assumptions were implicit in the EPR paper: (1) "The world can correctly be analyzed in terms of distinct and separately existing 'elements of reality,' " and (2) "Every one of these elements must be a counterpart of a *precisely* defined mathematical quantity appearing in a *complete* theory." The latter is essentially the explicit definition of completeness given by EPR, but Bohm felt that a stronger statement was implied and added that the elements of physical reality in a complete theory must be precise mathematical entities.

Bohm notes that these assumptions are at the root of classical theory, but are still a comparatively new idea historically, arising from the great success of the mathematical analysis of mechanics and electrodynamics in the three centuries after Newton. Bohm calls this "the hypothesis that reality is built upon a mathematical plan." He comments that while this assumption seems natural to us, it is by no means inescapable.[27] In a later chapter I will discuss how this neo-Platonic view of reality is part of the new metaphysics, despite the fact that quantum mechanics seems to argue against it.

In quantum theory, as described by Bohm, we make quite different but equally plausible hypotheses. In his 1951 book Bohm cites the conventional wisdom that, "At the quantum level, the mathematical description provided by the wave function is certainly not in a one-to-one correspondence with the actual behavior of the system under description, but only in a statistical correspondence."[28]

In the EPR view of reality, all the components of electron spin, like momentum and position, are part of physical reality and so must be included in any complete theory, even when they are not specifically measured. To the extent that this is not done in quantum mechanics, which includes in its description of particle states only those quantities that can be simultaneously measured, quantum mechanics is incomplete.

As we will see later, a variation of the Bohm experiment that was analyzed by John Bell provided a crucial empirical test of quantum mechanics, a test it passed with flying colors. But first we need to explore further the notion of the quantum mechanical state. As we see from the above discussion, the wave function does not seem to tell us everything about an individual quantum system that we would like to know.

The State of the Individual

The question of whether quantum mechanics can still be used to unambiguously describe the state of an individual quantum system such as a single particle is a controversial one that remains unsettled. Several points of view exist on the issue, but all agree with the general conclusion that the results of measurements on individual particles are not in general predictable by conventional quantum mechanics.

Leslie Ballentine has identified two differing views of the quantum mechanical state. According to Ballentine, the state represented by the wave function Ψ is either (1) a pure state that provides a complete and exhaustive description of an *individual* system, or (2) a pure (or mixed) state that describes the statistical properties of an ensemble of similarly prepared systems.[29] Ballentine argues that an application of the standard formalism of quantum mechanics leads to contradictory results for the first interpretation and that a major revision of quantum mechanics is needed to provide for the description of individual states. He claims that the second interpretation, the **ensemble interpretation**, is viable in the standard formalism, although whether this provides an adequate description of quantum mechanics remains in dispute.[30]

But Ballentine makes a very important point that is not disputed. The wave function is defined by the Born postulate to be associated with the probability for finding the system in a certain state. Probabilities are defined mathematically in terms of ensembles of objects, at least in most physics applications. (Technically, the ensemble must have an infinite number of members, but for our purposes let us just imagine the number to be very large.) To know with certainty that an individual object is in a specific state, all the objects in the ensemble must be in that state.

For example, given a large number of shuffled decks of ordinary playing cards, the probability that you will randomly select an ace is 1/13. Only if you prepare your decks ahead of time by selecting out all the aces can you say, with certainty, that the "state" of the card you chose will be aces.

James L. Park has similarly argued that the jargon of modern physics induces us to regard the phrase "an electron in the state Ψ," as merely the quantum analog to the classical expression "an electron in state (q,p)," where q is the position and p the momentum of the electron. This jargon is used "in spite of the fact that the former refers physically to statistics of measurement results upon an ensemble of identically prepared electrons whereas the latter just means that a single (classical) electron *has* position q and momentum p."[31]

Park cautioned that we cannot relate wave functions to single systems analogous to the classical manner. He argues, "The linguistic extension of Ψ from its role in describing ensembles to its further function as the state of a single system has given birth to monumental barriers to the understanding of quantum theory as a rational branch of natural philosophy."

In a reversal of the usual derivation of classical statistical mechanics from Newtonian particle physics, Park shows how the classical description of individual particle motion follows unambiguously from classical statistical mechanics. The key idea is to treat the individual particle as a subensemble of the original ensemble.

However, Park proved it was not always possible to move unambiguously from the quantum ensemble to a subensemble that describes an individual state. Park summarized his results:

> Although classical statistical mechanics admits of an unambiguous assignment of individual states, quantum theory fails to satisfy the necessary criteria. Hence, the simplest, and most natural, conclusion is that the pure "state" vector Ψ of quantum theory must not be interpreted—even theoretically—as referring to the physical state of a single system at a single time.

Park does allow that an unambiguous interpretation of the state of a single particle can be made in some cases. If a particle is a member of an ensemble that has already been prepared to have a sharply defined position, then its position will be predictable with essentially complete certainty. If its momentum has been similarly prepared, with no attempt to measure its position, then that momentum will be predictable with essentially complete certainty.

If Ψ does not always represent the state of an individual system, then it would be nice to have something else that does. James Hartle suggests one possibility.[32] Hartle points out that the state of an individual quantum system can be specified uniquely, if one incorporates indefiniteness as part of the definition.

For a classical system, the state is defined by a list of all the possible propositions that can be made about the system and whether each is true or false.[33] For an individual quantum system, however, it is impossible to simultaneously specify the truth values of all possible propositions about the system; some propositions are necessarily indefinite. Nevertheless, we can still specify a unique quantum state by listing each proposition as true, false, or indefinite.[34]

To see how this works, recall the example discussed above in which a beam of electrons is sent through perpendicular spin-analyzing magnets. We saw that we could not specify the spin states of the electrons as, for example, UR (up-right) or DL (down-left), because the measurement of the spin component along one axis makes the spin component along a perpendicular axis indefinite.

Using "I" for indefinite, if we measure the vertical spin component and find it to be up, then that state of the electron in the Hartle scheme is UI. When we take that beam and measure its horizontal spin component to be left, then the state has become IL. We now predict that if we proceed to measure whether the spin is up or down we will have a fifty-fifty chance of getting either. If instead we remeasure its horizontal spin component, we will get L with unit probability and R with zero probability.

While this provides a simple way to define an individual state, it does not really change anything. As Hartle points out, his state is not an *objective* property of an individual system. We see this in the above example, where the state depends on what measurements we perform and their sequence. However, the state contains useful information that can be used for making predictions about future measurements.

So, while Ballentine, Park, and Hartle do not see quantum mechanics in quite the same light, the practical effect of their different approaches (and that of the many others who have made similar attempts to define quantum states) is the same: however we define our quantum states, whether in terms of individual systems or ensembles, the theory that utilizes them only enables us to compute probabilities. Since the wave function Ψ is defined in terms of probabilities, it is difficult to interpret it as representing the unique state of an individual system.

Polarized Light

Polarized light provides another, perhaps more familiar example to illustrate many of the concepts we have developed. It also has been utilized in most of the EPR experiments that have been actually carried out in the laboratory.

Light polarization phenomena were first studied in the eighteenth century and have long been known to exhibit curious effects that are counterintuitive at first glance. Light can be plane-polarized by passing it through certain materials, like the common polarizing lenses in sunglasses that help cut down the glare from reflected sunlight. Suppose unpolarized light from a normal incandescent lamp is incident on a sheet of polarizer oriented so its polarization axis is vertical. The light emerges with vertical polarization; that is (in classical language), the electric field vector associated with the electromagnetic wave oscillates in a vertical plane. If we now take this light and try to send it through a horizontal polarizer, nothing will get through.

However, if we take a third piece of polarizer, orient its axis at 45° with the vertical, and place it in *between* the two polarizers, light will now pass through the horizontal polarizer. Thus, placing some material between the two polarizers actually allows light to go all the way through. Without that material, the light is blocked. This is counter-intuitive, since you would expect more material to block more light from passing through.

Yet it happens exactly this way. The experiment can be easily performed and is a common classroom demonstration. The classical explanation is that the vertically oriented electric field is projected at 45° by the middle polarizer. After passing through this polarizer, the field then has a horizontal component able to pass through the final polarizing sheet. Without the middle polarizer, the field has no horizontal component.

Light can also be circularly polarized. Circular polarization is classically described by the rotation of the electric field vector around the direction of propagation. Looking toward an incoming beam of light, right-circular polarization corresponds to clockwise rotation and left-circular polarization to counterclockwise rotation. In classical electromagnetic theory, linear polarization occurs when the electric field vectors of two oppositely rotating, circularly polarized beams with the same amplitude add up to a net electric vector that remains fixed in a plane.

In quantum mechanics, light is composed of corpuscular photons. Like electrons, photons also have spins and these can only be aligned along or opposite their direction of motion.[35] The two photon alignments correspond to the two states of circular polarization, as illustrated in figure 3.8. We can picture a linearly polarized beam of photons as being composed of half right- and half left-circular photons.

Polarization is a property of the classical electromagnetic fields that are associated with light. In carrying this concept over to quantum mechanics, we must treat polarization as a property of the wave function that describes the ensemble of photons that forms a beam of light. When we say an individual photon is in a state of right-circular polarization R, we are precisely stating that it is a member of a beam of right-circularly polarized light. Photons in a state L are members of a beam of left-circular light. We can visualize these beams as streams of photons moving in a certain direction with

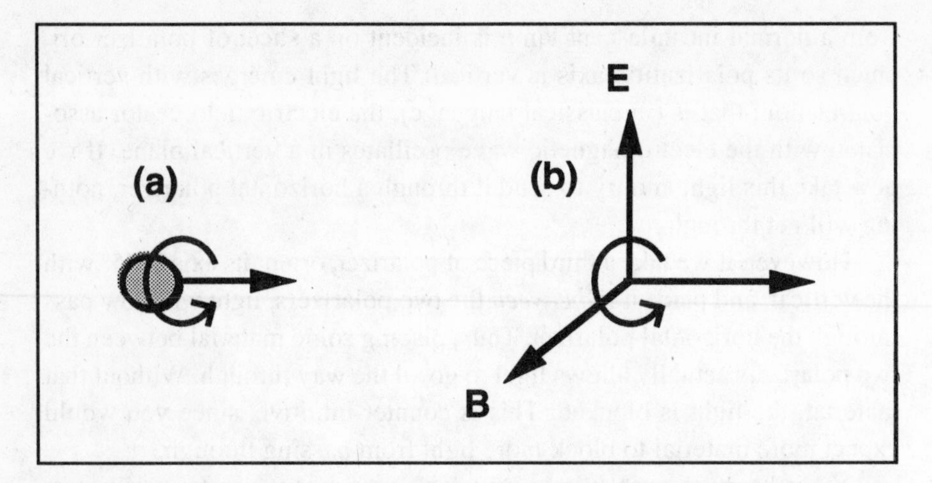

Fig. 3.8. The right-handed photon shown in (a) is spinning around an axis along its direction of motion. This corresponds to a left-circularly polarized electromagnetic wave in which the classical electromagnetic field vectors **E** and **B** rotate around the direction of propagation of the wave, as in (b).

their spins all pointing opposite to that direction in the first case, and the same direction in the second case.

In the case of linear polarization, we again define state vectors that correspond to a beam with, say, either horizontal polarization H or vertical polarization V. An individual photon is said to be in a state H when it is a member of a beam of light that is linearly polarized along a horizontal axis at right angles to the direction of the beam, state V when so polarized along the vertical axis.

Since photons can only be modeled as spinning along or opposite their directions of motion, you cannot draw any simple picture of how an individual photon achieves this state. However, the fact that we cannot picture a quantum system in familiar terms does not imply that this system is unreal or its description is less precise. Within the framework of quantum mechanics, H or V states of photons can be defined as precisely as can L and R states.

We can use the Hartle quantum state scheme to specify the state of individual polarized photons in a manner similar to that done above for electrons. We label the state with two symbols, the first indicating the circular polarization and the second the linear polarization. Again we use "I" for indefinite. Thus, a right circular photon state would be RI, where the linear polarization is indefinite. Sending it through a horizontal polarizer, the photon comes out in a state IH. Note how the state is not an objective property of the photon, since it depends on the measurements made. It simply contains information about past measurements performed on the photon, and the prediction of future measurement. In the cases of an IH photon, we predict that it will have 100 percent probability of being H if sent through a horizontal polarizer, 0 percent for a vertical polarizer, and 50 percent for either a right or left circular polarizer.

These same results can be obtained in the statistical view, where the wave function refers only to ensembles of photons. So, it appears, we can take our pick of either an individual or ensemble interpretation of the states of polarized photons.

EPR with Photons

Now let us see how the EPR experiment can be performed with polarized light instead of electrons. The dual nature of light, classical waves and quantum particles, originally led to the discovery of quantum mechanics. Furthermore, the most important class of actual EPR experiments has been performed with polarized photons.

By now the process should sound very familiar. As with the electron experiment, we start with a singlet (spin zero) atomic system that decays into two photons. As seen in figure 3.9, this implies that the two photons emitted have opposite spins, but the same circular polarizations (by definition). The correlation between photons follows from angular momentum conservation, and no ambiguity results when we describe each photon as a member of an ensemble having a given circular polarization, left or right, which is defined by looking toward an incoming beam.

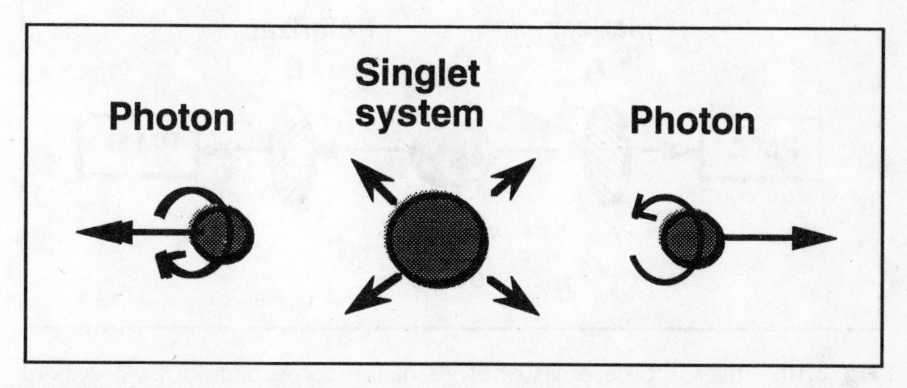

Fig. 3.9. A spin zero (singlet) system decays into two photons that go off in opposite directions with spins that cancel. Thus the photons have the same polarization. In the case above, they each have right-circular polarization.

In any given decay of a spinless atom, we do not know ahead of time whether the emerging photon polarizations will be left or right. Each is equally likely. But once we measure one polarization to be right, then we can predict with 100 percent certainty that the other also will be right. Likewise, if one is found to be left, the other is predicted to also be left. This presents no paradox. Angular momentum conservation will guarantee that the polarizations are the same, and any given measurement has a fifty-fifty chance of either result.

Now we can also do the experiment with linear polarizers, as seen in figure 3.10. We can use electronic coincidence on the photodetector outputs so that we know we are detecting the photons from the same pair emitted by the source. After the photons are emitted from the source, we align the axis of the linear polarizer A at a random angle around the beam line. If we then orient polarizer B in the same direction, we predict that a photon will be detected by photodetector B with 100 percent probability (less the corrected-for inefficiencies mentioned above). These correlations are analogous to those found previously for the Bohm experiment with electron spins. Indeed, they can be viewed as arising from the same source, namely angular momentum conservation. They can also be viewed as a rather obvious consequence of classical light polarization.

We can imagine a transmitting station sending out light beams in which signals are coded by the polarization in the beams. Receivers at various places could intercept the beams and decode the signal. This is a feasible means of communication, from transmitter to receivers.

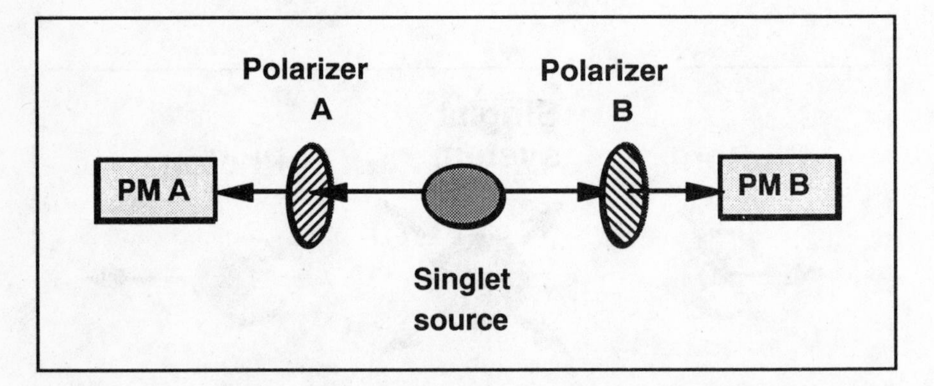

Fig. 3.10. The source S emits photons of the same circular polarization, which pass through linear polarizers A and B. The photons are detected by photodetectors PM A and B. The orientation of polarizer A is chosen randomly after the photons are emitted from the source. If polarizer B has the same orientation, the photon will be detected by PM B.

Some have fantasized that an experiment of this type could also be used to communicate between receivers—at speeds greater than the speed of light! The idea is to code a message into a pattern of changes in the polarization axis of one receiver. Another receiver located at a different position at the same distance from the transmitter should then be able to *instantaneously* decode the message by moving around its polarization axis.

At first glance this seems possible, since all receivers at the same distance from the transmitter that have the same orientation of their polarization axes will detect a photon. However, it does not work. In the next chapter, I will analyze this type of superluminal communicator in detail and show how no information can be transmitted by this means. Basically, whatever message is sent, the receiver will obtain random noise.

We will return to the EPR experiments in the next chapter, when we consider the hidden variables interpretations of quantum mechanics. There we will see how these experiments have been actually performed and how they have ruled out the existence of the most important class of such variables. First, however, let us move on to the other paradoxes of Copenhagen quantum mechanics that were raised after EPR.

Schrödinger's Cat

Among the pioneers of early twentieth-century physics, only Einstein and Schrödinger were not won over by Bohr and the Copenhagen school. Even de Broglie was converted to what became the orthodox view, although in his later years he changed his mind yet again. Schrödinger, however, stood firm. Enthused and inspired by the EPR paper in 1935, he dreamed up a colorful example of his own that has become even more famous, the paradox of Schrödinger's cat.

Schrödinger's cat has been discussed widely in the popular and semipopular literature of quantum mechanics and the reader has most likely encountered it before.[36] A cat is placed in a chamber that contains a phial of poisonous gas that is under the control of a Geiger counter. The phial may or may not be opened depending on the random emission of a particle from a radioactive substance. If a particle is emitted, the Geiger counter will detect it and open the phial. In a certain time interval, there is a fifty-fifty chance of the phial being opened and the cat killed. Schrödinger, with tongue in cheek, asked whether the cat was dead or alive before the chamber was opened and the cat was observed. Conventional quantum mechanics would seem to imply that the cat is neither dead nor alive but in some kind of limbo. Then, when someone looks inside, the cat's wave function collapses and it is either dead or alive. As someone has said, "Curiosity kills the cat."

Think of the cat as having two possible states, analogous to the spin of an electron along a vertical axis. Spin "up" corresponds to the cat being alive. Spin "down" corresponds to the cat being dead. The original live cat is in a pure state up. But after the prescribed time interval, the cat has a fifty-fifty chance of being up or down, and until we open the chamber we do not know which. In the Copenhagen interpretation, the wave function or state vector of a system does not collapse until the state of the system is observed. Thus, the cat is in neither a pure up (alive) nor a pure down (dead) state, but is an absurd mixture of the two until such a time as its state is observed.

One objection that is often raised about the Schrödinger cat paradox is that a macroscopic object such as a cat cannot be considered to be an individual quantum system describable by a wave function. Like almost all macroscopic systems, it is an incoherent mixture of quantum states. Certainly a cat is not a pure quantum state, but Schrödinger did not intend his example to be taken literally. He was just pointing out, in a dramatic way, what he perceived as an apparent absurdity within quantum mechanics.

Schrödinger's argument could just as well have been made with a simple system replacing the cat. Consider the ensemble of radioactive nuclei that trigger the poison. If similarly prepared, each nucleus can be considered to be in a pure quantum state. Schrödinger's argument then is that, according to Copenhagen, the nucleus exists in a state of limbo, half decayed and half not decayed, until the chamber is opened.

I have little doubt that Bohr thought anything other than that the nucleus decays, when and if it decays, and the cat is dead, when and if the phial breaks, whether we are looking or not. Most physicists would agree with Bohr's contention that we should not try to draw conclusions that are too profound about things we cannot observe directly or test empirically. A half-dead, half-alive cat has never been observed.

Statistical quantum mechanics simply tells us the obvious fact that, when we open the chamber, the cat will be dead or alive with equal probability. The wave function of the cat, in the statistical view, represents our knowledge of an ensemble of cats, not the "real" status of any given cat. Operationally this means that if we do the experiment with 20,000 cats in 20,000 identical chambers, we will be left with 10,000±100 dead cats and 10,000±100 live cats.

Because of the randomness of radioactive decay inherent in its quantum mechanical description, the fate of any given cat is unpredictable. Such a prediction requires a new, causal theory that goes beyond quantum mechanics. We cannot take it for granted that such a causal mechanism exists in nature. The Bohr-Heisenberg view is that quantum mechanics is complete, and such a mechanism does not exist. The Einstein-Schrödinger-Bohm view is that quantum mechanics is incomplete and we will someday find a better theory.

The Schrödinger cat paradox usefully supplements the EPR paradox in again raising, in a different and colorful way, the issue of whether quantum mechanics is complete. Einstein and Schrödinger accepted the calculational power of the formalism of quantum mechanics, and indeed each contributed immensely to its development. They simply felt that the formalism was not enough. Bohr and Heisenberg said: All we know is what we measure, so quantum mechanics is the best we can do. Einstein and Schrödinger argued that a deeper reality must exist and that quantum mechanics must either be modified to account for it, or we must seek a deeper theory and let quantum mechanics content itself with calculating probabilities.

Quanta and Consciousness

In 1978, John Archibald Wheeler suggested a variation on the double slit experiment that has been applied to other experiments as well, including tests of the EPR paradox.[37] Wheeler's idea was to delay the decision on whether or not to measure if a photon has passed though a particular slit until we know that the photon, traveling at the speed of light, has already passed through one or the other. This is how Wheeler put it: "We, by our last minute decision, determine whether the photon shall travel 'one of the two routes' or 'both routes'—to use a misleading and unacceptable language—*after* the photon has *already* accomplished that travel! Quantum mechanics seems at first sight to make 'what has already happened' in the past depend on a choice made in the present!"[38]

Recall my earlier discussion of Feynman's description of the double slit experiment. (Feynman was Wheeler's student at Princeton.) When you do not measure which slit the photon passes through, you get the classical Young's interference pattern expected from waves. When you place a photodetector in the path between the source and either slit, the interference pattern goes away and you get what you would expect for a particle moving in a straight line through the slit.

By delaying the decision until after the photon has passed though one or both slits, you demonstrate that the collapse of the wave function does not occur until the experimenter decides what to do. The photon is neither a particle nor a wave, until one or the other properties are measured.

Wheeler gives an example where the decision of an observer today can apparently affect what happened billions of years in the past: Astronomers have accumulated several examples of the light from a distant galaxy being split into two or more beams by passing near a large mass along the line of sight from earth, producing multiple images of the galaxy by a process known as gravitational lensing. Photodetectors located at each image position will register photons that travelled a particular path.

Wheeler argued that the two beams could be brought together by a mirrors so that they constructively interfere in the direction of one photodetector and destructively interfere in the direction of the other. Thus only the first detector gets a signal and we would say, incorrectly in Wheeler's view, that the photon travelled by each route.

Once more, we can make a last-minute decision whether or not to include the mirror that brings the beams together. By that time the photon has, again using a mistaken way of speaking, already travelled both routes. The mistake, according to Wheeler, is to assume that any definite path exists before the photon is detected. In an oft-repeated quotation, he has said, "No elementary quantum phenomenon is a phenomenon until it is a registered phenomenon. . . . In some strange sense, this is a participatory universe."[39]

I think Wheeler is now sorry he said this. His use of this language in describing the delayed choice experiments has encouraged quantum mystics to conclude that human consciousness is the agent of the participatory universe, the force that makes the possible actual. They often quote Wheeler to that effect, reinforcing the notion with another quotation from an equally distinguished physicist, Eugene Wigner, who said, "The laws of quantum mechanics itself cannot be formulated . . . without recourse to the concept of consciousness." [40]

David Albert describes Wigner's scenario as follows:

> All physical objects almost always evolve in strict accordance with the dynamical equations of motion. But every now and then, in the course of some such dynamical evolutions (in the course of *measurements* for example), the brain of a sentient being may enter a state wherein . . . states connected with various *conscious experiences* are superposed; and at such moments, the *mind* connected with that brain (as it were) *opens*

> *its inner eye,* and *gazes* on that brain, and that causes the entire system to collapse, with the usual quantum-mechanical probabilities, onto one or another of these states.[41]

Danar Zohar interprets these ideas in this way: "According to this view, proposed chiefly by quantum physicists Archibald Wheeler and Eugene Wigner, human consciousness is the crucial missing link between the bizarre world of electrons and everyday reality." That conclusion, she says, "ironically, is very close to my own."[42]

I am not sure what Zohar finds ironic. Perhaps the irony is that Wheeler also said, "Consciousness, we have been forced to recognize, has nothing whatsoever to do with the quantum process."[43] Wheeler is far from the New Age guru that the quantum mystics would like to claim. He has spoken out strongly against parapsychology, calling it pseudoscience and objecting to its affiliation with the American Association for the Advancement of Science.[44]

Commenting on how his statements have been misused, Wheeler has said, "No subject more attracts the devotees of the 'paranormal' than the quantum theory of measurement. To sort out what it takes to define an observation, to classify what it means to say 'no elementary phenomenon is a phenomenon until it is an observed phenomenon' is difficult enough without being surrounded by the buzz of 'telekinesis,' 'signals propagated faster than light' and 'parapsychology.' "[45]

Wheeler's true views are not shared by theoretical physicist Henry P. Stapp, whose review of the Copenhagen interpretation was discussed in chapter 2. Stapp has written copiously on theories that connect quantum and mind. He argues that consciousness has no place in classical, deterministic physics but that quantum mechanics has a "perfectly natural place for consciousness, a place that allows each conscious event, conditioned, but not bound, by any known law of nature, to grasp a possible large-scale metastable pattern of neuronal activity in the brain, and convert its status from 'possible' to 'actual.' "[46]

The conclusion I will reach in this book, however, is closer to that of philosopher Karl Popper, who writes, "I have often argued in favor of the evolutionary significance of consciousness, and its supreme biological role in grasping and criticizing ideas. But its intrusion into the probabilistic theory of quantum mechanics seems to me to be based on bad philosophy and on a few very simple mistakes."[47]

Notes

1. Letter from Einstein to P. S. Epstein dated November 10, 1945, Einstein estate, Princeton, N.J.
2. Schilpp 1949.
3. Einstein, A., letter to C. Lanczos, March 21, 1942.
4. Einstein in Przibam 1967.
5. Franck 1947, p. 216.
6. Jammer 1974, p. 127.
7. Stern, Otto, in a discussion with Res Jost taped on December 2, 1961, as quoted in Pais 1982, p. 445.
8. Essentially the same lecture appears in a series given at Cornell in 1965 and televised by the BBC. See Feynman 1965, chapter 6.
9. See Jammer 1974, pp. 132–36.
10. Pais 1979.
11. Pais 1982, p. 455.
12. Einstein 1935.
13. See d'Espagnat 1989 for a book-length discussion of the role of reality in physics.
14. Bohr 1935.
15. Pais 1982, p. 456.
16. Jammer 1974, pp. 187–88. At this writing, Nathan Rosen is still alive in Israel, and writing occasionally on the EPR paradox.
17. Unstable particles have uncertainties in their rest masses that result from their finite lifetimes, a consequence of the uncertainty principle.
18. The experiment is easier to perform when the particles are electrically neutral. We will not worry here about the extra complication of having charged electrons.
19. The angular momenta of macroscopic bodies are so much greater than $h/2\pi$ that the quantum effects described here are unobservable.
20. Goldstein 1980, p. 419.
21. Bohm 1980.
22. The story may be apocryphal, but Bohm was said to be the only person, besides the author himself, to have solved all the problems in a notorious textbook on electromagnetism. At least two generations of graduate students in physics and electrical engineering will appreciate the enormity of this accomplishment.
23. Bohm 1951.
24. Bohm 1951, p. 623.
25. Bohm 1952, pp. 166–79 and 180–93.
26. Bohm's final work has just been published at this writing: Bohm 1993.
27. Bohm 1951, p. 612.
28. Bohm 1951, pp. 620–21.
29. Ballentine 1988.
30. Ballentine 1970 proposes an ensemble interpretation of quantum mechanics that continues to be debated. See Hume 1992 for a recent detailed review of the issues.
31. Park 1968.
32. Hartle 1968.
33. Any quantitative measurement can be reduced to a set of propositions that are true or false. Just think of a computer, which uses binary arithmetic to represent numbers.

34. Perhaps quantum mechanics should be programmed on a tri-state computer rather than a binary one.

35. This does not imply that electrons and photons have the same spin. An electron has spin ½ that can be aligned along or opposite any chosen axis. A photon has spin 1 that can be aligned only along or opposite the photon's direction of motion, a consequence of moving at the speed of light. Spin 1 particles with finite rest mass move at less than the speed of light and can have three possible spring components, -1, 0, and +1, along any axis.

36. See, for example, Gribbon 1980.

37. Wheeler, J. A., in Marlow 1978. Wheeler actually preferred an experimental arrangement using half-silvered mirrors as beam splitters, but the effect is the same as in the double slit experiment.

38. Wheeler in Elvee 1982, p. 1.

39. Wheeler in Elvee 1982, p. 17.

40. Wigner, E. P., "The Probability of the Existence of a Self-Reproducing Unit," in Polanyi 1961, p. 232.

41. Albert 1992, pp. 81–82.

42. Zohar 1990, p. 43.

43. Wheeler in Elvee 1982, p. 21.

44. See the discussion in Gardner 1981, pp. 189–98.

45. Wheeler, John A. "Not Consciousness but the Distinction between the Probe and the Probed as Central to the Elemental Act of Observation," in Jahn 1981, pp. 87–111.

46. Stapp 1993.

47. Popper 1982.

4
Hidden Variables

When we invent worlds in physics we would have them to be mathematically consistent continuations of the visible world into the invisible . . . even when it is beyond human capacity to decide which, if any, of those worlds is the true one.

—John S. Bell[1]

Razor Sharp

As we have seen, the problems of quantum mechanics concern its philosophical, or metaphysical, interpretation and not its mathematical formalism or computational methodology. Most quantum mechanics textbooks do not make a clear distinction between interpretations and indeed hardly acknowledge the subject. Even high-level texts will generally incorporate only the briefest introductory philosophical discussion before moving on to the more important business of training the physics or chemistry student to make useful calculations. These calculations can be performed successfully without the slightest attention to imagined superluminal correlations or the other worries and "paradoxes" generated by various interpretations.

Textbooks correctly introduce the wave function Ψ with the Born hypothesis, defining $|\Psi|^2$ as the probability (density) of a particle being found at a particular position, or more generally in terms of the probability of a certain set of measurement outcomes for a quantum system. These texts refer to Ψ as the "state" of an individual system, implying that it is some objective property of that system. However, as we have seen, Ψ technically describes the statistical character of an ensemble of many separate individual systems and not, in general, the quantum state of an individual system unless we are very careful to prepare all the members of the ensemble in the same way.

While the quantum state of an individual system can be specified if one includes the fact that the outcomes of some measurements may be "indefinite," this alternate specification is likewise not an objective property of the system. It too depends on the array of measurements that have been performed, and are planned to be performed, on the system.

In popular books and articles, another misleading view of quantum mechanics is presented. In these media, quantum effects are described as a kind of blurring of reality. This is exemplified by the usual illustrations showing atoms surrounded by electron "clouds," contrasted with the precise circles or ellipses used to indicate the orbits of planets in the solar system. Again, many physicists contribute to this impression by belittling attempts to make the quantum world accessible to common sense.

In a similar vein, the uncertainty principle, $\Delta p \Delta x > h$, is commonly described as a law of nature that places fundamental limits on our knowledge of the properties of individual particles. However, the uncertainty principle is also a statistical statement, and indeed this is how it is used in practice. In experimental applications, the quantities Δp and Δx characterize the distributions of the momentum and position of particles within a data sample that approximates a statistical ensemble.

Despite the definition of the wave function in terms of probability, the consensus of physicists has taken quantum mechanics to apply to individual systems, governing their behavior in a way different from but analogous to the laws of classical mechanics. Furthermore, quantum mechanics has been widely regarded as final, "complete," the best we can be expected to do because of the uncertainty principle.

Einstein, on the other hand, was almost alone in limiting quantum mechanics to a partial, purely statistical role. This may surprise the reader, who has likely heard of Einstein's bemoaning that, "God does not play dice." However these views were not inconsistent, since Einstein also insisted that quantum mechanics was incomplete. Quantum mechanics may be statistical, but Einstein believed that the universe was fundamentally deterministic in a way not described by quantum mechanics.

During the years when Einstein debated Bohr on the nature of quantum mechanics, he was seeking a **unified field theory**, patterned after his general theory of relativity, that would account for all the forces in nature. He conjectured that such a theory would lie deeper than quantum mechanics and account for its apparent statistical nature.

Einstein never attained this goal. Today, forty years after his death, the classical, field theoretic approach to the unification of forces has been largely abandoned. Instead, unification is sought within the framework of quantum field theory. Though unfinished, this strategy has met with considerably more success. The current Standard Model of elementary particles and forces has found a common ground in which the electromagnetic, weak subnuclear, and strong subnuclear forces are all mediated by the exchange of particles of unit spin.

In the meantime, the search for deterministic mechanisms lying beyond quantum mechanics, so-called hidden variables theories, has moved to the fringe of conventional physics. Largely ignored by the bulk of the research community (though gaining increasing attention in the science and popular media), a few physicists, philosophers, and hangers-on have continued a low-budget attempt to revise the foundations of quantum physics.

Hidden Variables Theories

Philosopher Michael Redhead is one of many who have attempted to systematize the various interpretations of quantum mechanics.[2] Let me use his classifications to indicate how hidden variables enter the picture.

Redhead asks how a particular interpretation views the meaning of a physical observable x when that observable is unmeasured and the incompatible variable p has just been measured:

View A: x has a sharp but unknown value.
View B: x has a "fuzzy" value.
View C: the value of x is undefined or "meaningless."

View A is the assumption of counterfactual definiteness referred to in the previous chapter. View B differs from view A in that, while x exists, it is simply not sharply defined. Views A and B can be termed realist views in which the familiar properties of bodies like position and momentum represent aspects of reality just as they do in classical physics. Of course, the result of any measurement of x will give a single, sharp value. But until that measurement is actually performed, the position is still real but "fuzzy," the fuzziness indicating that we cannot predict the result of the measurement of x with certainty.

View B corresponds to the popularized view and represents what Heisenberg probably had in mind when he first developed the uncertainty principle, later being talked out of it by Bohr, who held view C.

View C is usually associated with Copenhagen, which says that we can only deal with what can be measured. The position of a particle is meaningless until it is measured. When the position is measured, the momentum of the particle becomes meaningless.

Sometimes this attitude is taken as implying that physical quantities somehow spring into existence when they are measured, and disappear when an incompatible quantity is measured, and thus depend on conscious acts. But as I have noted, the most common version of the Copenhagen interpretation is basically a pragmatic one. It holds that we must limit ourselves to concepts that correspond to actual observations. Even to say that an observable comes into existence upon measurement assumes we know that the quantity did not exist before measurement, which still ascribes meaning to the concept.

Although view C is often regarded as the orthodox perspective, most physicists still talk as if the position and momentum of a body exist as real, meaningful properties, before and after their measurement. Simply, and rather obviously, the values of these quantities are changed by the interference caused by the measurement process.

However real, the fuzzy properties of bodies in view B imply an indeterminacy in the behavior of those bodies. Since, by Newtonian mechanics, both momentum and position must be sharply defined at some time in order to use the equations of motion to accurately determine the future momentum and position of the body, the uncertainty principle prevents the behavior of physical systems from being completely predictable.

In view A, the position and momentum are not simultaneously measurable with infinite precision; nevertheless they exist as real, sharply defined properties of a body. That is, these quantities are not inherently fuzzy as in view B. This view is associated with hidden variables.

Since both the hidden variables and the laws they obey (if any) are currently unknown, we obviously have lots of room for speculation. This speculation, I must emphasize, is not justified by any strictly empirical considerations at the present time. Conventional quantum mechanics still works in every application, from the data gathered at the highest energy accelerator laboratory to observations made with most powerful telescopes. Most scientists would argue that this is sufficient and that speculation about the unobserved serves no useful purpose. Hidden variables that remain forever hidden are indistinguishable from hidden variables that are nonexistent.

However, while the critical examination of quantum metaphysics is a major purpose of this book, I cannot simply dismiss hidden variables as so much unproven and unnecessary speculation. Perhaps something useful can come out of the exercise. If a realistic, economical, nonmystical interpretation of quantum mechanics can be defined, even with no empirical data to prefer it over other alternatives, then much of the wind will have been removed from the sails of those who claim a profound connection between quantum, mind, and the universe. Some of my colleagues would respond that Copenhagen already fills this bill and that the metaphysical elements that are often injected have nothing to do with its practical application. However true, this position simply ignores rather than refutes the huge popular literature on quantum metaphysics that looks to the Copenhagen interpretation for its authority.

The undisputed statistical behavior of quantum systems can result as a natural consequence of underlying deterministic, or even indeterministic, that is, *stochastic* hidden variables. The observed uncertainties could be the manifestation of fluctuations in a deeper theory, the way classical statistical mechanics follows from the atomic theory of matter.

The theory of atoms as the basic constituents of matter extended our knowledge into new, deeper realms of the physical world. It made unique predictions that were not anticipated by the classical thermodynamics of the nineteenth century, today confirmed in experiment after experiment. The atomic theory is the foundation for chemistry and condensed matter physics and has provided us with many of the material benefits of modern society. Perhaps a subquantum level of matter awaits discovery.

Atoms and Hidden Variables

The atomic theory of matter was first proposed in ancient Greece. It became part of chemistry in the early eighteenth century. Despite this noble pedigree, direct evidence for the reality of atoms was not established until the early twentieth century. And, only in recent times have actual pictures of atoms been obtained by scanning tunneling microscopes and other modern instruments. During all those previous millennia, atoms could have been regarded as "hidden variables."

In the late nineteenth century, Ludwig Boltzmann showed how the principles of thermodynamics could be derived from the statistically averaged motions of the large numbers of the invisible atoms that were assumed to comprise matter. Statistical mechanics became one of the most powerful instruments in the theoretical physicist's toolbox.

It remains to be seen if hidden entities analogous to atoms will be found to explain quantum mechanics. By this date, they have not. We have had quantum mechanics for most of this century, yet no hint of subquantum hidden variables has been found in any of the myriad data collected over that time. To remain a viable alternative, subquantum hidden variables at some point will need to receive empirical support.

This eventually happened in the case of atoms. In fact, many hints, such as Brownian motion, existed in the data and excellent estimates of the scale of atoms had been made long before their direct observation. No such hints have been uncovered for subquantum variables.

Pilot Waves

As mentioned in chapter 3, de Broglie presented what was perhaps the first quantum hidden variables theory in a talk given at the fifth Solvay conference in Brussels in 1927. This was the same meeting that saw the Einstein-Bohr debate intensify and gain public attention. In de Broglie's "theory of the double solution,"[3] a particle has two waves associated with it: the normal quantum wave function Ψ, which accounts for interference effects, and a second wave U that possesses the localization that allows us to identify it with a particle.

De Broglie's theory avoids the problems associated with the interpretation of Ψ by maintaining it as a fictitious, mathematical object from which one calculates probabilities. On the other hand, the bunched U field is the "real" particle; that is, particle behavior is manifested in the localized U wave packet. In short, de Broglie's model postulates that the fluctuations of some yet-undiscovered underlying, subquantum dynamical system accounts for the observed statistical behavior.

The Ψ and U waves in de Broglie's theory are not themselves the hidden variables, since they are associated with observable wave and particle effects. They are no more hidden than electric or magnetic fields. But whereas electric and magnetic fields have visible electric charges and currents as their sources, the sources behind the pilot waves are still hidden from our view. These sources are the hidden variables.

I have mentioned that de Broglie did not pursue his pilot wave proposal after it was strongly criticized by Pauli and others in 1927, though Einstein reacted favorably. Stung by the criticism, de Broglie was eventually won over by the persuasive arguments of Bohr and the Copenhagen school, as were most physicists.

This consensus became almost frozen in cement after von Neumann appeared to prove that hidden variables were impossible. When von Neumann wrote his monumental 1932 book laying the axiomatic foundation of quantum mechanics, he included a section in which he claimed to disprove the possibility of hidden variables.[4] Von Neumann's proof was eventually rejected, not because it did not follow logically from its assumptions but because those assumptions were not all-inclusive.[5] After hearing that von Neumann's proof was not general, de Broglie regretted having given in so easily and once again started writing about pilot waves.

But this was a generation after von Neumann. In the intervening years, at least partially as a result of von Neumann's enormous stature, hidden variables did not receive much notice within the physics community. Two decades after the original work, they were revived by David Bohm. Then, as now, the reception was less than overwhelming.

David Bohm and the Quantum Potential

Bohm developed his hidden variables theory while in exile in Brazil after losing his Princeton job, and effectively his U.S. citizenship, to the anticommunist purges. The theory was published in 1952 in two consecutive articles in the *Physical Review.* [6]

Although amplified over the years by Bohm and various collaborators, the idea is basically very simple and follows from the same line of thinking as de Broglie used for his pilot waves.[7] A mathematical formulation essentially equivalent to Bohm's theory had been published in 1926 by E. Mandelung.[8] Bohm was apparently unaware of this work, although it was sufficiently well known to have appeared in at least one English-language textbook.[9]

By writing the wave function in terms of its amplitude and phase, and substituting it into Schrödinger's equation, Mandelung and Bohm obtained an equation that looked just like classical equations of motion, where a particle's velocity is simply related to its phase. The only difference with the classical equation was an additional potential energy term that was related to the mathematical form of the amplitude of the wave function. This Bohm dubbed the **quantum potential**.

In Bohm's notation, the wave function $\Psi = R \exp(-iS/\hbar)$. Substituting this in the nonrelativistic, time-independent Schrödinger equation and separating the real and imaginary parts gives the two equations:

$$-\frac{\partial S}{\partial t} = \frac{(\nabla S)^2}{2m} + V + Q \qquad (4.1)$$

and

$$\frac{\partial P}{\partial t} + \nabla \cdot \left(P\frac{\nabla S}{m}\right) = 0 \qquad (4.2)$$

where $P = R^2 = |\Psi|^2$ is the usual quantum probability density, V is the potential energy of the particle resulting from familiar forces, and the quantum potential is given by $Q = (-\hbar^2/2m)\nabla^2 R/R$.

Equation 4.1 is just the classical Hamilton-Jacobi equation of motion, where the momentum of the particle is given by $\mathbf{p} = \nabla S$ (a connection from classical physics that de Broglie also had used in his pilot wave theory). Equation 4.2 is the equation of continuity that occurs in classical physics for any conserved current like the electric current and the conserved "probability current" already familiar in standard quantum mechanics.

The quantum potential Q depends only on the mathematical form of the modulus (absolute value) of the wave function, $R = |\Psi|$. Thus it does not fall off with distance like the gravitational potential of a point mass or the electric potential of a point charge. Bohm likened the role of Q to that of a radio signal used to guide a ship, where the information to do the guiding does not depend on the strength of the signal but rather the signal's information content.

In this sense, he seems to agree with those who interpret the wave function as information, or a "propensity," rather than as a real field. Also, we recall from chapter 1 that a similar argument is used by promoters of ESP to explain the absence of a distance effect. Unsurprisingly, Bohm's quantum potential has provided another metaphor for their imagined psi field.

Bohm, as had Mandelung before him, succeeded in writing the Schrödinger equation in the same form as the classical equation of motion, with an additional potential energy term Q. The Bohm twist was one of interpretation: He restored the classical notion that particles occupy specific positions and momenta and move deterministically under the action of external forces. Bohm's theory deviates from classical mechanics with the addition of the quantum potential that exerts another force on the particle and causes it to deviate from classical behavior.

Like the classical electric and gravitational potentials, the quantum potential has a value everywhere in space, not just at the location of the particle. In college physics, the electric potential is computed from a knowledge of the electric charge distribution; the gravitational potential is similarly computed from the mass distribution. Bohm's quantum potential, on the other hand, has not yet been directly computed from any assumed source; indeed no evidence exists for such a source. Instead, Bohm incorporates the potential as part of the otherwise-standard mathematical solution of Schrödinger's equation, subject to boundary conditions that include the experimental setup.

Bohm imagined that the quantum potential, like de Broglie's pilot waves, is controlled by some more fundamental subquantum process to be determined by future scientific developments. When Q is ultimately calculated in this future theory, then the probability $P = |\Psi|^2$ for the particle's possible trajectories will have been determined from some more basic principle than conventional quantum mechanics.

In practice, the quantum potential has been computed—from the wave function, not subquantum processes—for specific problems like the double slit experiment. The resulting theoretical trajectories of the particle can be plotted and shown to exhibit the expected interference pattern.[10] Since that pattern went in, as the wave function, it should be no great surprise that it comes out.

Utilized in this fashion, the Bohm model is capable of giving results identical to those of conventional quantum mechanics, in particular the same statistical predictions. The interference pattern in the double slit experiment is still interpreted probabilistically, with bright bands where the plotted particle trajectories are dense and light bands where the trajectories are of lower density or absent.

But Copenhagen and Bohm differ on the nature of what takes place. In Bohm's picture, the particle is imagined to have a definite trajectory that passes though one slit or the other, while its quantum potential penetrates both slits to give the observed interference pattern. When a detector is placed behind a slit, the potential is suitably affected so that the interference pattern disappears.

Copenhagen, on the other hand, maintains that we cannot describe the particle as passing through one slit or the other since we have no measurement that provides that information.

Copenhagen and Bohm agree that the entire experimental apparatus must be considered in making quantum mechanical predictions about the values observables will have when they are measured. Like Copenhagen,

Bohm's viewpoint is contextual. In both cases, the motion of a particle depends on how it is eventually to be measured at a later time.

Copenhagen explains this incredible fact by asserting that the classical picture of predetermined particle motion must be discarded. The path of a particle cannot be decided until the entire experimental setup, including the locations of all detectors, is considered. Bohm's theory advances another idea: an unobservable, aethereal quantum potential pervades all space and time—past and future—and particles follow the spacetime paths laid out in that aether.

Solving Bohm's two equations is completely equivalent to solving Schrödinger's equation. Not surprisingly, the results are experimentally indistinguishable from any other viable interpretation of quantum mechanics.[11] It follows that no new empirical information is provided about the universe, such as the existence of a quantum aether.

So, most of my colleagues would now ask, "What's the point?" Bohm has claimed he was doing nothing more than demonstrating, by a counter example, the falsity of von Neumann's theorem on the impossibility of hidden variables. Here, the same results as standard indeterministic quantum mechanics are obtained in a deterministic-looking theory akin to classical Newtonian mechanics. But if the results are the same, how can any meaningful difference between the two approaches exist?

Reaction to Bohm's Theory of Hidden Variables

Did Bohm succeed in solving any interpretational problem of quantum mechanics? Einstein, who we have seen was about the only one sympathetic to de Broglie's original hidden variables ideas in 1927, was not so impressed by Bohm's proposal in 1952. In a letter to Born he said, "Have you noticed that Bohm believes (as de Broglie did, by the way, twenty-five years ago) that he was able to interpret the quantum theory in deterministic terms? That way seems too cheap to me."[12]

John S. Bell, whom we will meet shortly as the next important figure in this story, has remarked about the Bohm model, "This scheme reproduced completely, and rather trivially, the whole of nonrelativistic quantum mechanics. It had great value in illuminating certain features of the theory, and in putting in perspective various 'proofs' of the impossibility of a hidden variable interpretation. But Bohm himself did not think of it as in any way final."[13]

Several reasons have been put forward why Bohm's theory was not accepted when it first appeared in 1952, and remains unaccepted today, by the bulk of the physics community. Some have speculated that cold war politics played a role, since Bohm was regarded as a communist sympathizer while his critics, notably Pauli, were staunch conservatives.[14] However, the probable explanation then and now is a simpler one: Bohm's theory made no new, testable predictions. Few physicists are ready to abandon convention until the data demand it.

Contrary to the impression created by popularizer Gary Zukav[15] and others,[16] Bohm did not invent any deep new underlying theory of matter from which quantum mechanics emerges as statistical mechanics emerges from the atomic theory. Bohm's theory only showed that was possible; otherwise it is simply another way of doing conventional quantum mechanics.

There was value in providing an example where hidden variables are not impossible, as von Neumann thought he had demonstrated. Further, Bohm showed that it was still feasible to at least think in terms of individual particles with definite trajectories prior to their measurement, thus rendering unnecessary the mystical notion sometimes attributed to quantum mechanics that particles possess no reality until they are brought into existence by the conscious act of observing them.

In this light, it may appear ironic that Bohm later became the major spokesman for a new kind of science mysticism, the implicate order. This development was in fact connected with his hidden variables—not so much because they were hidden but because they were *nonlocal.*

Shortly we will see how EPR experiments of the type first suggested by Bohm in his 1951 book appear to have ruled out the existence of realistic, *local* hidden variables that control the motion of individual particles. If this is true, then the only remaining possibility for realistic, hidden variables is that they be nonlocal. But as we will also see, nonlocality is a heavy price to pay for a return to deterministic physics, which rationalists as well as New Agers find unappealing anyway.

Bell's Theorem

John Bell studied at the University of Birmingham, receiving his Ph.D. in 1955. Although interested in the foundations of quantum mechanics, he had read about von Neumann's proof of the nonexistence of hidden variables and "relegated the question to the back of my mind and got on with more practical things."[17] By this he meant that he went to work as a theorist at the CERN laboratory in Geneva, the center of European high-energy physics. We see a hint of Bell's famous humor here, where theoretical particle physics is labeled "practical" when compared with the esoteric realms of the foundations of quantum mechanics.

After reading Bohm's papers, Bell returned to his interest in quantum mechanics. In 1964, while visiting the Stanford Linear Accelerator in California, he wrote a paper on hidden variables that he submitted to *Reviews of Modern Physics.* The manuscript was unfortunately misfiled in the journal's offices and did not get published until 1966.[18]

In this paper, Bell pointed out that the assumptions made by von Neumann in "proving" the impossibility of hidden variables, and those in related papers by other authors, were insufficiently general for the conclusion to hold in all cases. That is, the proofs were mathematically correct, insofar as they followed from their assumptions, but the assumptions were too restricted. Like Bohm, Bell used a counter-example to prove his case.

While Bell laid the foundation, the nature of the assumptions in hidden variables impossibility proofs was made precise, at least to experts, in a highly technical 1967 paper by S. Kochen and E. P. Specker.[19] According to the **Kochen-Specker theorem**, "It is impossible to build a hidden-variables theory compatible with quantum mechanics in which the values of certain sets of non-compatible observables are simultaneously determined."[20] Two key assumptions are counterfactual definiteness and **non-contextuality**, where the value of an observable does not depend on the values of other compatible observables.

The same year, 1964, that Bell wrote the *Reviews* paper, he published an article on the EPR paradox in the first of the handful of volumes of a now-defunct journal called *Physics*.[21] In this paper, now cited as often as the 1935 EPR paper itself, Bell proved a remarkable and unanticipated theorem. We may be able to avoid some of the mythology that surrounds Bell's theorem by quoting directly from his conclusion:

> In a theory in which parameters are added to quantum mechanics to determine the results of *individual* measurements, without changing the statistical predictions, there must be a mechanism whereby the setting of one measuring device can influence the reading of another instrument no matter how remote. Moreover, the signal must propagate instantaneously. (emphasis added)

What Bell had succeeded in proving was that realistic, local hidden variables cannot provide a deterministic mechanism for quantum mechanics—if quantum mechanics is to apply to individual measurements.

I have emphasized the word *individual* in the above quotation because considerable misunderstanding exists on the conclusions that may be drawn from Bell's theorem. In the case of hidden variables determining the behavior of individual measurements, the mechanism must be nonlocal, that is, holistic with instantaneous connections between all points in space.

As speculative and metaphysical as it sounds, Bell's theorem was cut from pure scientific cloth because it was *testable.* Analyzing the version of the EPR experiment proposed by Bohm in 1951, Bell showed how a real experiment could be performed to compare the predictions of quantum mechanics with the predictions of any theory of local hidden variables.

Proof of Bell's Theorem

Bell's theorem provides a mathematical inequality that must hold in the Bohm-EPR experiment when the components of the electron's spin along all coordinate axes are assigned simultaneous physical reality. Here physical reality is used in the EPR sense: a measurable quantity surely has an aspect of physical reality when its value can be predicted with complete certainty.

For the sake of completeness, and in order to make a trivial point that nonetheless does not seem to be widely recognized, I will include here a mathematical proof of the theorem. Nonmathematical readers should be able to continue to skip the mathematical sections without losing the thread of the argument.

First I show that Bell's inequality holds when the spin components are treated as completely classical, objectively real angular momenta that can take on a continuous range of values. I have not seen this proof in any of the considerable literature I have personally encountered on the subject, perhaps because the conclusion is too simple and obvious.

Following this, I will repeat a simple proof given in Redhead[22] demonstrating that the same inequality holds when the spin components take on the discrete values allowed by quantum mechanics, but are otherwise as objectively real as the classical spins. Finally, I will derive the quantum result that violates Bell's inequality.

The Bell variation on the EPR experiment is conceptually illustrated in figure 4.1. The source emits pairs of electrons in singlet two-electron states; that is, the total spin of each emitted electron pair is zero. At the end of two beam lines are "spin meters" A and B, inhomogeneous magnets that can be rotated so that measurements can be made of the spins of the electrons along two different axes perpendicular to the beam.

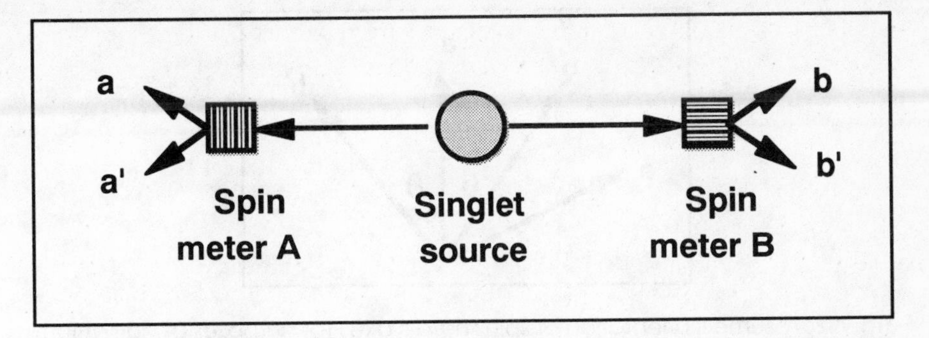

Fig. 4.1. Bell's EPR experiment. A singlet source emits electrons in opposite directions whose spin components along either of two axes can be measured by the spin meters A and B. The results for the two orientations of A are indicated by a, a′ while the results for B are given by b and b′.

Assume the relative orientations of the four spin meter axes shown in figure 4.2. Since we are free to rotate the whole system by any angle, no loss of generality occurs if we assume the electron moving along beam line A has a perpendicular spin component **a** of unit magnitude and points along the axis of A.[23] The spin component measured when the spin meter axis is **a'** will be $a' = \mathbf{a}\cdot\mathbf{a'} = \cos 2\theta$. At the other end of the beam line, the spin components must be such to cancel the spin a, since the total spin is zero. That is, $\mathbf{b} = -\mathbf{a}$ and the magnitude $b = -\mathbf{a}\cdot\mathbf{b} = -\cos\theta$ while $b' = -\mathbf{a}\cdot\mathbf{b'} = -\cos\theta$. For a given angle θ, the spin components a', b, and b' are thus determined and will be the same for every pair of electrons emitted from the source.

Following Bell, we define a function S that measures the correlation between the four spin components:

$$\begin{aligned} S &= ab + ab' + a'b - a'b' \\ &= -\cos\theta - \cos\theta - \cos 2\theta \cos\theta + \cos 2\theta \cos\theta \\ &= -2\cos\theta \end{aligned} \tag{4.3}$$

from which we note that $|S| \leq 2$.

Now let me derive the limits on the correlation S in the case where the spin components are quantized. Because of the quantum nature of electron spin, only two values are possible for each of these four measurements, so let us call those values ±1. Suppose we make N measurements in which all four quantities are determined. For each specific measurement n, we form the following quantity similar to S above:

$$\begin{aligned} g_n &= a_n b_n + a_n b'_n + a'_n b_n - a'_n b'_n \\ &= a_n (b_n + b'_n) + a'_n (b_n - b'_n) \end{aligned} \tag{4.4}$$

Now, either $b_n + b'_n$ or $b_n - b'_n$ must be zero. The combination that is not zero must be ±2. This can be seen from the following sequence of

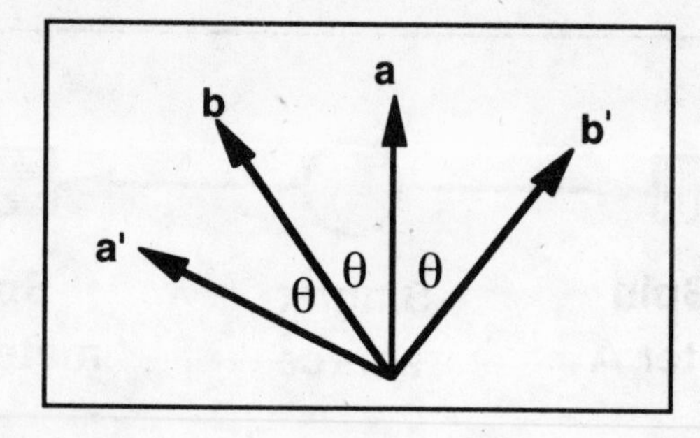

Fig. 4.2. Assumed orientation of spin meters axes for purposes of derivation of Bell's theorem.

possibilities: 1 + 1 = 2, -1 – 1 = -2, 1 – 1 = 0, -1 + 1 = 0. Thus we conclude that $|g_n| \leq 2$, that is, the magnitude of g_n is never greater than 2.

For the N measurements, made on N electron pairs, we can form the **correlation function**:

$$C(a,b) = \frac{1}{N}\sum_{n=1}^{N} a_n b_n \tag{4.5}$$

This quantity measures how well the results at the ends of the two beam lines correlate with one another. A random distribution of values of a and b would give C(a,b) = 0. If the axes of both spin meter A and spin meter B are aligned for all measurements, $a_n b_n$ will equal -1 for all n and C(a,b) = -1, since the total spin is zero. However, we need to consider all possible orientations of the spin meter axes.

We can now compute the quantity S, discussed above in the classical case, by combining correlation functions as follows:

$$S = C(a,b) + C(a,b') + C(a',b) - C(a',b') \tag{4.6}$$

Then, from equation 4.5,

$$S = \frac{1}{N}\sum_{n=1}^{N} g_n \tag{4.7}$$

Since $|g_n| \leq 2$ it follows that

$$|S| \leq 2 \tag{4.8}$$

which is Bell's inequality.

Bell's theorem says that the quantity S that measures the correlations of spins in the experiment must have a magnitude no greater than 2, provided that the components of the quantum spins of the electrons along each beam line are common-sense, objectively real properties that exist independent of any measurements at the end of either beam line. Again, note that this result applies to classical, continuous angular momenta as well as quantized spins.

What does conventional quantum mechanics yield as a prediction for the quantity S? The singlet state of two electrons is given by the state vector: $|\Psi> = (|+> - |->) /\sqrt{2}$, where "+" and "-" indicate whether the spin components are along or opposite the axis along which the spin is being measured.

Note the "entanglement" of the electron spin states. That is, individual electrons in the singlet, two-electron state are not in pure spin states—states with definite spin components—although the total spin is zero. This can happen with a given electron spinning in any direction, as long as the other is spinning opposite. In the jargon of quantum mechanics, we say that the two-electron system was "prepared" in such a way that the total spin was zero, but the individual electrons were not specially prepared to have any particular spin orientation.

From conventional quantum mechanics, the correlation function will be given by

$$C(a,b) = <\Psi \mid \sigma_1 \cdot \mathbf{a} \; \sigma_2 \cdot \mathbf{b} \mid \Psi> \qquad (4.8)$$

where σ_1 and σ_2 are the Pauli spin vectors for the two electrons and **a**, **b** are the unit vectors along the two spin meter axes. Choosing **a** as the z axis, x as the other axis in the plane of **a** and **b**, and θ as the angle between **a** and **b**, we can write,

$$C(a,b) = <\Psi \mid \sigma_{1z} (\sigma_{2z} \cos\theta + \sigma_{2x} \sin\theta) \mid \Psi> \qquad (4.9)$$

Using the singlet form for $|\Psi>$ given above, the Pauli spin matrices for σ_{1z}, σ_{2z}, and σ_{2x}, and doing a standard calculation, we get the simple result,

$$C(a,b) = -\cos\theta = -\mathbf{a} \cdot \mathbf{b} \qquad (4.10)$$

Note that this is not the same as the classical form $C(a,b) = -ab$, where a and b are the spin components along the two axes. To make this a bit clearer, note that the classical form can be written $C(a,b) = \sigma_1 \cdot \mathbf{a} \, \sigma_2 \cdot \mathbf{b}$, which is to be compared with equation 4.10. We see the usual difference between classical and quantum physics, with an exact classical value for a measurement replaced by an ensemble average, the so-called "expectation value."

We can now compute the quantum mechanical prediction for the quantity S defined in equation 4.6:

$$S = -\mathbf{a} \cdot \mathbf{b} - \mathbf{a} \cdot \mathbf{b}' - \mathbf{a}' \cdot \mathbf{b} + \mathbf{a}' \cdot \mathbf{b}' \qquad (4.11)$$

Let the angles between the vectors be shown as in figure 4.2. Then,

$$S = -3\cos\theta + \cos 3\theta \qquad (4.12)$$

Figure 4.3 shows how S varies with θ for the quantum and classical cases.

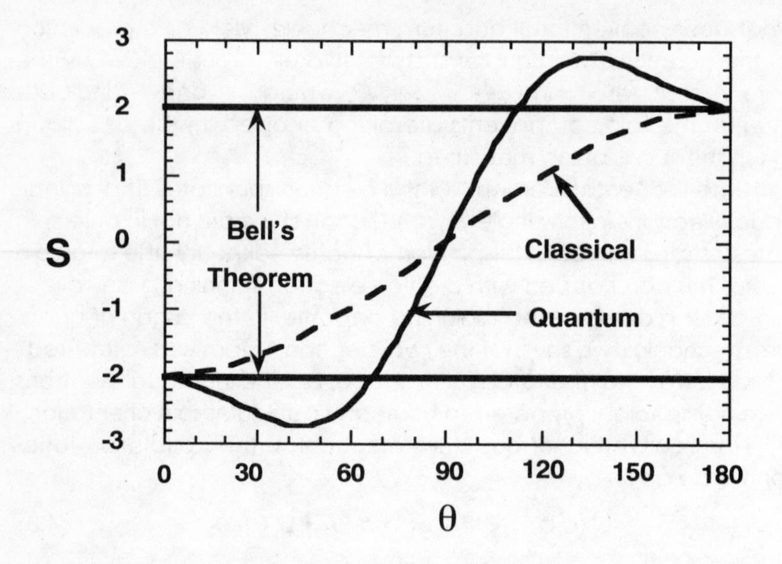

Fig. 4.3. Bell's correlation function S plotted as a function of the angle θ defined in figure 4.2. Shown are the classical and quantum predictions. Also shown is the range allowed by Bell's theorem for any theory of local hidden variables.

Now, in case you skipped all this technical stuff, and even if you waded through it, here's the bottom line: A quantity S computed from classical physics is given by equation 4.3. It has a maximum magnitude of 2. If we have quantized spins, but assume their components along all axes are simultaneously local and real, we get the limit given in equation 4.8, again a maximum of 2. Conventional quantum mechanics gives equation 4.12, which has a maximum value when $\theta = 45°$: $S = 2\sqrt{2}$. This is obviously greater than 2 and so violates Bell's inequality, $S \leq 2$.

Thus quantum mechanics predicts that measurements made at the ends of the two beam lines are more correlated with one another than they would be if the variables involved have the meaning one would expect from common sense, or more precisely, from locality and the EPR definition of physical reality.

> Note why it was necessary for Bell to introduce the complication of two orientations of the spin meters at the ends of the beam lines. If we had only one at each, the results, equations 4.3 and 4.11, would each be $S = -\cos\theta$ and no difference between the classical and quantum results would occur.

Thus Bell provided us with a seemingly straightforward empirical test to distinguish between conventional quantum mechanics and a hidden variables theory in which the still-quantized components of electron spin are objectively real.

Nonstatistical Version of Bell's Theorem

The above derivations for the two-electron system are based on the application of a statistical test, and indeed that is how they have been tested in experiment, as we will see shortly. This leaves Bell's theorem open to the challenge that it applies only to ensembles of systems and not to individual systems. More recently, another version of Bell's theorem has been derived in which the correlations are perfect and so statistics and inequalities are not needed.

Daniel Greenberger, Michael Horne, Abner Shimony, and Anton Zeilinger have analyzed systems of various configurations of three- and four-particle systems.[24] Using examples of spin correlations and spinless multiparticle interferometry, they show that the EPR premises are inconsistent. The incompatibility of EPR with standard quantum mechanics is stronger than that in the two-particle Bell's inequality discussed above.

These proposed experiments are more complicated than the two-particle situation discussed in detail above and their discussion would carry me beyond the technical limits I have set for this book. Furthermore, actual experiments have so far only tested the two-particle correlations. So permit me to limit my discussion to the simpler case, with the understanding that, while the multiparticle examples are undoubtedly interesting, they serve mainly to further confirm the incompatibility between the EPR hypotheses and conventional quantum mechanics.

Tests of Bell's Inequality

In a lengthy1978 paper, John Clauser and Abner Shimony reviewed the six experimental tests of Bell's theorem that had been performed up to that date.[25] All but one involved using photon pairs and the authors derived the formulas that apply in that special case. Recall from chapter 3 that photons have two states of circular polarization that correspond to their spins being in the direction of their motions, or opposite. The results are similar to those above for electrons.

In particular, Bell's inequality, $S \leq 2$, also holds for photons, although the quantum mechanical correlation function is $C(a,b) = -\cos 2\theta$, instead of $-\cos\theta$ as derived for electrons above. Thus the maximum of S occurs at 22.5° rather than 45°, but with the same value of $2\sqrt{2}$, well above Bell's limit.

These preliminary tests largely confirmed the quantum mechanical predictions. However, the definitive series of experiments was reported in 1982 by Alain Aspect and coworkers at the Institut d'Optique Théorique et Appliquée in Orsay, France.[26]

Their experimental setup is sketched in figure 4.4. A singlet (spin zero) source emitted two photons in opposite directions. The linear polarimeters were oriented in the directions indicated in figure 4.2, for various angles θ. The measured result for $\theta = 45°$ was $S = 2.697 \pm 0.015$, significantly above the maximum value of 2 allowed by Bell's inequality. The quantum mechanical prediction for this specific experiment was $S = 2.7 \pm 0.05$, some minor error resulting from a lack of perfect symmetry between the two beam lines. Clearly quantum mechanics agreed with the experiment.

In a paper published six months later, the same authors presented their results for a modified experiment in which time-varying analyzers were placed in the beam lines in order to make the delayed-choice test of the type suggested by Wheeler (see chapter 3). By using a fast switch, they caused their polarizers to randomly jump between two orientations in a time interval short compared to the light travel time between the ends of the beam lines. This guaranteed that any signal between these points would have to travel faster than the speed of light and any correlation observed would have to be nonlocal in the Einsteinian sense.

According to Einstein's relativity, all events that can be connected by a signal moving at less than the speed of light are local since an observer moving along with that velocity will see the two events at the same place. Such events are said to be *within the light cone.* The modification to the Orsay experiment described above was designed to guarantee that any signal between the beam ends be superluminal. And, since the results still violated Bell's inequality, correlations above those expected from locality and "physical reality" were reportedly shown to exist.

No experiment has the perfect efficiencies that are usually assumed in thought experiments, and the Aspect experiment is not an exception. A photodetector is capable of detecting no better than about 20 percent of the

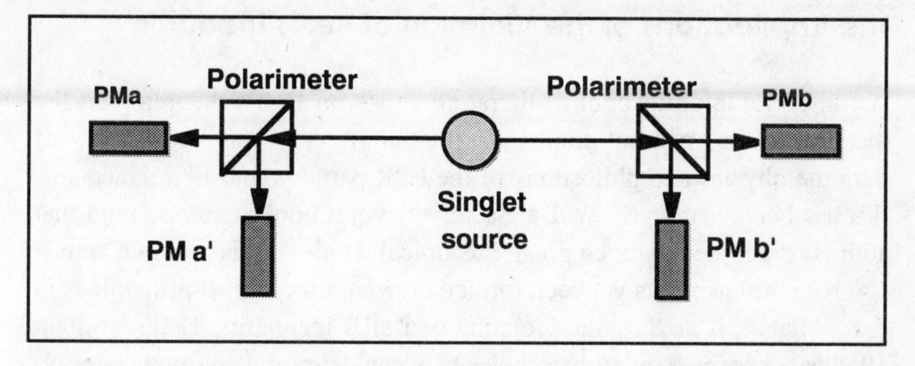

Fig. 4.4. The setup of the Orsay experiment. The source emits photons of the same, unknown polarization. The polarimeters at the end of the beam lines each have two possible orientations. The photons are detected by photodetectors PM.

photons that hit it. Two detectors operating in coincidence, as in the Aspect experiment, then have only about a 4 percent probability of triggering when each is hit by a photon. Other inefficiencies exist in the transmissivity of polarizers, the reflectivity of mirrors, and geometrical acceptance of the apparatus. So we are talking about an experiment that detects about 1 percent of the events it is seeking.

The experimenters were competent and made suitable corrections for inefficiencies, concluding that these could not account for their results. They are probably right. However, it is possible to find a local, realistic hidden variables model that gives the same apparent correlations measured by Aspect and his collaborators.[27] The model requires that the photon detection efficiency depend on the hidden variables. This has been regarded as implausible because it would imply, in some cases, that the insertion of a polarizer in the beam line actually enhances the signal. As I showed in chapter 3, however, this can in fact happen with polarizers; a diagonal polarizer placed between a horizontal and vertical polarizer will result in light being transmitted where previously it was not.

While most experts agree that the experiments likely confirm a violation of Bell's inequality, we should at least take note that this claim rests on the additional assumption that photon detection is independent of the hidden variables in a local, realistic theory. However, since the experimental results in fact agree with conventional theory, namely quantum mechanics, the experimenters are not making an extraordinary claim and so do not require extraordinary evidence. Unfortunately, others have made extraordinary claims about these experimental results, even going so far as to propose that they require a total overhaul of our conception of the universe.

The Implications of the Violation of Bell's Inequality

A considerable philosophical literature exists on Bell's theorem and the meaning of its apparent empirical violation. A very complete review of both the physics and philosophy of the EPR paradox and Bell's inequalities has been written by W. De Baere.[28] Several books can be found that address the issues in much greater technical detail than is possible here.[29]

No consensus has yet been formed on what ontological principle is in fact violated, if any, by the violation of Bell's inequality. Bell's original 1964 paper seemed to imply a choice between determinism (predictability) and locality. That is, violation of either principle would yield his inequality. The universe could be deterministic and nonlocal, nondeterministic and local, or nondeterministic and nonlocal.

A few years later, Bell produced a probabilistic derivation that made no explicit assumptions about determinism.[30] The derivation I gave above, due to Redhead, also does not assume determinism. It still requires something beyond locality, but exactly what?

Jon Jarrett has analyzed the various assumptions that can be made in deriving generalized Bell-type inequalities for Bohm-Bell-EPR experiments.[31] He finds that determinism and locality provide one route, while completeness and locality provide another.

Jarrett uses determinism in the operational sense that I have in this book—predictability with complete certainty. That is, we have a deterministic theory when the probability for a particular outcome is either zero or unity.

Locality is defined by Jarrett as the condition that predictions of the outcome of the observation at one beam end remain the same regardless of how the parameters of the detector at the other beam end are modified. When this is not the case, superluminal signalling is implied.

Jarrett defines completeness as the condition that predictions on the outcome of the observation at one beam end remain the same regardless of the *outcome* of the observation at the other. Although this sounds like another version of locality, and indeed he defines *strong locality* as the condition that prevails when both locality and completeness are operative, Jarrett claims an informal connection between his definition of completeness and the completeness issue raised in the EPR paper.

Don Howard suggests that Jarrett's completeness is equivalent to what he calls **separability**, the condition that each of two previously interacting physical states possesses its own physical state.[32] Indeed, the singlet state of the two-photon system in the Aspect experiment is nonseparable in precisely this manner.

Whatever the terminology, and you can see the verbal contortions some go to, these authors uniformly conclude that Bell's inequality can be broken by way of the violation of some principle other than locality. Nonlocality is a *sufficient* condition for a violation, but all these analyses indicate that it is not a *necessary* condition.

I have tried to summarize the logic in figure 4.5. Each circle represents a condition. The intersecting region of any one of the three circles of determinism, completeness, or separability with the circle of locality will lead to a system that obeys Bell's inequality. For example, any classical theory will obey Bell's inequality because it will be deterministic and local, complete and local, and separable and local. Local, complete hidden variables theories will also obey Bell's inequality, even if they are not deterministic or separable.

While nonlocality is sufficient to violate Bell's inequality, it is not necessary to do so. The inequality can be violated by local theories that are indeterministic, inseparable, and incomplete. This would seem to rule out a local, complete version of quantum mechanics, but this is by no means clear since Jarrett's definition of completeness is not precisely that of EPR.

Physicist Henry Stapp, whose 1972 review of the Copenhagen interpretation has been discussed earlier and whose current theory of quantum consciousness will be discussed later, has provided a derivation that purports to show that locality *alone* is sufficient to derive Bell's inequality.[33] This claim has been disputed by Michael Dickson[34] among others, although Stapp claims to have answered all objections. The proof is very complicated and probably still contains some additional hidden assumptions such as Jarrett completeness.

We will have to wait and see how the consensus develops on this. However, I am inclined to the view that any theoretical version of quantum mechanics that is necessarily nonlocal is likely to be constructed from more than the minimum set of hypotheses needed to describe the data, which remain local. A judicial application of Occam's razor may be able to reduce these hypotheses to a smaller, local set. If current quantum mechanics is, as Stapp insists, nonlocal, then perhaps we should seek another version that is local. If and when superluminal effects are seen in experiment, then we will have the justification required to re-introduce nonlocality into the theory.

I am sure this debate about the ontological significance of Bell's theorem will continue in the philosophical journals while continuing to be ignored by most physicists. Nonetheless, it remains important to clarify this issue. Most authors, including many of those who are physicists, have adopted the view that the empirical violation of Bell's inequality has

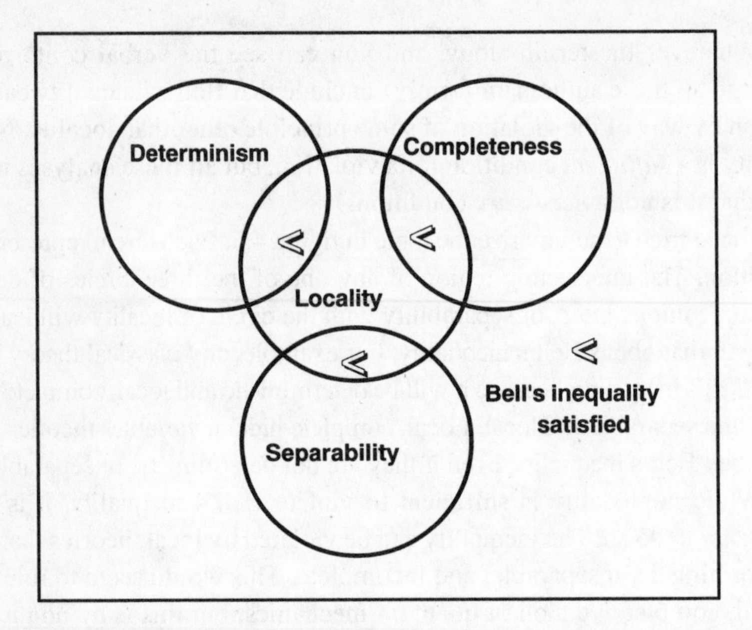

Fig. 4.5. Routes to the violation of Bell's inequality. The symbol "<" indicates the conditions required for Bell's inequality to be satisfied.

demonstrated that quantum mechanics is necessarily nonlocal. This is not in fact the case. In remains possible that a local version of quantum mechanics can be developed, as long as the data do not demand nonlocality. In the following chapters I will provide other, alternative ways to view the EPR paradox that are consistent with all the data and do not require superluminal connections.

Conventionally applied quantum mechanics, independent of interpretation, agrees with the observed violation of Bell's inequality. Until an experimental disagreement is found, no modification of conventional quantum mechanics is necessary. The issue is not the nonlocality of quantum mechanics but whether there exists some yet-undiscovered hidden variables theory that lies deeper than quantum mechanics and must necessarily violate locality.

If the subquantum theory is to be deterministic in the way classical mechanics is deterministic, then it will almost certainly have to be nonlocal, based on the EPR experiments. If the subquantum theory is not deterministic, but still retains those aspects of physical reality that are associated with the concepts of completeness or separability, then it also probably will have to be nonlocal. If we do not introduce any subquantum theory, but more or less stick with quantum theory as we now have it, then it is by no means demonstrated that we have a need for nonlocality.

As long as the possibility for a local route to Bell's inequality remains, whether its violation is manifested as experimental inefficiencies, a failure of determinism, completeness, separability, or whatever, then that local theory will remain the most economical alternative, since it will require the least modification of existing ideas. In the next chapter we will see why nonlocality is the alternative of last resort.

Bohm Revisited

As we have learned, Bohm's hidden variables theory is necessarily nonlocal. Papers written by Bohm and his collaborators after Bell's 1964 paper contain the hindsight that Bohm was thinking nonlocally in his 1952 papers on hidden variables. For example, in 1987 Bohm wrote, in recalling his early work, "It seems clear that at this stage I was anticipating the implicate order."[35] The "implicate order" is Bohm's holistic model of the universe that will be discussed in the next chapter.

However, a look at his original papers suggests that Bohm was (quite sensibly) writing at that time about far more conventional hidden variables, analogous to atoms and statistical mechanics, but operating at distances of nuclear dimensions or less. Bohm is explicit in viewing the positions and momenta of particles as simultaneously "real" but "hidden" because they cannot be measured simultaneously. Particle positions are local by definition.

As Bohm put it in the second of his 1952 papers:

> In our interpretation, however, we assert that the at present "hidden" precisely definable particle positions and momenta determine the results of each individual measurement process, but in a way whose precise details are so complicated and uncontrollable, and so little known, that one must for all practical purposes restrict oneself to a statistical description of the connection between the values of these variables and the directly observable results of measurements. Thus we are unable at present to obtain direct experimental evidence of precisely definable particle positions and momenta.[36]

Bohm never really specified the nature of his hidden variables. If he had, then they would no longer be hidden! Further, Bohm's model gave the same results as conventional quantum mechanics and indeed, as we have noted, provided essentially just another way of writing the Schrödinger equation. This is not to belittle Bohm's contribution, for his 1952 papers are sober and make no claim to be the final story. And, as Bell has emphasized, although Bohm's theory was in a sense "trivial," it helped break the orthodox attitude that the Copenhagen interpretation is the only way to comprehend the meaning of quantum mechanics.

Bohm at some point did fully realize that his theory was fundamentally nonlocal. For one, the quantum potential did not fall off with distance in any obvious way, depending on the form of the amplitude of the wave function and not its magnitude. For another, it depended on taking into account the whole experimental setup, including measuring apparatus.

As I have indicated, the reluctance of the physics community to discard a perfectly workable theory, not Bohm's leftist politics, were likely responsible for his model not being accepted into the mainstream of physics. Despite recent efforts to revive it, especially in the science media, the Bohm theory remains an outsider for much the same reasons.

Nevertheless, Bohm's hidden variables theory has two aspects that offer scant comfort to those asserting a connection between quantum and mind: the wave function does not collapse, and conscious observers need not participate.

> Our basic assumption has actually nothing to do with consciousness. Rather it is that the particles are the direct manifest reality, while the wave function can be "seen" only through its manifestations in the motions of particles. This is similar to what happened in ordinary field theories (e.g., the electromagnetic), on which the fields likewise manifest themselves only through the forces that they exert on particles. (The main difference is that the particles can be sources of fields, whereas, in the quantum theory, particles do not serve as sources of the wave function.) Moreover the conclusion that after an irreversible detection process has taken place, the unoccupied packets will never manifest themselves in the behavior of the particles follows, as we have seen, from the theory itself, and has nothing to do with our not being conscious of these packets.[37]

Just before his death, long after he took up mysticism, Bohm wrote, "It is difficult to believe that the evolution of the universe before the appearance of human beings depended fundamentally on the human mind. . . . Of course one could avoid this difficulty by assuming a universal mind. But if we know little about the human mind, we know a great deal less about a universal mind. Such an assumption replaces one mystery by an even greater one."[38]

Bell and the Paranormal

John Bell was not particularly enamored with the more mystical notions that have come to be associated with him. He once remarked,

> It is easy to understand the attraction of the three romantic worlds [of quantum mechanics] for journalists, trying to hold the attention of the man in the street. The opposite of truth is also truth! Scientists say matter is not possible without mind! All possible worlds are actual worlds! Wow! And the journalists can write these things in good conscience, for things like this have indeed been said . . . out of working hours . . . by great physicists. For my part, I never got the hang of complementarity, and remain unhappy about contradictions.[39]

I had the pleasure of meeting Bell in Italy in 1987. We discussed how promoters of paranormal ideas had jumped on the quantum bandwagon, claiming that quantum mechanics provided a mechanism for ESP, psychokinesis, and other psychic phenomena. Bell told me he received many letters on the subject and tried to be polite and open-minded, but could not offer the psychical crowd much comfort. Bell was a fine gentleman, and I was very saddened by his premature death a little later.

Notes

1. John S. Bell, "Six Possible Worlds of Quantum Mechanics," in Sture 1986 and in Bell 1987, p. 195.
2. Redhead 1987.
3. De Broglie 1964.
4. Von Neumann 1955.
5. Bell 1966. For a complete history of hidden variables impossibility "proofs" and "disproofs," see Jammer, pp. 265–78.
6. Bohm 1952, pp. 166–79 and 180–93.
7. For a recent popular discussion, see Albert 1994, p. 58.
8. Mandelung 1927.
9. Temple 1934, chapter 3.
10. J. P. Vigier et al. in Hiley 1987.
11. While the original theory was developed nonrelativistically, Bohm and others have shown that it can be made relativistic.
12. Born 1971, p. 192.
13. Bell 1987, p. 91.
14. Jammer 1974, p. 290; Cushing 1993, pp. 815–42. On page 292 Jammer mentions an acclamatory review of Bohm's theory written by Hans Freistadt in the Spring 1953 issue

of a Marxist quarterly, *Science in Society.* At the time, Freistadt was an assistant professor at Newark College of Engineering where I was a student, but I do not remember any discussion of either the political or scientific issues surrounding Bohm.

15. Zukav 1979.

16. Albert 1992, Cushing 1993, and Albert 1994.

17. Bell 1982. Bell's most important papers, as well as several interesting informal papers, can be found in Bell 1987.

18. Bell 1966.

19. Kochen 1967.

20. Cabello 1994, p. 179.

21. Bell 1964.

22. Redhead, pp. 83–85.

23. The spin of the electron is $\hbar/2 = h/4\pi$, and I refer to this as "unit magnitude," that is, one unit of electron spin.

24. D. M. Greenberger, M. A. Horne, and A. Zeilinger in Kafatos 1989, p. 73; Greenberger 1990. See also Mermin 1990.

25. Clauser 1978.

26. Aspect 1982.

27. Clauser 1974, Marshall 1983.

28. De Baere 1986.

29. Selleri 1988, Cushing 1989.

30. Bell 1971, reprinted in Bell 1987. See also Clauser 1978.

31. Jarrett 1984, and "Bell's Theorem: A Guide to the Implications," in Cushing 1989, p. 60.

32. Don Howard, in Cushing 1989, p. 230.

33. Stapp 1990.

34. Dickson 1993.

35. David Bohm, "Hidden Variables and the Implicate Order," in Hiley 1987, p. 35.

36. Bohm 1952, p. 183.

37. Bohm 1993, p. 336.

38. Bohm 1993.

39. Bell 1987, p. 181. The three "romantic" worlds are mind/matter duality, Bohr's complementarity, and the many worlds interpretation to be covered later. Three unromantic worlds, according to Bell, are pragmatism, introducing special elements such as nonlinearity or stochasticity, and de Broglie/Bohm hidden variables.

5

Nonlocality, Holism, and the Arrow of Time

> Quantum concepts imply that the world acts more like a single indivisible unit, in which even the "intrinsic" nature of each part (wave or particle) depends to some degree on its relationship to its surroundings.
>
> —David Bohm[1]

Bohm's Implicate Order

As we have seen, experimental evidence for the violation of Bell's inequality indicates that any theory seeking to explain the universe in terms of some yet-undiscovered subquantum entities that determine the motion of individual bodies must necessarily involve superluminal effects. That is, such entities are *nonlocal*. At some point, David Bohm realized that the quantum potential in his hidden variables model was indeed nonlocal, flowing simultaneously through two separated slits while the physical particle it controls passes through one or the other.

Bohm sought a metaphor for his nonlocal potential and found one while watching BBC television, where occasionally a program may be seen that is not brain-deadening. On the screen, Bohm saw a device in which an ink drop was placed in some glycerine that occupied the space between two concentric cylinders. When the inner cylinder was rotated by several turns, the ink spot disappeared as it spread throughout the glycerine. Nothing unusual there. The surprise came when the cylinder was rotated back the same number of turns and the spot reappeared. The information about the localized spot was not lost as the ink spread throughout the glycerine. As Bohm put it, the glycerine contained "a 'hidden' (i.e. non manifest) order that was revealed when it was reconstituted."[2]

Bohm made an analogy between this experiment and the hologram in which a three-dimensional picture can be at least crudely reconstructed from a small portion of the photographic plate where the interference pattern from the original object is stored. Unlike an ordinary photographic plate, where one region on the plate corresponds to one point on the object, each point on a hologram contains information about the whole object.

Bohm used the hologram and ink drop experiments as metaphors for what he called the *enfolded order*: a small region of space that contains information that results from the "enfolding" of an extended order that can then be "unfolded" into the original order. In this view, the universe is like a hologram in which each point in spacetime is not simply connected to every other, but *contains* every other. Bohm had the thought that a constant movement of enfolding and unfolding occurs in nature. Bohm dubbed this the *holomovement.* In interpreting this idea, he reversed the usual reductionist view that the whole is a result of its parts. Rather, Bohm proposed that the whole is fundamental and the parts, which he termed the *explicate order,* are the result of the enfolding of the *implicate order* that is the primary framework of existence.

However, it should be noted that the smaller the region on the plate from which a hologram is constructed, the less sharp the resulting image. Thus one should not imagine that *all* the information about the object is stored at each point in a hologram, just *some* information from all the object. Thus the hologram is not a complete metaphor for the idea that Bohm proposed, in which every point contains the whole.

Bohm's implicate or enfolded order has resonated with those who envision connections between quantum mechanics and Eastern mysticism. As Gary Zukav has put it, "All eastern religions (psychologies) are compatible in a very fundamental way with Bohm's physics and philosophy. All of them are based on the experience of a pure, undifferentiated reality which is 'that-which-is.' "[3]

Danah Zohar credits David Bohm's "picture of undivided wholeness" with "the general revival of Eastern mysticism and its emphasis on the oneness of all things."[4] As discussed in chapter 1, an unbroken wholeness of the universe imbedded in both quantum theory and Eastern mysticism was the central theme of Fritjov Capra's *Tao of Physics.* Capra says quantum theory "reveals an essential interconnectedness of the universe. It shows that we cannot decompose the world into independently existing small units."[5] The Hindu *Brahmin* is described as "the unifying thread in the cosmic web, the ultimate ground of all being."[6]

Quantum holism is not limited to the Eastern mystic-speak that has become so ubiquitous since the rediscovery of India in the hippie 1960s. Non-hip Christian theologians have also been inspired by the new metaphysics, as ever-flexible Christianity adapts to the New Age.

On the Western shores of the seas of New Age philosophy and theology, Robert John Russell has interpreted Bohm's ideas as transcendent features of nature that "could correspond to a divine presence."[7] David Trickett imagines humans as being images of the implicate order, and wonders if God is a projection of this image.[8] The holomovement has been suggested by Kevin J. Sharpe to be manifest in the personal, Christian God himself, putting a twentieth-century physics spin on Aristotle's notion of God as the Prime Mover: "The holomovement God is the source of all our objective and subjective experiences. Thus, God could relate to us personally. . . . Since the holomovement is more complex and internally connected than the brain, one could think of it as having the highest form of consciousness. It might even be pure consciousness. Thus God's consciousness transcends ours."[9]

And so modern theologians have finally solved their age-old problem of how many angels can dance on the head of a pin—an infinite number, because the head of a pin is a hologram!

The implicate order is the kind of idea that appeals to those who are less interested in the truth, whatever it may be, than in rationalizing their already-existing presumptions of truth. They use scientific argument not as a spade to probe for unknown answers to deep questions but as a brush to lay down a veneer of respectability over answers they have already determined.

Is there any depth of substance to the implicate order and the holomovement? What exactly is the implicate order? How does one utilize the concept in any practical way, test it, or link it with scientific methodology or anything else for that matter, including religion? How can any sense be made of the concept without testable consequences?

Bohm could only suggest vaguely that his quantum potential is the substance of the implicate order. J. P. Vigier and his collaborators have proposed that the quantum potential originates in "non-locally correlated stochastic fluctuations of an underlying covariant ether."[10] But we have seen that the quantum potential, as introduced by Bohm some four decades ago, has no unique empirical content.

Nothing of empirical significance has grown out of the development of Bohm's ideas over all that time. Undoubtedly Bohm, a highly competent physicist, thought hard about how to put some experimental meat into his holistic model. But he died before achieving that goal, and the rest of us probably will too.

Nonlocality

Quantum mysticism contains two aspects that are usually muddled together, as they were in the theological speculations about Bohm's holomovement. These aspects are consciousness and nonlocality (or holism). As we have seen, the connection between human consciousness and quantum mechanics appears in an unwarranted, mystical reading of the Copenhagen interpretation, at least in some writings.

Henry Stapp has argued, with the apparent assent of Heisenberg and Rosenfeld, that the founders of Copenhagen had a pragmatic intent that did not include mystical notions. However, we saw in chapter 2 that this seems to be disputed by the historical record.[11] Whatever the historical facts, however, pragmatism certainly characterizes the views of most practicing physicists today.

Consciousness and nonlocality surface within the theoretical frameworks of certain classes of interpretations that go beyond the pragmatic applications of conventional quantum mechanics. These interpretations regard mathematical entities, such as the wave function or quantum potential, as representing real, existing fields that directly control the behavior of individual particles and may be closely related to the nature of consciousness.

Even without the direct intervention of human consciousness, a new subquantum theory lying below conventional quantum mechanics may still provide a basis for a holistic universe that will vindicate those who promote a holistic model of reality. The holistic universe need have nothing to do with consciousness.

It must be emphasized, however, that the holistic solution is a deterministic one, which is not what most holists really have in mind. While no data exists to provide any hints at the nature of a subquantum theory, our experience with atoms warns us not to be to hasty in ruling it impossible.

An important question remains, however. How can the nonlocality of hidden variables of any form be reconciled with Einstein's assertion that there exists a limit on the speeds, namely the speed of light, at which signals can pass from one point in space to another?

The implications of a limitation on signal speed can be illustrated with the help of figure 5.1. This shows a spacetime diagram like those used in relativity discussions, with the time axis up and the three spatial axes reduced to two and indicated as lying in a place perpendicular to the time axis. "Events," indicated by black dots, are specified by their positions in space and their times of occurrence. For convenience, we can place one event at the origin—the event occurring "here" and "now" to a particular observer. Another event at a different position or time can then be located relative to the first event by its separation from the origin.

According to Einstein, only events within the **light cone** shown in the figure can communicate with an event at the origin, since only those spacetime points can be connected by a signal traveling at the speed of light or less. We call these "local," although they may be separated in space on any given diagram. Such events will occur at the same place to *some* observer, namely the one moving along with the connecting signal.

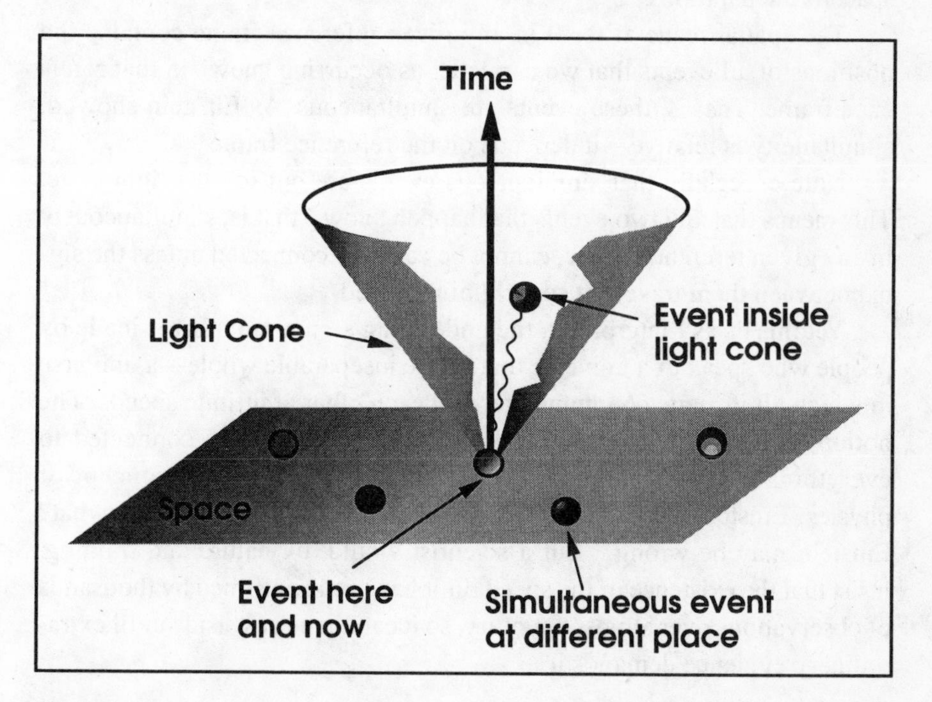

Fig. 5.1. Representation of events in space and time. For illustration, space is indicated as a two-dimensional plane, with the time axis vertical. An event occurring "here" and "now" is indicated at the base of the time axis. It can only be causally connected to events inside the light cone. To reach an event outside the light cone, a signal must travel faster than the speed of light. Simultaneous events at different places cannot be connected, since they lie outside the light cone.

For example, Paris and Lyon may be separated in a reference frame fixed to the earth, but they are at the same place in the reference frame of the high-speed train that runs between them. If you are riding on the train with your girlfriend, and kiss her once on leaving Lyon and again on arriving at Paris, the two events are local. They occurred at different times but at the same place—on the same lips.

The postulates of relativity require that all frames of reference be equivalent. So, if two events are local in any realizable frame of reference, they must be considered local in all frames of reference.

Physicists refer to the separation between two spacetime points within a light cone as **timelike**, since they will occur at the same place to some observer, but generally at different times. Two spacetime points outside the light cone are said to have **spacelike** separations, since they will occur at the same time to some observer, but generally at different positions. Nonlocally connected events lie outside one another's light cone and have spacelike separations.

The spatial plane at $t = 0$ in any given reference frame contains the positions of all events that we can label as occurring "now" in that reference frame. That is, these events are simultaneous. As Einstein showed, simultaneity is relative—it depends on the reference frame.

Note especially that simultaneous events are outside the light cone. This means that any two events that happen "now," that is, simultaneously in any given reference frame, cannot be causally connected unless the signal between them travels at superluminal speed.

Yet this is, as I interpret it, the underlying assumption that is made by people who speak of a universe that is one inseparable whole—a universe in which all its parts communicate with each other at infinite speeds. The notion that everything in the universe is instantaneously connected to everything else fundamentally conflicts with a very basic concept of physics: Einstein's speed limit. Now, a nonscientist might say, "So what? Einstein may be wrong." But a scientist would, by nature and training, insist that the existence of the speed limit has been confirmed by thousands of observations over almost a century, so it cannot be cast aside until extraordinary evidence demands it.

Holistic = Nonlocal

In this book, I often use the words "holistic" and "nonlocal" interchangeably. This is not common usage, so let me explain my rationale. Normally we should assume words to mean what most people take them to mean. However, in the case of holism, two meanings, one trivial, and one profound, are commonly mixed together in vague discourses on wholeness.

Profound implications, labeled "holistic," are often wrongly read into the trivial fact that different parts of the physical world are connected to one another. For example, promoters of holistic (sometimes "quantum") medicine assert that the brain is connected to the rest of the body and so mental states must be considered in the treatment of any body part, even the distant big toe. Of course the brain is connected to the body—but within the light cone. Signals travel from the brain to the big toe and back, but no faster than the speed of light.

Writers use holistic in this trivial sense to argue that a new paradigm should replace reductionist materialism. They may turn out to be right about the paradigm, but psychosomatic illness, the placebo effect, and other brain-body phenomena provide no evidence that the universe is one simultaneously connected whole. These phenomena have perfectly material, biological explanations that fall well within conventional, reductionist scientific methodology.

So I interpret the intent of the term **holism** to refer nontrivially to the simultaneous, or more generally superluminal, connection between spatially separate events. That is, the holistic universe allows for all points in spacetime to be connected, including those outside the light cone. By using the term holism in this way, we can make a clean distinction between the trifling examples that holists often give, which prove nothing, and their intention of expounding a deep, revolutionary concept.

In a universe with purely local connections, forces still are transmitted from one body to another body located at a different position; the carriers of these forces do not move faster than the speed of light, however. Indeed, this is the modern physical picture of the interactions between bodies. For example, an electron emits a photon, recoiling backward as it does so. The photon then travels at the speed of light to the place of a second electron, where it is absorbed by that electron. Thus energy and momentum are transferred from the first electron to the second, and each is repelled away from the other.

Put simply, subluminal connections are normal, reductionist physics. Superluminal connections are holistic.

The idea of the holistic universe is very serious in its implications. It says that electrons do not have to await collisions with photons sent out by other electrons. They, and all the other bodies in the universe, act in continuous concert with one another. In the holistic view, the bodies of the universe participate in a cosmic dance. The universe is one inseparable whole that cannot be completely understood by the traditional reduction to particles that move about independently until coming in contact with other moving particles.

Nonlocal in Theory

As I have indicated earlier, a holistic universe was not ruled out by nineteenth-century physics. It had become an inherent part of Newtonian physics with the introduction of the notion of fields. Indeed, we saw how Newton was puzzled by the apparent "action at a distance" implied by his theory of gravity.

Prior to the twentieth century, no limit on the speed of bodies was known to exist. So, it was conceivable that a signal travelling at superluminal or even infinite speed could move between events. Furthermore, the fundamental forces of gravity, electricity, and magnetism were well-described in terms of fields pervading all of space that seemed to be capable of affecting events at great distances. Those who find the notion that quantum mechanics is nonlocal so profound should pause to reflect on the fact that Newton's theory of gravity is also nonlocal. Those who claim that twentieth-century physics has "discovered" a holistic universe should pause to consider that the universe was holistic prior to twentieth-century physics. It was the greatest scientist of the twentieth century, Einstein, who destroyed holism.

Now I must admit that I have taken a somewhat narrow view of the role of nonlocality in physical theory, insisting that it be applied to those physical notions that have an aspect of reality to them. I believe this view is essential in delineating fundamental issues that have been severely obfuscated.

Certainly we can imagine purely mathematical entities that are nonlocal. In a prosaic example, consider a mathematical quantity that measures the wealth of an individual, say, her credit potential. Suppose our subject has entered a publishers' sweepstakes where the prize is $10 million and has the good fortune to win. At the instant when her number is chosen, she is a thousand miles distant from the drawing location. Nevertheless, her wealth instantaneously, nonlocally (if you assume it is something attached to her person) increases by $10 million.

We can call wealth a nonlocal variable since it truly represents a property of our subject that changes the moment her number comes up at the drawing. An economic theory based on this concept of wealth would indeed be a nonlocal theory. If a computer at the drawing site was programmed to make an investment the moment she won the lottery, that could have happened before any signal reached her. Nevertheless, the practical effect of the drawing must still await the transmission of a signal to the winner, and to the winner's bank account and credit agencies.

Similarly, a nonlocal theory such as Bohm's hidden variables can be conceived that describes what is observed in physics experiments. But the practical information flow still moves within the light cone, and this is all that counts when the time comes to discuss the practical effects of the theory. In my view, the ideas of nonlocality and holism that are used in discussion of quantum mechanics are inconsequential and superfluous, like my economic example. They do not by themselves justify the conclusion that quantum phenomena are fundamentally nonlocal. The main feature that distinguishes quantum from classical phenomena is their contextuality, their dependence on the full experimental arrangement. Nonlocality will not be demonstrated until some direct superluminal flow of particles or information is observed to take place.

Faster-Than-Light Communication?

Let us examine in some detail whether the correlations implied by the apparent empirical violation of Bell's inequality can be used to communicate superluminally, indeed instantaneously, between two separated points in space.

Such a possibility has been proposed in the past by physicists Jack Sarfatti and Nick Herbert, among others.[12] As mentioned in chapter 1, Sarfatti and Bohm had witnessed a demonstration of Uri Geller's alleged psychic abilities in London in 1974. Sarfatti came home from his London stint convinced that the mind is a quantum system, able to affect other quantum systems such as radioactive nuclei, as reported in Geller's London performance. Since then Sarfatti has been involved with Herbert, Fred Alan Wolf, and others in various initiatives that promote the quantum-consciousness connection, though Sarfatti told me he no longer believes in Geller's powers. Gary Zukav acknowledges Sarfatti as a "catalyst" for his book on science mysticism, *The Dancing Wu Li Masters.*[13]

Superluminal communicators based on the EPR experiment of the type proposed by Sarfatti utilize the notion that a decision made at one end of the beam line, such as the orientation of a polarizer, will instantaneously affect what is observed at the other end. This would seem to suggest that signals can be sent from one beam end to the other faster than the speed of light.

One version of such a device has been analyzed by David Mermin.[14] Mermin considered an apparatus similar to the Orsay experiments described in chapter 4, except that electrons rather than photons are used, as in Bohm's version of the EPR experiment.

The basic experimental setup is the same as illustrated in figure 3.7. At the center is a singlet (total spin zero) source S of pairs of electrons that are emitted along the two opposite beam lines. At the end of each beam line is a "spin meter," an inhomogeneous magnetic field that is used to measure electron spin. Each spin meter can be rotated about its axis so that the spin component of either electron can be measured along any desired axis. For simplicity, the beam lines are taken to be equally distant so the electrons from a given pair arrive simultaneously at each beam end (in the laboratory reference frame).

Also for simplicity, only three possible settings, 120° apart, as illustrated in figure 5.2, are considered. Electron detectors in each spin meter measure whether the electron is spinning along or opposite the selected axis. At spin meter A, a red light on the apparatus flashes on in the first case and a green light in the second. This arrangement is reversed at spin meter B so that the same color light will flash at both ends when the axes are in the same position and the electrons from a singlet pair pass through. Let us label the three directions 1, 2, and 3.

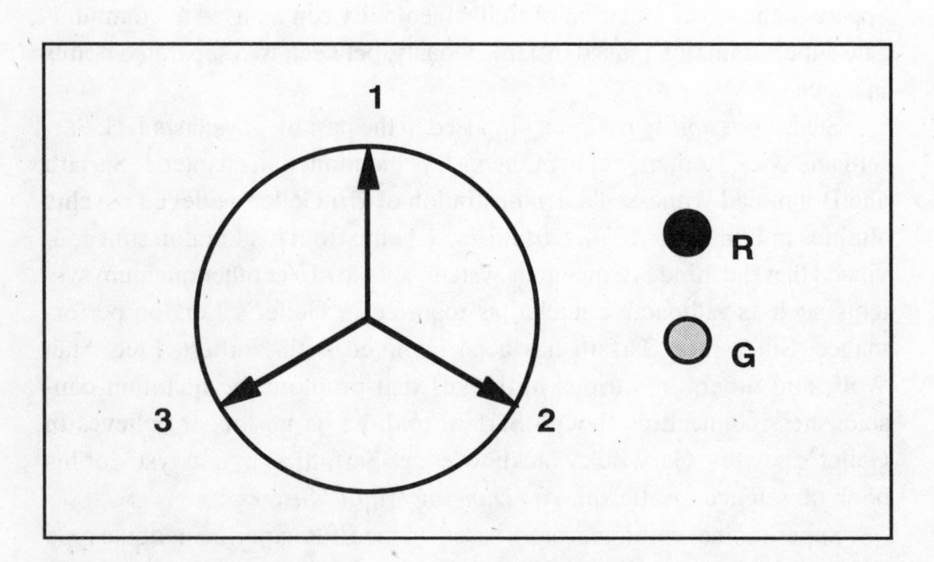

Fig. 5.2. Each spin meter in an EPR communicator can be rotated to one of three positions 1, 2, and 3 that are 120 degrees apart. When at a given setting, the spin meter measures whether the electron's spin is along or opposite the chosen axis. Lights flash red or green depending on which of the two results occur.

Now suppose that Alphonse at the end of beam line A wishes to send a message to Beth at the end of beam line B. He can attempt to do this by coding a message in terms of the three spin meter settings. Suppose he sets his spin meter to 1 and observes the red light flash, meaning that an electron spinning along the 1 axis was detected. At the end of beam line B, Beth sets her spin meter to 1 and detects the other electron in the singlet pair and sees her red light flash.

Suppose Alphonse next sets his spin meter to 2 and sees a green light. If Beth moves her meter to 2, she will also see a green light, and a message 1-2 will have been transmitted. Thus it would appear that a signal containing the information 1-2 has been transmitted instantaneously from Alphonse to Beth.

However, note that in the above example Beth had to know ahead of time to set her spin meter to 1 and 2 consecutively. Thus she already knew the message and no new information was transmitted. In order for information to be transmitted, Beth must try different settings. Still, one might think that information transfer is still possible since the spin measurements at the ends of the two beam lines are more correlated, according to quantum mechanics, than they would be if the measurements were independent.

Mermin tested this hypothesis by writing a simulation program on a computer and printing out the results. Rather than simply reprinting his results, I have written my own program to check for myself and add a trimming. Mermin's program sends a random message from A, but any message should exhibit the same results, namely, the lack of information transfer. In my simulation, I can allow Alphonse to send any arbitrary message and then see if the message is received.

For purposes of illustration, assume Alphonse tries to send the message 1-2-3, repeating it over and over again by setting his spin meter consecutively in each of the three orientations and detecting an electron in each case. The computer then randomly decides whether the spin of the electron detected by Alphonse is along or opposite the chosen axis, that is, whether Alphonse's light flashes red or green. He has no control over this.

The computer, acting for Beth, randomly chooses her spin meter orientations and determines, from the rules of quantum mechanics, what color light flashes at Beth's detector. Those rules are simple: (1) if spin meters A and B are at the same position, the lights will flash the same color; (2) if the spin meters are at different positions, they will flash colors randomly with the same color as A occurring on average one-quarter of the time ($\cos^2 120° = 1/4$) and the opposite color three-quarters of the time.

The program is given at the end of this chapter. It is only about forty lines and can easily be run on a personal computer. I encourage the reader to try it out.

Sample Results

Following is a typical output. Each run will be different in detail if started with a different random number. The first sixty trials are listed below, and the totals for the runs of six thousand trials are given at the end.

The three spin meter positions are indicated by 1, 2, and 3. R and G specify the two colors that represent electron spins being measured along or opposite the spin meter axes. Thus "1 G 3 R" means spin meter setting 1 for the message transmitted at A with a green light flash, and spin meter setting 3 at the receiver B seeing the red light flash. The message 1-2-3 is repeatedly transmitted, so each line below corresponds to two attempts to send this message. For example, in the first line the message being sent is 1-2-3-1-2-3 and the message received is 3-3-2-1-2-1.

I have indicated in boldface type those cases where the spin meter settings are the same. Note that whenever this occurs, the same colors flash. But as we see by the totals at the end, no information is transmitted.

1 G 3 R	2 G 3 G	3 G 2 R	**1 G 1 G**	**2 G 2 G**	3 R 1 R
1 R 1 R	**2 G 2 G**	3 R 2 R	1 R 3 R	2 G 1 R	**3 G 3 G**
1 R 2 R	2 R 3 R	3 R 2 R	**1 G 1 G**	**2 G 2 G**	3 G 1 G
1 G 2 R	2 R 3 R	**3 R 3 R**	1 G 2 R	**2 R 2 R**	3 G 1 R
1 G 3 R	**2 R 2 R**	3 G 2 R	1 G 2 R	2 R 1 R	3 R 2 R
1 R 2 R	2 G 3 R	3 G 1 R	**1 R 1 R**	**2 R 2 R**	**3 R 3 R**
1 R 2 R	2 R 3 R	3 R 1 R	**1 R 1 R**	2 G 1 R	3 R 1 R
1 G 2 G	2 G 3 R	3 R 2 R	1 R 3 R	2 G 3 R	3 R 2 R
1 G 1 G	2 G 1 R	3 R 2 R	1 R 2 R	**2 G 2 G**	**3 R 3 R**
1 R 2 R	2 G 3 R	**3 G 3 G**	1 G 3 G	2 R 3 R	3 G 1 R

Summary of results from six thousand trials:

Fraction of light flashes of the same color = 0.499
Fraction of light flashes of different color = 0.501
Expected statistical fluctuation: ±0.01

We see that the lights flash the same color half the time and different colors the other half. The reader is invited to run the program with different messages. You will find the same results for any message: equal numbers of red and green light flashes, within statistical fluctuations. You are also invited to try other strategies for Beth to select the orientation of spin meter B. Clearly if she simply leaves it in one position, she will still get half red and half green. No information is transmitted from the end of one beam line to the other, despite the apparent correlation that exists when the two spin meter settings are identical. We can conclude from this that no signals move at superluminal speeds in our EPR experiment.

Further, the application of conventional quantum mechanics to the experiment did not involve any unexpected statistical results. The $\cos^2\theta$ rule used to compute the probability that an electron will be detected with a spin axis making an angle θ with the spin axis of the other electron in the pair, 120° in our case, is exactly the same as given by Malus's law for polarized light, discovered by Etienne Louis Malus in 1809. This result is trivially derived in today's physics classes by simply projecting the field vector along the new axis and squaring to get the intensity. It applies equally well for electrons in this case.

Correlation without Communication

Correlation over spacelike separations does not require superluminal communication, nor quantum mechanics. As noted in chapter 1, detectors located at separated points along a wave front can simultaneously receive signals from a transmitting station and obtain whatever information is carried by that signal. For example, a television station sends out modulated electromagnetic waves that carry information that is reconstructed by thousands of television sets located all over the region. All the sets along a circumference of equal distance from the station receive the same picture simultaneously. Note that the television sets have to be tuned to the same channel to get the correlated message. They could be miles from one another, but this does not imply that the sets are communicating with each other at infinite speeds, telling each other what channels are selected.

We can imagine a wave function (thinking of it as some kind of real wave) being emitted from the singlet source in an EPR experiment analogous to the electromagnetic wave from the television station. It informs the spin meter detectors that the electrons they are receiving have a certain chance of being observed to deflect one way or the other as they pass though the magnetic fields of the meters, for various orientations of their field axes.

The orientations at both ends must be known in order to compute the correct probability. This is analogous to having to know when two television sets are tuned to the same channel to predict what is being viewed. However, a spooky action at a distance is implied when we interpret the setting at one end to instantaneously affect what is known at the other. In the television metaphor, it would be as if changing channels on my set immediately affected what you see on your set across town.

There are correlations between the individual measurements in our simulation of the EPR experiment. The table above, which lists simulated individual events, shows that the lights always flash the same color when the orientations are the same. Of course they should, or else angular momentum would not be conserved and we would have a severe violation of a long-established principle of physics. This principle alone, without quantum mechanics, requires at least some correlation between the results observed at the two beam lines.

Still, the experimental violation of Bell's inequality implies an additional correlation beyond this. My computer program uses a knowledge of the orientation of A in determining the outcome of the measurement at B. The relative angle between the two was assumed to be known in calculating the probability for the light to flash a given color.

The program simulates individual events in a statistical fashion that is consistent with the Born probability postulate $P = |\Psi|^2$. The probability follows from quantum mechanics.

> $P = \cos^2\theta$, really just Malus's law again, where θ is the relative orientation of the spin meters at the ends of the beam lines.

But, you might ask, doesn't a superluminal correlation exist in the way I implemented the quantum mechanical theory to simulate the outcome of individual particle detections? Does this not demonstrate that nonlocality is required after all? Actually, it nicely illustrates the point I have been trying to make that nonlocality can exist in theory, or, as in this case, computer simulation, without being required in reality.

A computer simulation, like a mathematical theory, can contain nonlocal correlations among the abstract entities it uses to represent an experiment. It matters not how we choose to manufacture our theoretical constructs, so long as they successfully describe the observed data and so long as people do not assume these constructs correspond to "real" elements of nature.

In the computer program, we simulate the experiment event by event. Each experiment has a given relative orientation of spin axes that must be assumed in simulating the outcome under those conditions. This is the contextual nature of quantum physics.

In the process we generate individual outcomes that are not specifically predicted by quantum mechanics. Quantum mechanics was only used to give the probability for a given outcome; the rest was determined by the computer's random number generator. We have done nothing different from what we might have done in simulating a coin toss by calling the random number generator and printing H when it returns a number between 0 and 0.499999 and T when it returns a number between 0.5 and .999999. The fact that we are simulating individual outcomes does not mean that we have programmed a theory that determines the outcome of an individual coin toss in the real world. In fact, we are probably simulating something close to what actually happens in a nondeterministic universe. Perhaps, as some have suggested, the computer program is a better metaphor for physical systems than mathematical solutions to the equations of physics.

Both our simulated experiment and conventional quantum theory agree on the observed outcome of an attempt at superluminal communication by means of an EPR apparatus. It would not work, and so Einstein's speed limit would remain inviolate.

Time's Arrow, Cause and Effect

The assertion is often made that Bohm's quantum mechanics, or other nonlocal interpretations, do not violate Einstein's relativity. Strictly speaking this is true, even though they may have superluminal connections. Superluminal motion is in fact not forbidden within the formal theory of special relativity. Einstein's basic equations can be satisfied by particles that travel faster than the speed of light c *but never slower.* These hypothetical particles, called **tachyons**, have a lower limit of c to their speeds, just as normal particles have an upper limit c. They are entities of a different type than normal, subluminal **tardyons**.

Tachyons cannot be produced by accelerating tardyons. They have a world all their own. Tachyons have strange properties, such as gaining speed as they lose energy. Thus they move at infinite speed when they have zero kinetic energy. They would indeed be exotic particles if searches for tachyons could reveal their existence. So far they have not.

In his original development of special relativity, Einstein had made an additional postulate that ruled out superluminal motion. He applied the common notion of a cause-and-effect connection between events, what I will call **causal precedence** (sometimes this is called **Einstein causality**) in order to distinguish the concept from the more generic *causality*, which suggests other meanings, such as a simple connection between events.

Given two connected events, and assuming some time direction, the earlier event is conventionally labeled as the "cause" and the later event is labeled as the "effect." Einstein asserted that this relationship cannot be one way when an observer is moving at one speed with respect to the events being observed, and the opposite way for an observer moving at another speed.

The time sequence can be reversed, however, when the relative speed is superluminal. Einstein rejected this possibility, insisting that what you call the cause and what you call the effect cannot depend on your frame of reference. Since then it has become standard folklore that nothing can move faster than the speed of light.

Einstein's causal precedence postulate did not mean that cause and effect cannot be interchanged. Indeed, they often are. More precisely, cause and effect are not always distinguishable, especially at the level of elementary interactions.

All chemical, nuclear and most elementary particle reactions are reversible. That is, if A+B → C+D, then the time-reversed reaction C+D → A+B occurs with the same basic probability.[15] Thus A+B can "cause" C+D, or C+D can "cause" A+B. But given a specific time sequence for a reaction, that reaction will appear in the same time sequence in all reference frames moving at subluminal relative speeds. This is the meaning of causal precedence.

Causal precedence, however, does not seem to be required at the level of elementary interactions. Although a notion of causality as a connection between separated events is retained at this level, the labels "cause" and "effect" here are arbitrary. Two electrons recoiling from one another under the exchange of a photon can be equally well described by the photon being emitted by either electron and absorbed by the other. Cause and effect are indistinguishable for such processes.

Certainly the notion that cause must precede effect is common sense. It agrees with everyday experience, and indeed is built into most scientific theories. It would be difficult to imagine biology or psychology without a concept of prior cause, subsequent effect. But causal precedence may be an *emergent* property. Emergent properties are those that arise out of interactions in complex material systems. They do not necessarily correspond to principles that exist at the elementary level.

We are gradually converging on the view that much of our difficulty in interpreting quantum phenomena comes from prejudicially assuming deep significance to familiar, common-sense concepts that are in fact nothing more than arbitrary conventions. The notion of causal precedence may be such a concept. Another very closely related one is the **arrow of time**.

Ludwig Boltzmann proposed that the arrow of time of common experience is a purely statistical phenomenon, meaningful only for systems of large numbers of particles. Basically, we define the arrow of time as the direction of most probable occurrences, which in the case of macroscopic systems leads to an apparent directionality to time.[16]

Consider the experience of aging, which is the clearest, most personal testimony supporting a specific direction for time. Technically, the atoms of our body can reassemble by chance into a more youthful-looking configuration. However, it is far more likely for the erratic movements of these atoms to degrade our once smooth-skinned and slim-waisted structures. We define the arrow of time as the direction in which we age and assume that the time-reversed process, getting younger, is impossible. In fact it is not impossible, just highly improbable. At the quantum level, no such consensus on the direction of time is possible.

If we lived in a world with few particles, we would not have any basis for assigning a direction to time. Time-reversed processes would have about the same probabilities as processes in the original time direction. However, systems of large numbers of particles, like our macroscopic world, are characterized by probability distributions that are very sharply peaked about the most probable outcome. This is sometimes called the *law of large numbers.*

From very general principles, the fractional width of the probability distribution, which measures the scatter of outcomes that result from statistical fluctuations, goes like $1/\sqrt{N}$, where N is the number of particles in the sample. For example, a liter of hydrogen gas contains about 10^{26} particles. The pressure of the gas will fluctuate by one part in ten trillion ($1/10^{13}$).

Because of such tiny fluctuations, many observations that we make on the macroscopic scale are smooth and predictable with such a high probability that we tend to think of them as certain. We assign the time direction of that high predictability as the direction of time.

Boltzmann connected the arrow of time to the *second law of thermodynamics,* in which a quantity called the *entropy* of a closed system is required to increase or at best remain constant for all physical processes. We will discuss the second law in more detail in chapter 8. For now, suffice it to say that entropy is associated with the disorder of a system of many particles. If time's arrow is an arbitrary selection we make for the direction of most probable occurrences, then this will coincide with the direction of increased entropy.

Clearly the absence of an arrow of time at the elementary level precludes any distinction between cause and effect. If that distinction cannot be made, then no basis exists for the causal precedence postulate that rules out superluminal motion. Consequently, superluminal motion would seem to be possible at the elementary level, since it is not relativity per se, but rather causal precedence, that rules it out. Still, tachyons have not been observed in any domain, microscopic or macroscopic. Experimental fact continues to support Einstein's speed limit, even if theory does not. And in science, experiment rules over theory.

If tachyons are ever seen, then the causal precedence postulate will have to be discarded. But even so, this could still apply only to elementary interactions and not to the macroscopic world. Indeed, the violation of causal precedence at the macroscopic scale would have fantastic implications.

If cause and effect were relative at the everyday or astronomical scales, then time travel and all the logical paradoxes that attend it would have to be confronted. If the time sequence of events is different to different observers, then we are the parents of our parents, the children of our children, and so on up and down our ancestral lines. In one reference frame, Lee Harvey Oswald shoots John F. Kennedy. In another, JFK shoots Oswald. In some reference frame, the dinosaurs are still to come and the first land creature is yet to crawl out of the sea. One way out of these absurdities is the introduction of parallel universes, but that leads us down a line of uneconomical speculation that we need not follow until forced to do so by the observation of a tachyon.

The implications of nonlocality, at least if extended to macroscopic scales, are profound. Those who continue to write rather blithely about quantum mechanics and nonlocality should give pause to consider them. The universe might indeed be holistic, with everything everywhere and everytime connected, with the past determining the future, the future determining the past, and no room for chance or choice. However, we should not be too hasty in concluding that this is indeed the case until empirical evidence, stripped clean of all possible mundane explanations, forces us to do so.

If superluminal motion is possible at all scales, then surely all that will happen will have already happened. No wonder New Agers have that glazed look in their eyes! No wonder most physicists' eyes glaze over when considering the consequences of superluminal motion! Still, nonlocality at the microscopic scale could happen with no requirement that it also happen at the scale of humans and their everyday interests.

Zigzagging through the Vacuum Aether

Some think quantum mechanics can provide nonlocality without glazed eyes. J. P. Vigier has pointed out that an aether consistent with causal precedence and relativity can theoretically exist.[17] This has been known for some time and not regarded as troublesome. Vigier refers back to a 1951 paper in which Dirac argued that a quantum aether would not have a definite velocity at certain spacetime points, because of the uncertainty principle.[18] Consequently, an aether in which all velocities are equally probable becomes possible. The resulting state is a vacuum, but at least it does not violate relativity. As mentioned above, Vigier and others have suggested that the vacuum aether may correspond to Bohm's quantum potential.

The vacuum is capable of producing interactions between particles at effectively spacelike separations. This occurs when quantum fluctuations in the vacuum cause a particle to zigzag backward and forward through spacetime. Let me explain.

No doubt the idea of motion backward in time makes a grievous assault on common sense. The world just does not seem to operate that way, as our ever-aging bodies testify. However, to a particle physicist raised on a diet of Feynman diagrams, motion backward in time is not all that disturbing. All fundamental particle interactions work backward as well as forward and, with rare exceptions, do not distinguish between directions of time.[19]

Feynman used the idea of motion backward in time when he invented his famous diagrams in the late 1940s. Dirac had developed his fully relativistic quantum theory of the electron in 1928, and discovered that it contained negative energy solutions. These solutions were identified as anti-electrons or positrons. Positrons were observed as predicted in 1932. Following Stückelberg[20] and Wheeler, Feynman reinterpreted positrons as electrons moving backward in time.[21]

Feynman's idea grew out of his earlier work at Princeton as a graduate student of John Wheeler. Together they had developed a theory of electromagnetic waves involving solutions of Maxwell's equations that travel both ways in time, the so-called **retarded** and **advanced** waves. The advanced waves traveled *backward in time,* that is, they arrived at the detector before they left their source. Despite their presence as valid solutions to Maxwell's equations, advanced waves had been previously ignored by less bold thinkers.[22] Feynman later extended the idea to quantum field theory, in which waves are particles and vice versa, associating *antiparticles* with the advanced waves.[23]

Feynman noted that whether you say you have a particle moving forward in time with negative energy, or its antiparticle moving backward in time with positive energy, is really quite arbitrary at the fundamental level. Energy conservation and the other laws of physics remain intact. By reversing the charges and momenta of the backward particles, charge and momentum conservation are unaffected.

The violation of causal precedence by tachyons, if they are ever found, will result not from their motion backward in time but from their superluminal motion. In the case of the known elementary particles, whether or not they move backward or forward in time they still remain within the light cone and retain causal precedence. That is, they do not exchange cause and effect from one reference frame to another. And, as I will now show, the apparent nonlocality proposed by Vigier is simply an artifact that can be understood without superluminal motion.

The Feynman diagram for the zigzag process is illustrated in figure 5.3.[24] As usual, the time axis is up and a single spatial axis is indicated to the right. An electron starts at point A and follows a path through spacetime at constant velocity, changing its position as time progresses. At point B, a fluctuation in the vacuum results in a momentum transfer to the electron, which turns it around so that it goes backward in time. At point C, another vacuum fluctuation causes the electron to turn around again and resume its forward course in time, passing point D at the same time as it had point B, but at a point separated by the distance BD. Thus it appears that the particle has made an *instantaneous* jump from B to D.

Actually, it is possible to view this nonlocal artifact without introducing motion backward in time, as illustrated in figure 5.4. Note that all the particles are moving in one time direction. At time C an electron-positron pair is created by a vacuum fluctuation. The positron goes to the left and collides with the original electron at B, where they annihilate each other, the annihilation energy disappearing into the fluctuating vacuum. In the meantime, the electron from the pair created at C continues on and is interpreted as the original electron from A transported instantaneously from B to D.

The net result, in either view, is an effectively instantaneous jump of the electron over the spacelike separation BD. At time B the electron disappears and reappears at D some distance away. A quantum jump, a

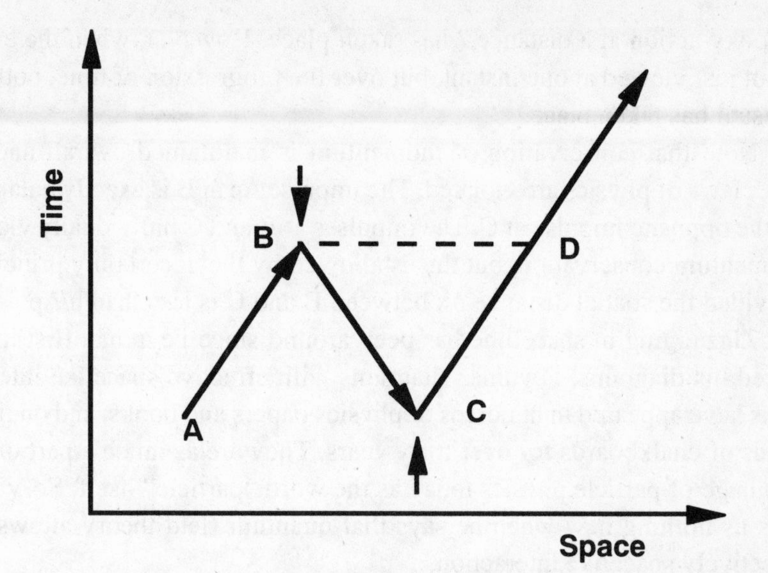

Fig. 5.3. How an apparent instantaneous jump in space can result at the quantum level. An electron starts at A. At B it receives an impulse from the vacuum that sends it backward in time to C. At C it receives another impulse sending it again forward in time. Thus it appears to jump instantaneously from B to D. This is allowed by the uncertainty principle, provided the instance BD is less than Planck's constant divided by the impulse.

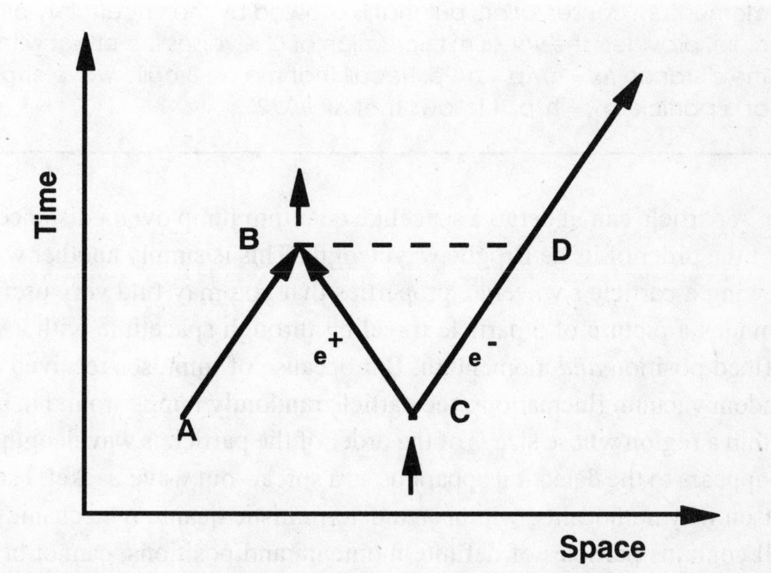

Fig. 5.4. An alternate way to view figure 5.3. Here an electron-positron pair are created by a vacuum fluctuation at C. The positron annihilates the original electron at B. The other electron from the pair created at C continues on, taking the place of the original electron.

"spooky action at a distance," has taken place. However, when the event is not just viewed at one instant, but over the progression of time, nothing unusual has taken place.

Note that conservation of momentum is maintained overall and no other laws of physics are violated. The impulse Δp at B is exactly balanced by the opposite impulse at C. The impulses at B and C individually violate momentum conservation, but this is allowed by the uncertainty principle, provided the spatial distance Δx between B and C is less than $h/\Delta p$.

Zigzagging in spacetime has been around since Feynman first introduced his diagrams. Feynman diagrams with effective spacelike interactions have appeared in hundreds of physics papers and books, and on thousands of chalkboards for over forty years. They are as much a part of the language of particle physics today as the word "particle" itself. So Vigier tells us nothing new when he says that quantum field theory allows for effectively spacelike interactions.

But can this be used to predict nonlocality across *macroscopic* distances? Here we need to get quantitative.

> Suppose the electron's initial momentum is p. To turn it around, the vacuum fluctuation at B must provide an impulse $\Delta p = 2p$. This violates momentum conservation, but that is allowed by the uncertainty principle, provided the vacuum fluctuation at C reverses the effect within the distance $\Delta x = h/\Delta p = h/2p$. Recall that the de Broglie wavelength of a particle is $\lambda = h/p$. It follows that $\Delta x = \lambda/2$.

A particle can undergo a spacelike quantum jump over a distance that is of the order of its de Broglie wavelength. This is simply another way of viewing a particle's wavelike properties that you may find very useful. It provides a picture of a particle traveling through spacetime with a well-defined position *and* momentum. But because of impulses received from random vacuum fluctuations, the particle randomly jumps around in space within a region whose size is of the order of the particle's wavelength, and so appears to the detection apparatus as a spread-out wave packet. I see no reason why nonlocality, within an indeterministic quantum mechanics that still contains particles of definite momenta and positions, cannot be formulated in this fashion.

An ensemble of similarly prepared electrons will have measured positions whose distribution is given by $|\Psi|^2$. Could vacuum fluctuations be the hidden variables? If so, they do not provide for nonlocal connections across the universe, or even across the room, for the material bodies of normal experience whose de Broglie wavelengths are infinitesimal.

> Effectively spacelike motion is possible over a distance of the order of the de Broglie wavelength λ. Consider an electron moving at, say, $v = 100$ meters per second. Its de Broglie wavelength is $\lambda = h/mv = (6.63 \times 10^{-34})/(9.1 \times 10^{-31} \times 100) = 7 \times 10^{-6}$ meter, where $m = 9.1 \times 10^{-31}$ kg is the mass of the electron and we have used the fact that momentum $p = mv$. The value of λ is comparable to the distances between atoms in solids.

Macroscopic objects do not produce measurable wavelike effects. If a one-kilogram object is moving at 10^{-10} meters per second, surely very close to being at rest (nothing is ever exactly at rest), it will have a de Broglie wavelength of 7×10^{-24} meter, far smaller than the size of a nucleus. Its zigging and zagging would never be noticed.

This illustrates why quantum effects are not observable in everyday life, at least for the familiar objects that we think of as material "bodies," and it demonstrates why Vigier's idea, while qualitatively correct, does not provide quantitatively for holistic connections over macroscopic distances, certainly not the whole universe.

Light and other electromagnetic waves, however, do exhibit quantum effects on the macroscopic scale. The wavelength of visible light is in the range 4×10^{-7} to 6×10^{-7} meter. Though this is small by macroscopic standards, light diffraction effects are observable to the naked eye. Radio waves can be of macroscopic and even planetary dimensions. Long-wavelength radio photons appear instantaneously at widely separated receivers. In the vacuum fluctuation picture being considered here, individual radio photons hit all receivers at once by zigzagging through spacetime, not by some superluminal transfer of energy.

The vacuum is thought to be alive with particles and antiparticles that are constantly being created and destroyed, or zigging and zagging through spacetime, if you prefer. Measurable effects have been calculated by quantum field theorists and checked to great accuracy against experiments for decades, with no violations of fundamental laws of physics evident or implied. Zigzagging in spacetime can produce what appears to be superluminal motion, but only when the wavelengths of the particles are of comparable dimensions. And even this is the result of random quantum fluctuations, and so cannot be perceived as transmitting superluminal "signals."

The measurable effects referred to above are precisely those quantum effects that physicists infer from observations in the laboratory, almost exclusively involving atomic and subatomic phenomena. The objects emitting and absorbing these zigzagging particles have sizes that are comparable to the wavelengths involved. For particles to similarly zigzag across the universe, the wavelengths would have to be of extragalactic extent. Such waves could not be emitted or detected by anything of human dimensions, like a brain or scientific instrument, by any conceivable application of existing knowledge.

One cannot simply speculate about possibilities. One must check the numbers. Much of pseudoscience is qualitative hand-waving. Until a concept can be made quantitative, or at least put on a firm logical foundation, it is not science.[25] Certainly, spacelike correlations across the universe, making the universe one "interconnected whole," are not possible unless you imagine particles of infinite wavelength. In short, the vacuum aether does not provide a quantitatively feasible metaphor for the holistic universe.

And what about the paradoxes of superluminal motion discussed earlier? Do they not exist for the effective superluminal motion produced by zigzagging in spacetime? No, since, as we have seen, no distinction between cause and effect is made at the elementary level. Only with complex systems, such as macroscopic bodies, do causal paradoxes present interpretational problems.

Local EPR in Reverse Time

As long as superluminal effects are not observed in experiments, any interpretation of quantum mechanics that requires nonlocal effects is not parsimonious. If, as many seem to think, conventional quantum mechanics is nonlocal, then proper scientific method demands that we seek alternative, local interpretations.

Now I would like to show how the EPR "paradox" can in fact be almost trivially resolved by interpreting the experiment in reverse time. As far back as 1953, French physicist Olivier Costa de Beauregard argued that the EPR paradox could be resolved by including the action of advanced waves.[26] He pointed out that the exclusion of advanced waves is a classical prejudice that has no *a priori* justification. If they are present as solutions of Maxwell's equations, we make an added hypothesis in ruling them out, namely the hypothesis that I have called causal precedence. (Note that this is the same hypothesis used by Einstein to rule out superluminal motion.) The following explanation of the EPR experiment goes along similar lines, but uses Feynman's association of antiparticles with the advanced waves.

Let us again consider the Bohm-EPR experiment in which a singlet (total spin zero) system decays into two electrons that go off down opposite beam lines A and B. At the ends of the beam lines are the usual spin meters that can be oriented in any direction perpendicular to the beams. Nonlocality is implied when the decision on what orientation to use at A is made just before the detection, so no time is left for a signal to reach B without travelling faster than light.

Now view the EPR experiment from the frame of reference that is time-reversed from the normal, familiar one, as illustrated in figure 5.5. The detectors at A and B then become polarized positron emitters. Suppose emitter A is set so that it gives a positron with its spin aligned along an axis x perpendicular to the beam line. Emitter B generally can emit a positron of any arbitrary spin axis orientation.

Let us first examine the special case in which the axis of emitter B happens to be the same as A and emits a positron whose spin is opposite to that of A. Then the total spin of the system of two positrons will be zero. When the two positrons moving *backward* in normal time along the beam lines come together they will form a two-positron state that, from angular momentum conservation, will have total spin zero. That is, a singlet state will be *locally* produced.

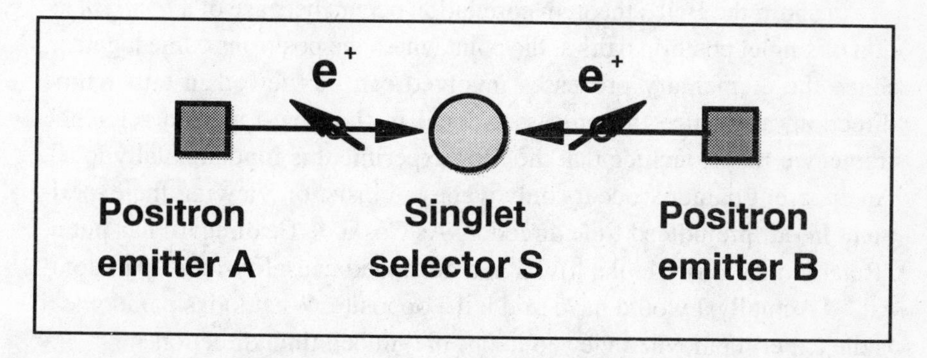

Fig. 5.5. Time-reversed EPR experiment. Polarized positrons e^+ with arbitrary spin axes are emitted toward each other from A and B. They collide in the center and form a two-positron state that will in general be a singlet (total spin zero) or triplet (total spin one). However, only singlet states are accepted (locally) by the singlet selector S. This results in a correlation that is the same as calculated by quantum mechanics in the normal EPR experiment in which electrons are emitted at S and detected at A and B.

If instead the spin of B were in the same direction as A, then a triplet state would be formed. However, the experiment, when viewed in the normal time sequence, was designed to include only singlet states as the electron source. Viewing this in reverse time, the triplet states that are formed are discarded (locally) from the sample.[27]

It is precisely this selection that produces the correlation that is observed in the experiment. Putting it another way, no correlations will be observed in a Bohm-EPR experiment if triplet states are included in the (normal time sequence) source. By using only singlets, we force a correlation.

If emitter B emits a positron with a spin along some other arbitrary axis, say the y axis, then it is a matter of chance (with calculable probabilities) whether a singlet or triplet is formed when the positrons collide. But once again a correlation is enforced by *locally* tossing out the triplet states. The equations for all this are the same as in standard quantum mechanics, and do not fundamentally distinguish between the two directions of time. Thus the quantitative correlation will be the same as that calculated assuming the macroscopic time direction.

This way of viewing the EPR experiment may also shed some light on why a deterministic theory is necessarily nonlocal. (Logicians note that I am not saying determinism is the only means for nonlocality.) If you insist on producing a specific state, then you must know the orientation of one positron emitter relative to the other at the moment of emission. But this is unnecessary when you are willing to take your chances on what state is produced when the two positrons collide.

In short, the Bell's theorem correlation occurs because of a *local* selection of singlet positron pairs at the point where the positrons come together. Since the elementary processes involved can be viewed in either time direction, and since the process is local in the time-reversed reference frame, we may conclude that the EPR experiment is fundamentally local. An apparent paradox occurs only when we insist on viewing the experiment in our prejudiced time direction. As Costa de Beauregard has put it, "Retarded causality looks trivial and advanced causality looks paradoxical."[28] Actually, I would have said it the opposite. What looks paradoxical is the experiment when viewed in our prejudiced time direction.

Costa de Beauregard does not conclude, however, that the consequences of time symmetry are trivial. On the contrary, he takes the directionlessness of time and causality at the elementary level to be so profound as to imply "the existence of subtle phenomena termed 'psychic' in a broad sense, inside the human, the animal, and possibly the vegetal kingdoms."[29]

I could not disagree more. The behavior of the microworld appears paradoxical only when we insist on applying to it concepts from the macroworld that have no meaning at the elementary level.[30] The fact that our common-sense prejudices do not apply cannot be taken to mean that the microworld possesses mysterious properties. On the contrary, we have found that the microworld is far simpler than the macroworld and can be understood with a minimum set of physical ideas that do not have to be supplemented by emergent qualities such as a direction of time and causal precedence.

Time Symmetry in Quantum Mechanics

The fact that the basic laws of physics do not contain inherent time asymmetry continues to bother modern thinkers. Several have taken the view that since time asymmetry is such an obvious, common experience, our formulation of the laws of physics will not be correct until they demonstrate time asymmetry at their deepest levels.[31] Others have proposed that the absence of directionality of time in elementary particle physics demonstrates that we should look to macroscopic physics, not elementary particles, for the fundamental laws of nature.[32]

My view agrees with what I sense is the developing interdisciplinary consensus: two sets of natural laws exist, one at the elementary level of fundamental particles that possesses a high degree of symmetry, and another that emerges at the levels of many particles where the elementary symmetries are accidentally broken and new laws appear to describe the structures that thereby evolve.

In the usual application of classical physics, the equations that govern the evolution of a physical system must be solved subject to certain **boundary conditions**. Because of our normal conception of time flow, these boundary conditions are usually taken to be *initial* conditions—that state of the system at some time $t = 0$ when we arbitrarily start our clock. Then the equations predict the future motion of the system, which is usually what we want to know. Prediction is the most common application of science, and its greatest power.

However, the equations computed with *final* conditions can also be used to postdict the past. We can use celestial mechanics to precisely date the past appearances of solar eclipses and comets, verifying certain historical events. For example, an eclipse occurred on March 28, 585 B.C.E. that may have been the one reportedly predicted by Thales of Miletus, perhaps triggering the development of Greek science and philosophy.

Nothing forces us to chose either initial or final boundary conditions. In fact, the most general methods of classical mechanics make no distinction between initial and final conditions.

In quantum mechanics, the situation appears at first glance to be fundamentally different. Conventional quantum mechanical formulations incorporate a distinction between past and future, despite the fact that the Schrödinger equation and all relativistic formulations of quantum mechanics are time-symmetric.[33]

In the Copenhagen description of the measurement process, the act of measurement selects the state of a system from among all its possible states. This is a nonreversible process, performed in the reference system in which the arrow of time is selected by the prejudice of everyday experience.

An important subtlety should be noted, however. The arrow of time, we have seen, is determined by the direction in which entropy increases. If we imagine a local system being organized by outside energy, it will have a decreasing entropy with reference to the arrow of time of the outside system. Should we not define its local time arrow in the opposite direction, and describe measurements in this system with reference to this time direction?

Issues of this sort have led quantum cosmologists, notably Penrose,[34] Page,[35] and Hawking,[36] to investigate ways in which time asymmetry can be built into cosmology. However, Gell-Mann and Hartle have shown that a time-symmetric quantum cosmology can be developed using a time-neutral, generalized quantum mechanics of closed systems in which initial and final boundary conditions are related by time reflection symmetry.[37] Thus even the quantum universe appears to be time-symmetric, despite our psychological perception of a unique direction of time.

In an electronically disseminated paper,[38] Paul Sommers has shown how time-symmetric quantum mechanics provides the natural way to view the contextuality of quantum mechanics. In classical physics, as I have noted, the normal procedure is to predict the future paths of particles using a set of initial conditions and solving the appropriate equations of motion. However, in quantum mechanics initial conditions alone do not suffice to determine the future. Each possible outcome is not predetermined, but occurs with some probability. Furthermore, the set of possible outcomes differs for different arrangements of the detectors.

Sommers suggests instead that quantum probabilities must be calculated using final conditions as well as initial conditions. The universe is presumed to be subject to a final boundary condition that limits the set of possible final states, just as the possible final states for a laboratory experiment are limited by a particular arrangement of detectors. Sommers further explores how particular types of final boundary conditions might account for the classical nature of the universe.

A quantum system can thus be viewed as being influenced by its future as well as its past. The final condition defines all the possible outcomes, with a quantum mechanical probability calculated for each. One of these outcomes happens in accordance with these probabilities. As long as the dice are being tossed to determine the outcome, that is, as long as we do not have deterministic hidden variables, then the macroworld can develop with a future that is not already written in the stars.

The Transactional Interpretation

At least one time-symmetric interpretation of quantum mechanics has been worked out in some detail: the transactional interpretation of John Cramer.[39]

Cramer applies the notion of advanced and retarded waves to the wave function itself. The problem that keeps raising its head in quantum mechanics is the manner in which experimental results depend on the complete arrangement of the experiment being conducted on that system. Changing the detector configuration seems to change the nature of the object being observed, even after the object is well on its way to the detector. That is, quantum phenomena are contextual.

For example, in the double slit experiment, photons or other particles from a source somehow travel both paths and interfere with one another when a detector directly behind one slit is turned off; but they travel only one path when that detector is turned on. When the decision to turn the detector on or off is made after the particle has had time to pass through either slit, we are led to the bizarre conclusion that the particles know ahead of time what choice of detector arrangement will later be made.

Cramer's solution to this conceptual problem uses the basic time symmetry of elementary processes described above. A measurement in quantum mechanics is viewed as a "handshake" between a source of particles and the apparatus set up to detect these particles. The source sends out **offer waves** in all directions that correspond to the wave function Ψ of conventional quantum mechanics. In the meantime, the particle detectors at their various positions in the experimental arrangement continually send out **confirmation waves**. The confirmation waves correspond to the complex conjugate wave function Ψ^*. The two act together to produce a kind of traveling wave pattern.

The confirmation waves are analogous to the advanced waves of

Feynman and Wheeler, traveling backward in time Thus the confirmation wave from a given detector reaches the source at the same time that the offer wave is emitted. The interference between the two sets of waves then produces the traveling wave we associate with the particle that moves from the source to the detector.

Cramer notes that the relativistic versions of Schrödinger's equation (the Klein-Gordon and Dirac equations) allow for solutions that propagate in either time direction. When the latter are taken to the nonrelativistic limit ($v << c$), two equations result, the normal Schrödinger equation and its complex conjugate equation. The latter is usually ignored because its solution propagates backward in time. Again we see the hypothesis of causal precedence being implicitly inserted into our theories so that they may retain our common-sense prejudices.

One solution of Schrödinger's equation for a free particle, that is, a particle not acted on by any external forces, is a complex wave moving forward in time. This can be written $\Psi = A \exp[i (kx - \omega t)]$, where A is the amplitude, k is the wave number and ω is the angular frequency. A solution of the complex conjugate of Schrödinger's equation is a wave moving backward in time and in the opposite direction with the same amplitude, wave number, and frequency. This can be written $\Psi^* = A \exp[-i (kx - \omega t)]$. The sum is then $\Psi = \Psi + \Psi^* = 2A \cos(kx - \omega t)$, which, we note, is a real function.

In classical wave theory, this function describes a traveling wave of wavelength $\lambda = 2\pi/k$ and period $T = 2\pi/\omega$ propagating in the x direction. In standard quantum mechanics, this is interpreted as a particle moving in the x direction with a momentum $p = \hbar k$ and energy $E = \hbar \omega$.

In an electronic mail message to me, Cramer commented that the transactional interpretation was suggested to him by the mathematical structure of quantum mechanics itself. He said he was amazed that no one had thought of it before he did.

All quantum predictions involve both Ψ and Ψ^*, so nothing new has really been added to existing theory. In fact, when the two complex wave functions are added, a simpler, real wave function results. (Here, "complex" and "real" refer to mathematical, not metaphysical, properties.) The handshake between source and detector is necessary for a particle moving between the two to have conventional meaning, that is, to exhibit measurable properties such as momentum and energy that we associate with particles.

The double slit, Schrödinger's cat, EPR, and other apparently paradoxical experiments are easily described in the transactional interpretation. In the double slit, a handshake occurs for offer and confirmation waves passing though both slits and an interference pattern results. When a detector behind one slit is turned on, its confirmation wave travels back through only that particular slit and not the other; thus the handshake only happens for the path through the first slit. Variations, such as with Wheeler's delayed choice option, do not change the conclusion.

As for Schrödinger's cat, Cramer suggests that the apparatus sends out a "weak" offer wave that has only a fifty-fifty chance of being confirmed by the Geiger counter that detects the radioactive particle and triggers the poison gas. No half-dead, half-alive paradox exists.

Similarly, in the Bohm version of the EPR experiment, the source sends out offer waves corresponding to the two electrons. The spin meters at the beam ends send confirmation waves backward in time to the source. These confirmation waves carry the characteristics of the spin-meter settings, which can be selected after the electrons leave the source. Remember, the confirmation waves travel backward in time and so shake hands with the offer waves at the source to produce electrons of exactly the right properties to register on the spin meters.

Cramer notes that, as in all time-symmetric interpretations, his interpretation demotes the role of the observer compared to the Copenhagen interpretation. Thus the observer has no special role as the source of wave function collapse and this metaphysical aspect of Copenhagen is absent.

Mystical Transactions

And now for some comic relief. Like most other interpretations of quantum mechanics, the transactional interpretation has not escaped being used to support mystical delusions, serious or nonserious. Fred Alan Wolf entertainingly suggests that Cramer's transaction between emitter and absorber takes place on the "psychic plane." The emitter does not emit a photon until it "senses" the presence of the absorber. This sensing is by means of the exchange of "psychons" that result from the "transformation of morphenes into feelings." He relates this to the "inner light" that mystics speak of. Ordinary "outer light" is then the physical manifestation of inner light. Wolf further relates this to the mystical concept of fate, that anything that happens "has already happened." As Wolf puts it, "The psychic emitter emits and the absorber absorbs the photon, provided that the 'moving finger' on the psychic plane 'wills' that it be so; or that there is a transaction of psychons between the two so that no movement will occur unless there is a mutual benefit to both, a synergistic effort of emitting and absorbing."[40] Cramer tells me that this is not quite what he had in mind.

Double Slit Again

We do not need to talk about offer and confirmation waves to resolve the paradoxes of quantum mechanics within the framework of time symmetry. I described above how the EPR paradox is removed when one views the process in time-reversed fashion and does not require determinism. Similarly, the double slit paradox can be easily explained. Simply imagine an electron passing through one slit to the detector, going back in time to the source through the other slit, and then forward in time again through either slit to the detector, as illustrated in figure 5.6. This will give an interference pattern when both slits are open and no pattern when one is closed, as is observed. It also provides us with an intuitive picture by which we can visualize a single electron passing through both slits, one of the long-standing bugbears of quantum mechanics.

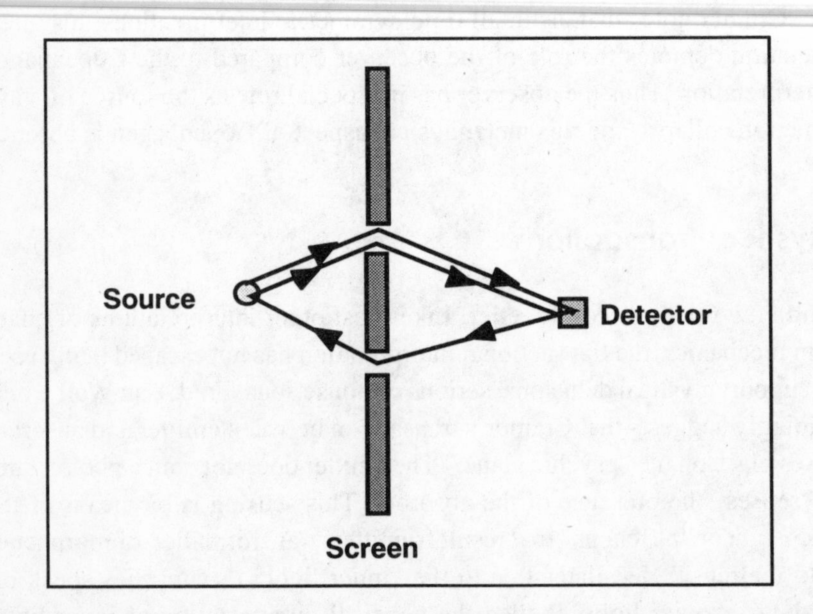

Fig. 5.6. The double-slit experiment, as visualized in a time-symmetric picture. An electron from the source goes through the top slit to the detector. Then it travels backward in time through the bottom slit to the source, and then back again through the top slit. (The path has been displaced for illustration.)

Nonlinear Quantum Mechanics

While I feel that time-symmetric quantum mechanics holds the key to resolving the nonlocality problem, I am probably in a minority on this issue at this time. The notion that time has a fundamental direction is too deeply embedded in our culture and psyche. So let me reluctantly return to the conventional picture in which time flows in one well-defined direction.

I have previously demonstrated that no faster-than-light communication occurs for the specific, EPR-type experiment that is simulated by the program included at the end of this chapter. As a matter of fact, the impossibility of superluminal communication has been proved generally within the framework of conventional relativistic quantum field theory.[41] This conclusion is noncontroversial and now accepted by Jack Sarfatti and others who originally hoped that a superluminal communicator might be built by means consistent with existing theory.

Thus, if superluminal communication exists, a new theory that goes beyond standard quantum theory will have to be found to explain it. This might be a subquantum hidden variable theory of the Bohm or de Broglie type. However, one of these models would have to be greatly fleshed out to provide some unique, testable features that are different from conventional quantum mechanics. Currently they do not provide such features.

One somewhat-fleshed-out substitute for conventional quantum mechanics is the **nonlinear quantum mechanics** proposed (for purposes other than superluminal communication) by theoretical physicist Steven Weinberg.[42]

Conventional quantum theory is a **linear** theory. This essentially means that when a quantum system is displaced from equilibrium, the change in the wave function of the system is proportional to that displacement.

Many classical systems, such as a stiff spring or a simple pendulum limited to small displacements from the vertical, have a similar linear response. But many others, such as the pendulum with large displacements, are nonlinear. As we will see in a later chapter, nonlinearity is a necessary (but not sufficient) condition for the phenomenon of *chaos* in classical systems.

Classical chaos has been a trendy field of study in recent years. Because of its built-in linearity, conventional quantum mechanics will not exhibit classical chaos, which should not be confused with the chaos associated with the uncertainty principle.[43]

Weinberg asked what might happen if quantum mechanics were slightly nonlinear. He looked for possible experimental consequences and found that these would be exceedingly small, at least in the examples he considered. Furthermore, nonlinearity was shown to lead to superluminal communication, which Weinberg found unacceptable.[44] For these reasons, and other technical ones, Weinberg decided to give up on nonlinear quantum mechanics.[45]

Others, however, have not given up. Recently, in a paper controversially published in the *Physical Review,* Henry Stapp has proposed a slight variation on Weinberg's nonlinear quantum mechanics.[46] He suggests that this might account for the so-called "causal anomalies" reported by Helmut Schmidt[47] and by Robert Jahn and his collaborators at the Princeton Engineering Anomalies Research group[48] in the random event generator (REG) experiments that were briefly described in chapter 1.

These experiments claim evidence that human actions can affect the outcome of random number generators, even when their results are controlled by radioactive decay and recorded months earlier. Thus the random number generators are alleged to exhibit both the consciousness connection and nonlocality, with the human mind affecting events that presumably happened in the past. Stapp explains that "the various possibilities in regard to the detection of radioactive decay remain in a state of 'possibility' or 'potentiality,' even after the results are recorded on magnetic tape: no reduction of the wave packet occurs until some pertinent mental event occurs."[49]

Since a deviation from the predictions of orthodox quantum mechanics is claimed, the simple Copenhagen linear variety of wave function collapse will not suffice. Thus Stapp finds it necessary to make three assumptions to explain the causal anomalies that the REG experiments purportedly demonstrate: (1) the human mind causes wave function collapse, (2) the wave function responds nonlinearly to mentally induced displacements, and (3) the effect is transmitted backward in time.

However, as remarked in chapter 1, the REG experiments are far from convincing, both in ruling out experimental artifacts and in possessing adequate statistical significance. Furthermore, they have not been independently replicated. The Schmidt and Jahn results do not quantitatively agree, and none of the hundreds of other REG experiments that have been performed confirm either group's results. Considerably more independent evidence will have to be garnered before the bulk of the scientific community will be prepared to discard linearity in quantum mechanics.

Appendix

The following program, which can run on any personal computer with True BASIC™, simulates an EPR "superluminal" communicator. The program can easily be modified for other versions of the BASIC language, or rewritten in other languages. Following the program listing is a sample of the results from one run of six thousand experiments.

```
! Superluminal Communicator?
! Program written in True BasicTM on the MacintoshTM
! By Victor J. Stenger (based on the idea of David Mermin)

DIM c$(2)
OPEN #1: name"SuperCom", create newold
ERASE #1
RANDOMIZE

LET c$(1)="R"                    !Color red
LET c$(2)="G"                    !Color green
LET sum1=0
LET sum2=0
LET sum3=0

FOR n = 1 to 6000                !Send message 6,000 times

    ! a is the position of spin meter A that represents
    ! the message to be transmitted. Here the message is
    ! 1,2, or 3 sequentially, simulating the repeated message 123.
    ! Change to any message you like.

    LET a = mod(n-1,3)+1

        ! b is 1,2, or 3 randomly, simulating the receiver at B
        ! trying the three spin meter positions at random.
        ! Change to any strategy you wish.

    LET b = int(3*rnd+1)

        ! k is 1 or 2 randomly, simulating the red and green lights
        ! that correspond to the two random spin orientations.
```

```
    LET k = int(2*rnd+1)

    IF a=b then                    ! The two settings are the same.
        LET j=k                    ! Set the color of B to the color of A.
        LET sum1=sum1+1            ! Count the times they are the same.
    ELSE IF rnd < 0.25 then        ! When the settings are different,
        LET j=k                    ! set color of B the same ¼ of
                                   ! the time.
        LET sum2=sum2+1            ! Count these.
    ELSE
        LET j=mod(k+1,1)+1         ! Set the color of B opposite to A
        LET sum3=sum3+1            ! Count
    END IF

    IF n <=60 then PRINT #1: a;c$(k);b;c$(j)
NEXT n

LET same=(sum1+sum2)/n
LET diff=sum3/n

PRINT #1: " Fraction of light flashes of the same color = ";same
PRINT #1: " Fraction of light flashes of different color = ";diff

END
```

Notes

1. Bohm 1951, p. iv.
2. Bohm, "Hidden Variables and the Implicate Order," in Hiley 1987, p. 40.
3. Zukav 1979, p. 326.
4. Zohar 1990, p. 74.
5. Capra 1975, p. 137.
6. Capra, p. 139.
7. Russell 1985.
8. Trickett 1982.
9. Sharpe 1993.
10. As quoted in B. J. Hiley and F. David Peat, "General introduction: The Development of David Bohm's Ideas from the Plasma to the Implicate Order," in Hiley 1987, p. 12.
11. Jammer 1974, pp. 160–66. See also Wilber 1984.
12. Sarfatti 1977, Herbert 1982, Hegerfeldt 1985, Datta 1987.
13. Zukav 1979, p. 7.

14. Mermin 1985, p. 38.

15. Reaction rates in one direction and its reverse can differ as the result of statistical factors.

16. See Davies 1974 for a complete discussion of time asymmetry in physics and cosmology. Also see Born 1949 for a classic discussion of the role of chance in physics.

17. A. Kypriandis and J. P. Vigier, "Action-at-a-Distance: The Mystery of Einstein-Podolsky-Rosen Correlations," in Selleri 1988, p. 273.

18. Dirac 1951.

19. The rare exceptions occur in so-called CP-violating interactions involving short-lived particles called K^0 and B mesons, which are thought to have different probabilities in time-reversed directions. These can be ignored in the current discussion since they play only a very indirect role in the structure of normal matter. Furthermore, within the standard model the fundamental forces are still CP, and T, invariant, with Cp violation a "broken symmetry."

20. Stückelberg 1941, 1942.

21. Feynman 1948, 1949a, 1949b, 1965b.

22. For an amusing anecdote concerning Feynman's first talk on the subject, given before Einstein, Pauli, and other physics greats, see Feynman 1986, pp. 77–80.

23. Feynman 1948. See also Stückelberg 1941.

24. Purists will object that the Feynman diagram is generally drawn in terms of four-momenta rather than space and time. However, the spacetime diagrams I show are an equivalent way of describing the same ideas. Even purists must admit that one can go from a momentum space to a spacetime description by a canonical transformation.

25. Some people have proposed that nonlocal effects are occurring in the alleged cold fusion process, so that energy is transferred holistically to a lattice without the telltale gamma rays or neutrons expected from nuclear processes. But this is also impossible for the same quantitative reason described here. The interatomic spacings in a material lattice are far greater than the distances at which spacelike interactions involving nuclear energies can take place. The wavelengths of nuclear particles are comparable to nuclear dimensions.

26. Costa de Beauregard 1953.

27. Actually, a triplet state can also be formed from oppositely spinning electrons, but this will also be discarded from the sample.

28. Costa de Beauregard 1987, p. 263.

29. Costa de Beauregard 1987, p. 284.

30. It might be argued that the EPR experiment, being conducted in a normal-sized laboratory, is part of the "macroworld." However, as I have noted in several places, the distinction between quantum and nonquantum effects is not one of scale. Quantum effects, including anything to do with photons, can appear on any scale. The EPR experiment, whether performed with electrons or photons, involves elementary interactions and so must be viewed in those terms.

31. Penrose 1988, pp. 302–47.

32. Prigogine 1984.

33. While the nonrelativistic Schrödinger equation does not appear, at first glance, to be time-symmetric, it becomes so if you change Ψ to its complex conjugate Ψ^*. See the later discussion on the transactional interpretation.

34. Penrose 1979.

35. Page 1985.
36. Hawking 1985.
37. Gell-Mann 1991.
38. Sommers 1994.
39. Cramer 1986, 1988.
40. Wolf 1984, p. 324. I suspect, although I have no evidence of it, that this is a spoof.
41. Eberhard 1989.
42. Weinberg 1989.
43. Here considerable confusion results from the use of the term "chaos" to refer to classical systems that behave in a very complex but still completely deterministic manner. This chaos is not connected in any direct way to the random quantum fluctuations that are often called "quantum chaos."
44. Polchinski 1991.
45. Weinberg 1993, p. 88.
46. Stapp 1994. See the letter by Jonathan Dowling and Stapp's response in *Physics Today* (July 1995): 78–79.
47. Schmidt 1993.
48. Jahn 1991, 1993.
49. Stapp 1994, p. 19.

6

One or Many?

> One arrives at very implausible theoretical conceptions, if one attempts to maintain the thesis that the statistical quantum theory is in principle capable of producing a complete description of an individual physical system. On the other hand, those difficulties of theoretical interpretation disappear, if one views the quantum-mechanical description as the description of ensembles of systems.
>
> —Albert Einstein[1]

Measuring the Wave Function

Given what appears to be the empirical violation of Bell's theorem, the conscious and holistic universes may stand or fall on the meaning of the wave function. In philosophical terms, the issue is whether the wave function is *ontological* or simply *epistemological.*

If the wave function is epistemological, an abstract, mathematical object of the theory that represents only our knowledge of the physical system under study, then it can possess whatever magical properties we choose to give it. The only requirement is that the results of any calculations we make with the wave function agree with observations. If, on the other hand, it is ontological—a fundamental element of reality—then the mystics may be right after all. The wave function is an aether that pervades the universe, perhaps a cosmic mind that controls events throughout space and time.

The only criterion for the scientific merit of a theory is that it provide a useful, economical description of our observations about the real world. As we have found, observational data provide no evidence for objectively real nonlocality nor the power of the human mind to overcome the limitations of physical law simply by an act of will. Such beliefs are based more on wishful fantasies than scientific fact.

I have indicated that certain physical quantities, such as rest mass, charge, and spin, have a predictability that may reasonably be associated with our common-sense notion of objective reality. If the wave function, or some related quantity such as Bohm's quantum potential, is a "real" field associated with individual particles, something carried along by a particle, then its apparent nonlocal behavior implies a holistic universe. And if this ontological wave function can instantaneously collapse throughout the universe upon some conscious act, then indeed our scientific notions of objective reality will have to be replaced by something more akin to Bishop Berkeley's idealism, in which human observation determines and sustains the universe.

How might we establish the wave function's reality? Let us approach the problem pragmatically. When we do so, the answer comes quickly to mind: measure the wave function! By that I mean, perform an experiment whose outcome depends unambiguously and predictably on the wave function. Show that the wave function cannot be sensibly regarded as simply an abstract, mathematical quantity descriptive only of the statistical character of an ensemble of similarly prepared systems or some other purely abstract representation of the likelihood of the various outcomes of experiments.

Mathematical quantities can carry important knowledge about the world without necessarily being classified themselves as real, concrete entities of the same nature as rocks, trees, and electrons. The volume of a ball is not a ball. The ratio of the circumference of a pie to its diameter is pi, which is not a pie. Though not real, these are meaningful quantities.

Similarly, the probabilities for the various outcomes of Las Vegas games of chance are mathematical quantities that no one associates directly with objects or fields in space and time. A slot machine has odds, but those odds are not some aura that radiates out from the machine. If it was, then we should be able to detect the emanations and perhaps even predict the outcome of a pull of the lever. We can take our clue from casino owners. They don't fret about whether or not their odds are real. They do not concern themselves with the possibility that someone might detect the auras of their slot machines. They set the odds for winning on their machines just slightly below the break-even levels calculated by their mathematical consultants and laugh all the way to the bank. Their bank deposits, numbers on paper and bits in a computer, are real enough to suit them.

For the wave function to be real, it must be more than a tool for calculating probabilities.

In the quantum theory of radiation, the wave function for an ensemble of photons, such as in a light beam, is associated with the potentials of classical electrodynamics, suitably modified so that they are consistent with the formalities of quantum theory (that is, the potentials are *quantized*).[2] One might think that these familiar, well-understood classical potentials are observable, in the sense that they produce measurable effects, such as the deflection of the pointer of a voltmeter. This would seem to provide a simple example of a measurable wave function.

However, potentials in classical electrodynamics are merely mathematical tools introduced to aid in the calculation of presumed real electric and magnetic fields, which in turn provide the forces on charged particles and deflections of pointers. Unlike the electric and magnetic fields, which can be directly measured by unambiguous procedures,[3] electromagnetic potentials have a certain arbitrariness of definition. No experiment, at least in classical physics, can measure potentials directly.

In the simplest example, the static electric potential is simply the "voltage" measured with a voltmeter at some point in an electrical circuit. This voltage is only determined relative to some other point, namely, the point where you have attached the other lead of the voltmeter. The voltmeter might read zero, but you still could get shocked touching either point. Only electric potential differences are physically meaningful, and you get the shock, according to the classical view, because an electric field exists between your hand and the point touched, resulting in an electron flow between the two.

It follows that the wave function of a photon at a point in space cannot simply be observed by measuring the electromagnetic potential at that point. That potential is arbitrary. However, in 1959 Yakir Aharonov, an Israeli student working in London with David Bohm, came up with a proposal for an experiment that could test whether the electric potential, and thus the wave function, are real fields. This has become known as the **Aharonov-Bohm effect**.[4]

The idea is shown in figure 6.1. We again have the double slit experiment so important in physics, in this case with a source of electrons. A magnetic solenoid, a tightly wound cylindrical coil of wire indicated by the circle, is placed behind the slits. When an electric current is applied to the coil, a magnetic field similar to that of a permanent bar magnet and directed

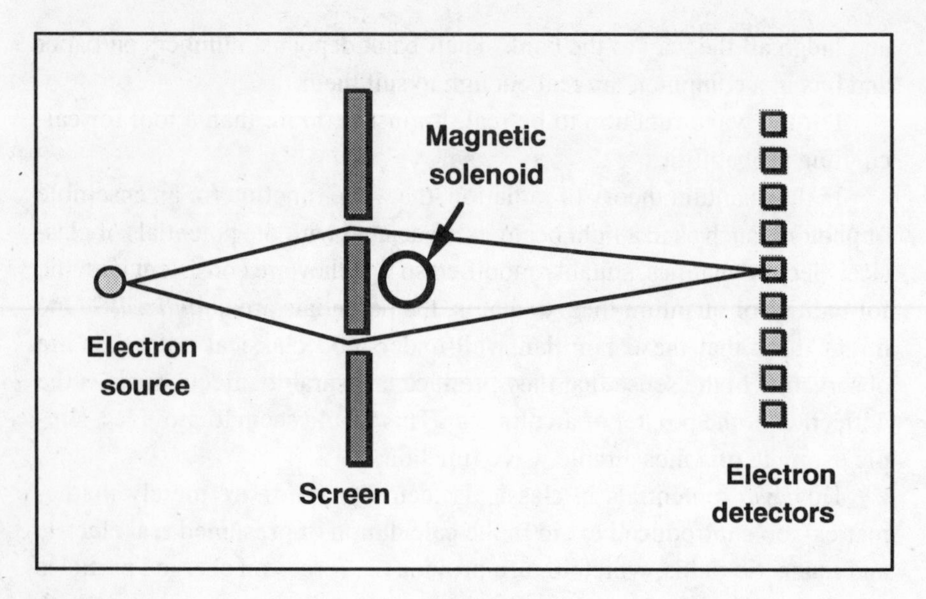

Fig. 6.1. The Aharonov-Bohm effect. Although the two paths of the electrons shown do not pass through the magnetic field inside the solenoid, the wave functions of the electrons undergo a phase change that results in a measured interference effect at the detector plane.

perpendicular to the page is produced. If the solenoid has many closely packed turns of wire, the return field outside the cylinder will be very small compared to the field inside.

Electrons from the source are split by the slits into two beams. Shown in the figure are two paths that pass outside the solenoid and come together at a wall of electron detectors. Since the magnetic field along the two paths is essentially zero, the electrons experience negligible magnetic forces, at least according to a naive application of classical electrodynamics.

However, Aharonov showed that according to quantum mechanics the wave functions of the electron beams along the two paths will experience a relative phase shift that depends on the magnetic flux through the solenoid. This phase shift will result in a change in the observed interference pattern of the two beams when they come together at the detector wall.

Although the magnetic field outside the solenoid is zero, the electromagnetic potential is not necessarily zero. This follows from classical electrodynamics, where the potential around a closed loop is determined by the magnetic flux passing though the loop. So, by measuring the shift in the electron interference pattern when the solenoid current is turned on, we have observed an effect of the electromagnetic potential where no magnetic field exists. By inference, we have thereby measured the wave function of a photon.

The Aharonov-Bohm experiment has been performed and indeed a change in interference pattern observed when the current in the solenoid is turned on.[5] The effect is now routinely applied in semiconductor quantum interference devices (SQUIDs). The phenomenon is surely real. But what does it mean? Aharonov has interpreted the result as demonstrating that the electromagnetic potential has a physical reality in quantum mechanics that it lacks in classical mechanics. The Aharonov-Bohm effect also is widely believed to demonstrate a quantum mechanical deviation from the local interactions between changes and fields in standard, classical electrodynamics.

This interpretation of the Aharonov-Bohm effect stood for thirty years, undoubtedly contributing to the common impression that the wave function has some aspect of objective reality that goes beyond its conventional interpretation in terms of probability. However, in 1991, an alternate, local, classical explanation for the Aharonov-Bohm effect was provided by L. O'Raifeartaigh, N. Straumann, and A. Wipf.[6] They showed that the same effect results from the purely classical interaction of the magnetic field produced by the electron current and the field inside the solenoid.

Any moving, charged particle constitutes an electric current, and currents generate magnetic fields that (speaking classically) can reach out and interact with other fields distant from the particle. Since this follows from Maxwell's equations, and Maxwell's equations are completely consistent with Einstein's relativity, no spooky action at a distance or other miraculous occurrence is involved.

Thus, while the Aharonov-Bohm effect is interesting, it does not seem to provide the empirical basis needed to conclude that the wave function is observable. Whatever the reality of the electromagnetic potential, it is no different in quantum physics than in classical physics.

In a series of more recent papers that approach the subject differently, Aharonov and his collaborators have attempted to show that the wave function of a single system can in fact be directly observed.[7] They argue that the usual situation in which the wave function supposedly collapses under the act of observation can be obviated by what they call a "protective" measurement. That is, when a measurement of a particular observable is performed that would ordinarily cause the collapse of the wave function, you add another interaction that compensates for that collapse. Then, Aharonov says, you have "observed" the wave function.

For example, suppose we have an electron beam that has been prepared so that all the electrons are spinning along, say, the z axis. The electron wave function (technically, state vector) is consequently known. We then try to measure the spin component along the perpendicular x axis. In normal quantum jargon, we say that this act of measurement destroys our knowledge that the electrons are spinning along the z axis and the wave function collapses to a different one corresponding to the spins all being aligned along the x axis. Remember that observations of spins along different axes are incompatible; they cannot be simultaneously performed.

Aharonov and coworkers suggest that while performing the measurement of S_x you also apply a strong magnetic field in the z direction that brings the electron's spin back to point along z. Thus, they say, an observation is performed on a noncompatible variable and the electron remains in its original state. This they interpret as meaning that the wave function of the individual electron is real, since it can be observed.

I am not sure what they have proved. As Aharonov and Vaidman admit, "The conceptual disadvantage of measurement with artificial protection is that we have to know the state in order to arrange proper protection. One might object, therefore, that we obtain no new information." They refuse to accept failure, though: "However, we can separate the protection procedure and the measuring procedure: one experimentalist provides protection and the other measures the Schrödinger wave itself. Then the second experimentalist does obtain new information."[8]

This strikes me as more than a bit contrived. How does holding back information from an experimenter that is already known by a coworker constitute a determination of new information when the experiment is performed? If I tell you my phone number, no new information is generated; already-existing information is simply added to your particular knowledge bank.[9] However, Aharonov collaborator Jeeva Anandan argues that, "Protective measurements still shows the manifestation of the wave function of a single particle, which has not been done before."[10] Since no protective measurement has yet been actually performed in the laboratory, we can suspend judgment until it is.

In summary, the Aharonov-Bohm effect has been tested in the laboratory, and is used in practical applications, but has a plausible, alternate, classical explanation. Incidentally, the experimental test involved statistical measurements only and so provided no basis for an interpretation in terms of individual electrons or photons. Finally, Aharonov's latest ideas of "protective measurements," claiming to demonstrate how the wave function of an individual particle may be "observed," require prior knowledge of the particle's state. Whether or not this constitutes an actual measurement that adds to our knowledge of the system being observed remains to be empirically demonstrated. No specific empirical effect, different from the predictions of normal quantum mechanics, has been suggested in these papers.

Copenhagen Revisited

We have seen how Einstein's original objections to quantum mechanics, the EPR paradox, and Schrödinger's cat are all related. Einstein initially expressed discomfort with the notion that the wave function collapses immediately upon the act of observation. Because of the implication of superluminal motion, he called this a "spooky action at a distance."

The connection that many imagine between the quantum and holism is flimsily constructed from the notion of nonlocality in wave function collapse. The connection that others imagine between quantum and consciousness is likewise flimsily constructed from the notion that the act of human observation causes the wave function to collapse.

Now, as I have tried to make clear, the alleged paradoxes of quantum mechanics are confined to purely conceptual entities that are not directly identified with those entities that we can safely associate with objective reality. Objectively real quantities that serve as properties of physical systems can be consistently defined. They have no paradoxes attached to them. The paradoxes of quantum mechanics attach to abstract entities such as the wave function.

Physical properties, such as mass and electric charge, are unambiguously imbedded in quantum theory. These properties are operationally defined in terms of well-prescribed procedures that are carried out in the laboratory under carefully controlled conditions that minimize spurious backgrounds and systematic errors. Unlike momentum and position, these *invariant* observables do not interfere with the determination of other properties of the system being studied. Momentum and position are examples of noninvariant observables. However, for either invariant or noninvariant observables, the numerical readings supplied by the laboratory apparatus are entered as real numbers into the theoretical equations used to make predictions that can be tested against further measurements.

Our theoretical equations contain symbols that represent observables, such as m for the invariant mass and x for the noninvariant position. These equations also contain other symbols that are not associated with actual observables. This second set of symbols may include special numbers such as π or $\sqrt{-1}$, or mathematical functions such as Ψ, that exist only within the formalism of the theory. The observables are real numbers; they correspond directly to instrument readings conventionally expressed as real numbers. However, the mathematical objects in the theory that do not correspond directly to measurements can be imaginary or complex numbers, matrices, operators, tensors, or any of a shopping list of available creations of the mathematical mind. These objects are bound only by the mathematical axioms that define them and the theorems that are derived from these axioms.

With this understanding, the Copenhagen interpretation remains fully viable as a way to conceptualize the actual formal procedures that a physicist carries out in applying quantum mechanics in the laboratory. However, Copenhagen still suffers from a conceptual difficulty. In the Copenhagen view, the measuring apparatus itself is considered a classical system separate from the quantum system being observed. Thus an artificial distinction is made between quantum and classical systems. This leads to several logical objections. Any classical system is fundamentally quantum mechanical at the atomic level. How and when does a quantum system become classical? Where does this transition take place? And what is the mechanism by which the measuring apparatus collapses the wave function? The Schrödinger equation does not collapse wave functions.

The Copenhagen interpretation offers no answers to these questions, and this failing has encouraged the ill-founded speculation that human consciousness itself causes the collapse. Attempts have been made to provide ad hoc physical mechanisms for wave function collapse, but these have so far failed to gain wide acceptance.[11]

Another logical objection to Copenhagen arises from the more esoteric consideration of the wave function of the universe as a whole. (Based on my prior quibbles, the correct terminology perhaps should be the wave function of an ensemble of similarly prepared universes, but let me use conventional language here, understood in these terms.[12]) If the universe is a quantum system, and we believe all systems are ultimately quantum, it must be describable by a wave function.[13] But the Copenhagen wave function is only defined by measurements performed on the system from outside the system. How can we perform a measurement on the universe from outside the universe?

Many Worlds

An important step toward resolving this dilemma was made in 1957 by Hugh Everett III. In his Ph.D. thesis from Princeton University, Everett introduced the *many worlds* interpretation of quantum mechanics.[14] It has slowly attracted a cult of followers. Originally called, less grandiosely, the "Relative State Formulation of Quantum Mechanics," the theory provided a formalism that reproduced the statistical results of quantum mechanics without the need for a separate set of classical measuring instruments or external observers. In the many worlds interpretation, measuring instruments and observers are all part of the quantum system that is being described by wave functions and the other mathematical apparatus of quantum mechanics.

In Everett's model, the wave function of the universe is a huge mathematical object containing the information needed to compute the probabilities of every possible measurement in the universe. When an actual measurement is made, the wave function does not collapse. Rather, each piece of the original wave function that contained the result of an alternative measurement outcome remains intact, describing the universe that would have existed if that particular branch had been taken.

In the common description of the Everett view, the universe is imagined to branch into as many parallel universes as there are possible results of the measurement. This connotation has given the many worlds interpretation the same pop science attention that we have found with Copenhagen consciousness and Bohm holism, often with all three in the same breath.[15]

To examine the many worlds idea in more detail, consider an experiment measuring the spin component of an electron. Recall that it has two possible outcomes. Suppose that, prior to the measurement, the wave function can be written $\Psi = \Psi_1 + \Psi_2$, where $|\Psi_1|^2$ is the probability for outcome 1 and $|\Psi_2|^2$ the probability for outcome 2. A measurement is performed and, say, the result "2" is obtained. In the Copenhagen view, the act of measurement collapses Ψ to Ψ_2. On the other hand, in the many worlds view, the act of measurement causes the universe to split into two universes, one where the measurement yields outcome 2 and another where the measurement yields outcome 1.

Everett solves the Schrödinger's cat paradox by having two universes—one where the cat is dead and another where it is alive. But at least the unfortunate cat's death occurs when the phial of poison is broken, where common sense demands it, rather than afterwards, when the box is opened by a conscious observer, as quantum mysticism might lead you to believe.

The EPR experiment is explained by many worlds as follows: At the instant that the choice of spin meter orientation is made at the end of beam line A, our universe takes the branch associated with that choice, while other parallel universes appear where all other possible choices are made. Another split occurs when the spin is measured, with one universe giving spin "up" along that axis and the other giving spin "down." This is illustrated in figure 6.2.

The result observed at the end of beam line B is simply the one expected for that particular universe where the given choice and measurement applies. If the observer at B decides at the last moment to change axis orientation, again the universe splits in two, one where the choice is made and one where it is not. This could happen, for example, under the control of a computer random number generator, so no special role for human consciousness should be implied in the process. In all cases, the results from the various choices and measurements occur in the universes where these choices and measurements are made. No superluminal communication is required within any of these universes. Each universe is simply a different experiment requiring a different calculation for the outcome.

The other so-called quantum paradoxes have similar explanations. They all quickly go away when we do away with wave function collapse. Without wave function collapse, Einstein rests more easily since no spooky

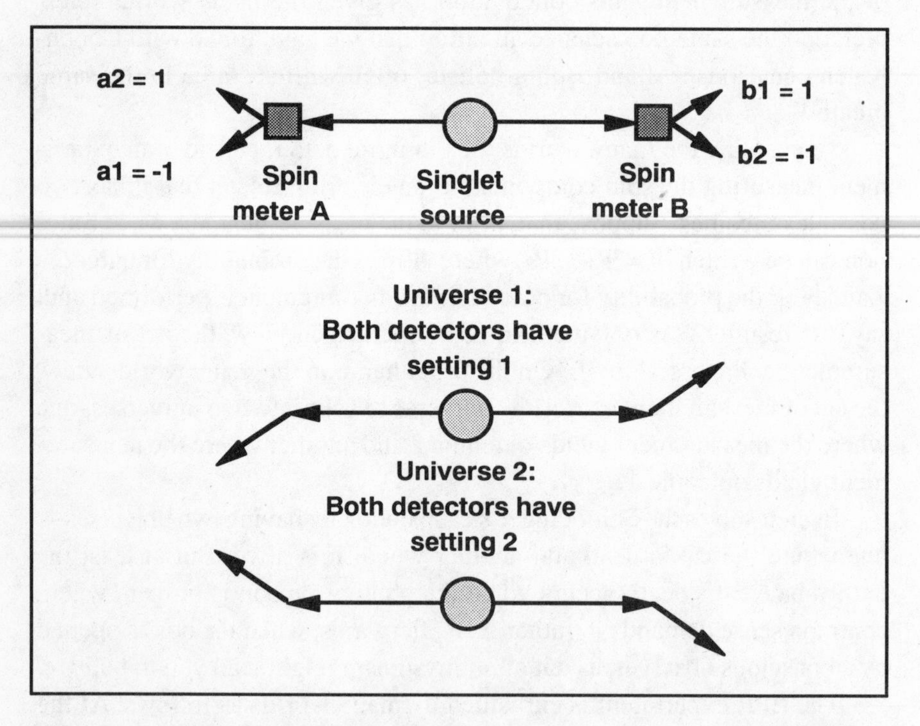

Fig. 6.2. The Bohm-Bell EPR experiment as viewed in the many worlds interpretation of quantum mechanics. When the spin meter settings are the same at each beam end, then the spin measured at one end will be opposite the spin measured at the other. In the many worlds view, the universe spits into two universes, one for each spin setting. Since in either case the particles originated from the same source, nonlocality is not required. Other universes, not shown, exist for the other choices of spin axes.

actions at a distance take place. We have no nonlocality, no basis for holistic cosmic mind.

Is many worlds really local? Author Alstair Rae says not. "If a measurement performed on a photon at one place causes an immediate splitting of the whole universe out to the furthest stars, a non-locality is implied that is certainly as radical as, and probably more so than, any other version of quantum theory."[16]

However, this is not what is implied by many worlds; the branches occur locally. No change occurs that requires instantaneous correlations across the universe, unless we again assume a deterministic reality to the wave function or some subquantum process. In other words, many worlds becomes nonlocal, as do all other interpretations, when you go beyond the statistical hypothesis and postulate some deterministic control over individual events. Indeterministic many worlds quantum mechanics does not lead to nonlocal effects. Any signals that pass between points in any given universe move no faster than the speed of light.

To many quantum theorists, Everett's theory offers advantages over Copenhagen. In particular, the measurement apparatus is not treated classically and remains part of the system. This means that many worlds can be used in cosmology, a point emphasized by Bryce DeWitt, who became the major promoter of Everett's idea.[17] Many worlds also eliminates some mysticism by not implying any central role for consciousness. While it seemingly introduces a new metaphysics of parallel universes into the debate, this can be extracted from the discussion, as we will see later, without discarding any of the benefits of the Everett formalism.

Since the parallel universes in the many worlds interpretation are presumably inaccessible to one another, and since the theory does not tell us which particular branch our personal world will take, the many worlds theory contains no empirical content that has allowed it, so far, to be tested against other interpretations. As with the other interpretations we have discussed, many worlds can only be judged by a comparison of its formal simplicity and convenience in application. If the Everett formalism were to lead to a viable quantum cosmology that entertains unique observational consequences, then many worlds would have a firm basis. But such an event would only represent a victory for the formalism and not imply the actual existence of parallel universes.

Where does the many worlds interpretation stand among the consensus of scientists? An informal poll of seventy-two leading physicists and cosmologists was conducted in 1991 by political scientist L. David Raub. Although not published at this writing (to my knowledge), Raub has reported that 58 percent agreed with the statement, "Yes, I think the MWI is true."[18] Supposedly included in this group were Stephen Hawking, Murray Gell-Mann, and the late Richard Feynman.

Based on their own published writings, however, I doubt very much that these luminaries specifically supported the notion that the parallel universes in the many worlds interpretation all simultaneously exist. In fact, as we will see, Gell-Mann has explicitly rejected that view. I think it is fair to say that most physicists dismiss as rather extreme a solution to the problems of quantum mechanics that requires the continual splitting of our universe into billions upon billions of new parallel universes, with probably no way of ever determining whether these universes exist.

Still, the Everett formulation has merit in finding a place for the observer inside the system being analyzed and doing away with the troublesome notion of wave function collapse. These features have been retained in a new class of interpretations that appears to solve all of the problems we have discussed in a manner that does violence to neither physics nor common sense.

Beyond Many Worlds

Gell-Mann and collaborator James Hartle,[19] along with a score of others, have been working to develop a more palatable interpretation of quantum mechanics that is free of the problems that plague all the interpretations we have considered so far.[20] This new interpretation is called, in its various incarnations, **post-Everett quantum mechanics**, alternate histories, consistent histories, or decoherent histories.[21] I will not be overly concerned with the detailed differences between these characterizations and will use the terms more or less interchangeably.

Post-Everett quantum mechanics grows out of a version of quantum mechanics proposed by Feynman almost fifty years ago, the many worlds formalism of Everett, and some insightful recent work by Robert B. Griffiths.[22] Much of the new interpretation remains under development and intensely debated.[23] However, it appears to be heading toward a very sensible solution to the interpretation problem that does not require the inclusion of invisible holistic fields, superluminal signals, parallel universes, or the intervention of consciousness, human or cosmic.

Feynman's Path Integrals

The Feynman path integral formulation of quantum mechanics was the early precursor of the post-Everett interpretation.[24] Around since 1946, this method has always been regarded as a particularly elegant idea representing a natural, quantum extension of an equally elegant approach to classical physics. One of the great virtues of post-Everett quantum mechanics, we will see, is to make the quantum-to-classical transition a smooth one.

In classical physics, the path followed by a particle between two points a and b is determined by an equation of motion. The most familiar equation of motion is Newton's second law, F = Ma, where F is the force on a particle of mass m and a is its acceleration.

In the general formulation of classical mechanics developed in the centuries after Newton, the path given by the equation of motion corresponds to the particular path in which a quantity called the **action** is an extremum (maximum or minimum). For historical reasons, this is somewhat inaccurately termed the **principle of least action**, or **Hamilton's principle**.

According to Hamilton's principle, the path followed by a particle from a to b is that for which the action integral along the path

$$S = \int L(x,v,t)\, dt \qquad (6.1)$$

is an extremum, where $L(x,v,t) = mv^2/2 - V(x,t)$ is the Lagrangian of the particle (kinetic energy minus potential energy). By using the calculus of variations, this can be used to derive the familiar classical equations of motion, as well as the geodesic equation that is the jumping-off point for general relativity.[25]

In optics, the path of a light ray through media of varying indices of refraction can be computed in an analogous manner using *Fermat's principle.* Thus, in classical physics, particles and light are interpreted as following a kind of path of least resistance among all the possible paths that they may follow through space.

In his path integral formulation of quantum mechanics, Feynman adopted a similar approach. In place of the principle of least action determining a definite path, Feynman asserted that we must calculate the probability for all alternate paths. The path taken is not necessarily the most probable but, as in the usual statistical view, an ensemble of similarly prepared particles will be observed to follow the frequency distribution given by this probability, with paths of probability 0.3333 followed one-third of the time and so forth.

One common rule of probabilities is that they add for the combined effect of mutually exclusive alternatives. For example, the chance of drawing a picture card from a shuffled deck of playing cards is $3/13$, and the chance of drawing an ace is $1/13$, so the chance of drawing a picture card or an ace is $3/13 + 1/13 = 4/13$.

Feynman noted, however, that the probability for a result in quantum mechanics will not always be the simple sum of probabilities for the apparent alternatives. Rather, in certain situations, the probability must be calculated as the absolute square of the sum of **probability amplitudes** over all imaginable paths between a and b. In this case, each path contributes a different phase to the amplitude, and interference effects occur when the relative phases over different paths result in their respective amplitudes summing constructively or destructively.

> In the Feynman path integral method, the probability for a particle to go from a to b is given by
>
> $$P(b,a) = |A(b,a)|^2 \quad (6.2)$$
>
> where A(b,a) is the probability amplitude of going from a to b computed by
>
> $$A(b,a) = \sum R \exp(iS/\hbar) \quad (6.3)$$
>
> where R is a normalization constant, S is the classical action defined in equation 6.1, and the sum is over all paths between a and b. Note that the action determines the phase of the amplitude, which is a complex number. The probability amplitude is the sum of amplitudes, or wave functions, for the various paths. Interference effects will result from cross terms in squaring the sum.

The classical principle of least action follows from the Feynman method when, as is the case for most macroscopic systems, the phases along all but the path of least action interfere destructively.

In their 1965 book on path integral quantum mechanics, Feynman and Hibbs suggested a rather unconventional statement of the uncertainty principle: "Any determination of the alternatives taken by a process capable of following more than one alternative destroys the interference between alternatives."[26] What they meant was that all possible alternatives for the path of a particle will interfere, which is why we must add their amplitudes and not simply their probabilities. However, determining which path a particle actually follows, say by placing a detector behind a slit, will eliminate the interference.

Clearly the probabilities for two alternative paths do not always add. Consider the double slit experiment illustrated in figure 3.2. If A_1 is the probability amplitude for a path through one slit to a particular detector, and A_2 is the amplitude for a path through the other slit to the same detector, then the probability P that the particle hits the detector is $|A_1 + A_2|^2$. This is confirmed by the interference pattern seen in the double slit experiment.

As Feynman and Hibbs put it, "From a physical standpoint the two routes are independent alternatives, yet the implication that the probability is the sum P_1+P_2 is false. This means that either the premise or the reasoning which leads to such a conclusion must be false. Since our habits of thought are very strong, many physicists find that it is much more convenient to deny the premise than to deny the reasoning." And so we hear bizarre statements like, "When you watch, you find that it [the electron] goes through either one hole or the other hole; but if you are not looking, you cannot say that it goes either way!"[27]

Feynman and Hibbs rejected this, denying the reasoning rather than the premise. Probabilities are not to be computed by adding the probabilities of the alternatives. Instead two kinds of alternatives are considered: those that interfere, requiring the addition of amplitudes, and those that do not interfere (or whose interference washes out) and allow the addition of probabilities.

Consistent Histories

A firm mathematical and logical basis for the interpretation of Feynman paths was provided in 1984 by Griffiths.[28] As we learned from the above discussion, not all conceivable paths through spacetime are meaningful in the sense that we can talk about them in terms of conventional additive probabilities. Griffiths established a set of conditions that are required for a path to be logically consistent in this sense. He called these consistent histories. For our purposes, we can think of a Griffiths' consistent history as a series of spacetime points for which we can calculate a probability.

Only those histories for which a probability can be meaningfully calculated are "consistent." A consistent history can be made up of other consistent histories, where the probabilities add. Only consistent histories are ever measured.

Now, in quantum mechanics, not all imaginable paths, such as the two slit paths of the double slit experiment when no detector exists behind either slit, are consistent histories. Griffiths, following Feynman, tells us we can think about such paths if we want, as long as we add their amplitudes before squaring to make a probability. But we are to remember that such paths can never be measured. In Copenhagen positivism, they simply do not exist. Consistent histories also puts such paths in a state of limbo, in some sense saying that they can exist in your mind as you sum up their amplitudes, but only consistent histories ever appear in any measurement.

This point of view, incidentally, is precisely that implicitly held by particle theorists when they use Feynman diagrams to compute the probability for some fundamental process such as electron-positron annihilation into photons. The theorists draw all possible diagrams and add their amplitudes. Normally experiment does not single out a particular diagram, so no one knows which one "really happened" in a given observation. All possible diagrams must be included in the calculation, just as all possible paths must be considered in a multiple slit experiment. Nevertheless, the individual diagrams are intuitively powerful and imagined as being real, in some sense, by the people making the calculations.

In a recent book, Roland Omnès argues that Griffiths' consistent histories is not just one interpretation of quantum mechanics but "the interpretation" that represents the orthodox view as inferred from the writings of Bohr. However, as we shall see, consistent histories seems to solve the problems of Copenhagen, in particular by including the observer in the quantum system and providing for a smooth quantum-to-classical transition. Thus it would appear to represent a major step beyond Copenhagen.

Alternate Histories

Gell-Mann and Hartle have followed a closely related line of reasoning in what they term the *alternate histories* approach. They have commented that Everett and other proponents of the many worlds theory caused unnecessary confusion by calling all the branches taken by the universe under the act of measurement "equally real." According to Gell-Mann and Hartle, what the many worlders should have said was that quantum mechanics makes no preference for one branch or another, except in so far as one has a higher or lower probability.

Gell-Mann and Hartle interpret the various branches of Everett's many worlds as being simply alternate histories for which quantum mechanics computes probabilities, as for example by Feynman path integrals. Which branch the universe takes is then a matter of chance, based on those probabilities. Thus, an infinity of independent parallel universes is not required to spring into existence at every blink of an eye or click of a Geiger counter.

In the alternate histories scenario, a universal wave function describes an ensemble of similarly prepared universes. That wave function includes all possible histories, not just those that actually happen. Following many worlds, alternate histories does not require the Copenhagen division of the world into quantum objects and classical measuring instruments, using the Everett formalism to incorporate the measuring apparatus into the system. Quantum mechanics then simply tells one how to calculate the likelihood of each possible history. What history happens, happens by chance.

Using alternate or consistent histories, we can interpret the double slit experiment in the following way, as illustrated in figure 6.3: Three histories exist for the electron, corresponding to two different experimental arrangements. In one experimental arrangement, the detector D1 behind the

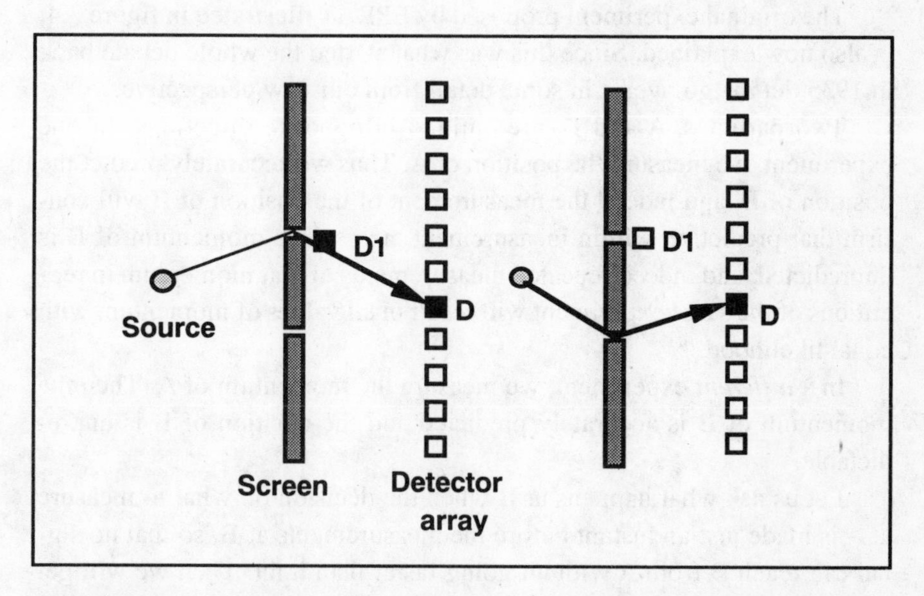

Fig. 6.3. Alternate paths of electrons to a detector D in the double slit experiment. On the left, the electron goes through the top slit. On the right, it goes through the bottom slit. The detectors D1 behind the upper slit can be on or off. Quantum mechanics computes the probabilities for the two paths shown, which can be checked by counting electrons in the detectors. When the detector behind the slit is on, the probabilities add. When it is off, the probability amplitudes add and an interference pattern is observed as D is moved to other positions in the detector array.

top slit is turned off. The probability for the electron hitting D is given, as in the previous discussion, by $P = |A_1 + A_2|^2$. This is one alternate history.

In the second experimental arrangement, D1 is turned on. If we get hits in both D1 and D, we know the electron went through the top slit. The probability for this happening is $P_1 = |A_1|^2$. This is the second alternate history. If we get a hit only in D, but not in D1, then we know the electron went through the bottom slit. The probability for this happening is $P_2 = |A_2|^2$. This is the third alternate history. The probability of a hit at D by either path is then $P = P_1 + P_2$, probabilities adding for these alternate histories.

In the Bohm-EPR experiment, the explanation is similar to that for many worlds illustrated in figure 6.2. However, the separate universes do not exist "in reality" but only as potentialities, with our universe being a *chance* selection among the various possibilities. Only if we insist on some underlying, determinate hidden variables forcing the choice does nonlocality enter the picture. Thus we once again see the strong connection between nonlocality and determinism. As Bell showed, it is impossible to escape nonlocality in deterministic theories.

The original experiment proposed by EPR, as illustrated in figure 3.4, is also now explained. Since this was what started the whole debate back in 1935, let me go over it in some detail from our new perspective.

Two particles, A and B, are emitted in opposite directions. In one experiment, we measure the position of A. Thus we accurately predict the position of B, and indeed the measurement of the position of B will confirm that prediction within measurement errors. The momentum of B is unpredicted, and indeed repeated measurements of that momentum in repetitions of the same experiment will result in all values of momentum with equal likelihood.[29]

In a *different* experiment, we measure the momentum of A. Then the momentum of B is accurately predicted and the position of B is unpredictable.

Let us ask what happens at B when the decision on what to measure at A is made just an instant before the measurements at B, so that no signal can reach B from A without going faster than light. Then we will get the result that corresponds to the particular choice at A. As in the experiment with spins, we find different experimental arrangements leading to different predictions.

Now, people will still insist that this result is "spooky," that some kind of nonlocality is implied. But the predictions correspond precisely to what is in fact measured. What more can one ask of a theory? Once again, I must emphasize that no superluminal signal has been transmitted.

As before, we witness that *contextuality* is the distinguishing feature of quantum phenomena. Results depend on the experimental setup; a different setup is a different experiment. In the double slit we have different experiments, depending on whether the detector behind a slit is on or off. In EPR, we have different experiments depending on the choice of what we measure at the end of one beam line. Each different experiment demands a different calculation. Alternate or consistent histories only calculates probabilities for the various experiments; it does not reveal which path is followed by an individual particle.

Griffiths has discussed the EPR experiment from his consistent histories viewpoint.[30] He insists it provides a realistic, objective, but indeterministic explanation without action at a distance or other "spooky" influences.

Griffiths shows that measurements carried out at A have absolutely no effect on the physical system at B. What is affected by a measurement at A is our *ability* to make a prediction of the result of a measurement at B. While this is still distinct from what one expects classically, Griffiths argues that this distinction is of the same nature as we find in trying to predict the outcome of the measurements on a single particle. For example, measuring the spin component of an electron along one axis affects our ability to predict a later measurement of the spin of the same particle along another axis. A history in which two different spin components are predictable is inconsistent and thus meaningless. This is not the case in classical physics and represents the primary feature of quantum mechanics. As Griffiths points out, no nonlocal effects are implied in the single particle case and the two-particle EPR experiment has a similar, local character.

The Observer

Gell-Mann and Hartle provide a conceptual basis for distinguishing between meaningful and meaningless histories as defined by Griffiths. They contend that only those histories for which probabilities can be assigned produce an effect that is observable by our macroscopic instruments. As they explain it, "An 'observer' (or information gathering and utilizing system) is a complex adaptive system that has evolved to exploit the relative predictability of a semiclassical domain."[31] This "semiclassical domain" is the domain of events that occur with some level of regularity and predictability. In other words, we may have evolved in the near-classical, predictable, macroscopic world because it was adaptive to do so. Life has enough uncertainties as it is, without having to deal with quantum uncertainties. By its very nature, life may have little direct connection with quantum phenomena and the fact that our world is so well approximated by classical physics could have been selected by evolution. Mystics can prate all they want about the bizarre quantum world, but it may have little to do with the bodily and mental processes of human beings because life evolved to avoid quantum mechanics.

Alternate or consistent histories differs from the many worlds interpretation in that every allowed history does not occur. What actually happens is selected by chance from a set of allowed possibilities. That set includes only those events for which probabilities can be logically defined.

Is the selection process by which one of the alternate paths is chosen random or causal? We are so accustomed to thinking causally that we feel compelled to require a causal explanation for every event that happens. However, decades of experiments described by indeterministic quantum mechanics suggest that the universe may not work this way at its deepest levels, even if it seems to do so on the level of human experience.

The lack of a distinction between cause and effect at the elementary level also suggests this. As I have conjectured earlier, perhaps the whole notion of causal precedence is an emergent principle of complex, macroscopic, nonquantum systems. At the quantum level, individual events just happen. The statistical "laws" they seem to obey, such as the exponential decay law for radioactive decay, argue for at least some noncausal, random processes.

Members of the school of quantum consciousness insist that we must seek an underlying causal explanation for all phenomena, regardless of their apparent statistical nature. They propose that an entity they label as "mind" is the creative force that chooses between alternative paths, acting to "actualize" events.[32] I will return later to this issue. First I need to develop a crucial aspect of post-Everett quantum mechanics—the idea of *decoherence*.

Decoherent Histories

The double slit experiment can be generalized in a way that provides a useful model for the consistent histories approach.[33] Imagine that, in the place of the single screen with two slits, we have multiple parallel screens, each with many slits (figure 6.4). Histories can then be visualized as the various possible paths followed by a particle as it works its way from the source to the final detector. The only consistent histories are those paths passing

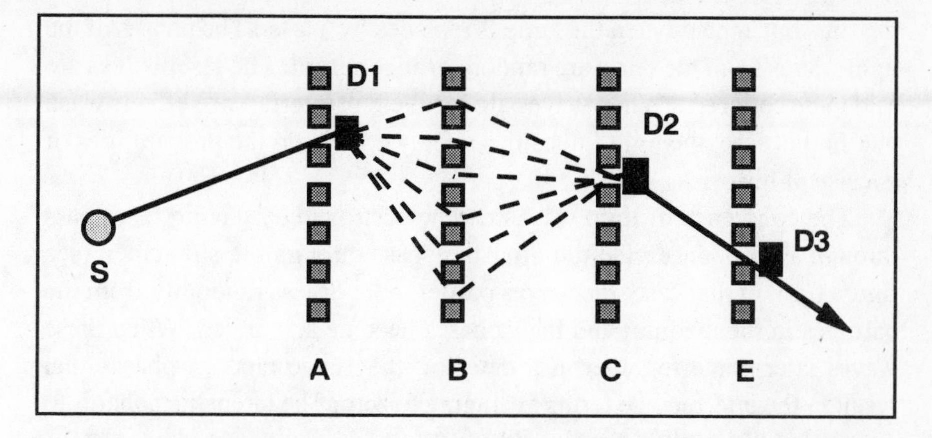

Fig. 6.4. The path of a particle from a source S through a series of four screens with multiple slits. The path is definitive (heavy line) through screens A, C, and E since detectors D1, D2, and D3 are located behind specific slits in these cases. These detectors cause the particle's wave function to decohere at their positions. Since screen B contains no detector, nor environmental particles, to cause decoherence, the wave functions for all paths (light lines) must be summed to determine the probability at screen C.

through slits behind which exist other particle detectors, because only they have additive probabilities. Where no detector exists behind a screen, all paths must be combined coherently, that is, their amplitudes must add. These are inconsistent histories. We can imagine them if we want, but they are never observed.

This may seem to continue to assign a special role to observing devices such as particle detectors, which many regard as a major undesirable feature of Copenhagen. However, Everett already showed that we can use his scheme to include the detector as part of the system rather than as some independent, external agent, as in Copenhagen. In the new post-Everett formulations, this feature is retained.

Furthermore, "detectors" need not be limited to the sensory apparatus of human beings or their scientific instruments. The term "detector" can, in this context, also encompass the particles in the environment surrounding the system. Consider the role of a detector placed behind a particular slit. It serves the purpose of removing the interference term in the probability calculation so that the probability for passing though the given slit can be added to the probabilities for other allowable paths.

In optics, interference effects occur when two light waves of the same frequency have fixed relative phases. For example, interference bands in the double slit are not seen when an ordinary incandescent lamp illuminates the slits, but appear when the lamp is replaced by a laser. The phases of the light waves from the lamp are random or **incoherent**. The laser waves are said to be **coherent**. When waves are incoherent, interference effects wash out. In this case, the light intensities simply add, as do the probabilities of consistent histories.

The coherence of light waves can be destroyed by having them pass through a semidense medium after they pass through the slit screen (see figure 6.5). In that case, the waves scatter more or less randomly from the particles in the medium and their phases become scrambled. When these waves later come together at a detector, they have random phases that result in the interference terms averaging to zero. The intensities, that is to say, probabilities, then simply add.

Thus any interaction of the particles in a system, whether with their environment or the detection apparatus, is capable of producing decoherence. This decoherence need not be perfect, just sufficient to reduce the interference terms in the wave function to a negligible amount.

Decoherence is precisely the property that distinguishes quantum systems from classical ones, providing the transition from quantum to classical, from the ephemeral to the familiar. In the macroscopic world of our experience, systems behave decoherently and predictably. Our concepts of space and time, mass and energy, grew from our observations of that

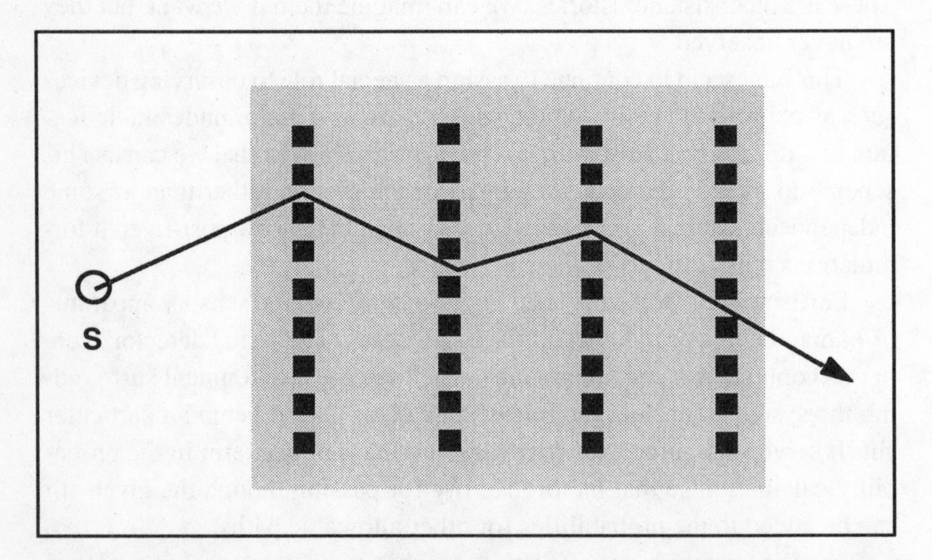

Fig. 6.5. The environment can act as a decohering medium. Here we imagine the screens immersed in a gas that scatters the particles from S. In this case, all possible paths are allowed, as in classical physics. Only one is shown.

world. Probabilities add in the macroscopic world, and macroscopic bodies obey Newtonian mechanics.

In the decoherent histories view, decoherence arises naturally by the interaction of physical bodies with their environment. The bodies of common experience interact with the other bodies they rub against, and against the air or water of their environments.

Even deep space is not empty, but contains dust and the photons of the 2.7°K cosmic microwave background left over from the big bang. Calculations indicate that interactions of visible photons with this background alone can produce decoherence during an attempt to measure the position of a massive body such as a planet.[34] This explains why planets and the objects of our everyday experience do not appear in one place at one moment, and another place an instant later. On the other hand, electrons in an atom do demonstrate such strange behavior, since the histories corresponding to different positions do not decohere.

A **coherence length** that allows a quantitative estimate of the boundary between localized, classical particle paths and coherent wavelike paths can be defined by $L = 1/n\sigma$, where n is the number density of environmental particles and σ is the interaction cross-section between those particles and the particles whose paths we are tracing. In the case of low-energy photons, σ is typically 10^{-30} cm^2. The earth's atmosphere has a number density n of about 4×10^{22} cm^{-3}. Thus the coherence length for photons in air is of the order of a few hundred kilometers. As a result, visible light and other low-energy photons appear to us as wavelike phenomena.

On the other hand, the cross section for a typical atom interacting with air molecules is of the order of $\sigma = 10^{-20}$ cm^2. The coherence length in this case is tens of microns, and so atoms appear localized on about that scale. By the same token, high-energy photons, such as gamma rays, behave like particles in passing through normal matter.

Information and the Arrow of Time

Decoherence can also be used to throw more light on the arrow of time, which I have argued is a macroscopic (or, more precisely, many-body) emergent concept. The unidirectional flow of time is deeply embedded in our experience, yet we have seen that this apparent process cannot be found in observations at the elementary particle level. All fundamental interactions can be run forward or backward. Decoherence, on the other hand, is (for all practical purposes) irreversible. In thermodynamic language, decoherence increases the entropy of the system, thus producing precisely the effect that we associate with the direction of time. Time's arrow is, by definition, the arrow of increasing entropy of a system plus its environment.

A localized system can decrease its entropy by donating some to another system or the environment. Now, in information theory, the quantitative measure of **information** is equated with negative entropy. So we can think of the communication between two systems as a flow of information in the direction opposite to the flow of entropy.

As illustrated in figure 6.6, when we perform a measurement on a system, we can imagine information flowing from the system to the measuring apparatus, which then becomes more ordered. The observed system becomes more disordered by the decoherence introduced by the measurement. Additional disordering results from the interaction of the system and the environment. Measurement errors, which often result from unwanted environmental effects, can be viewed in these terms.

Information also flows into the environment. This information is dif-

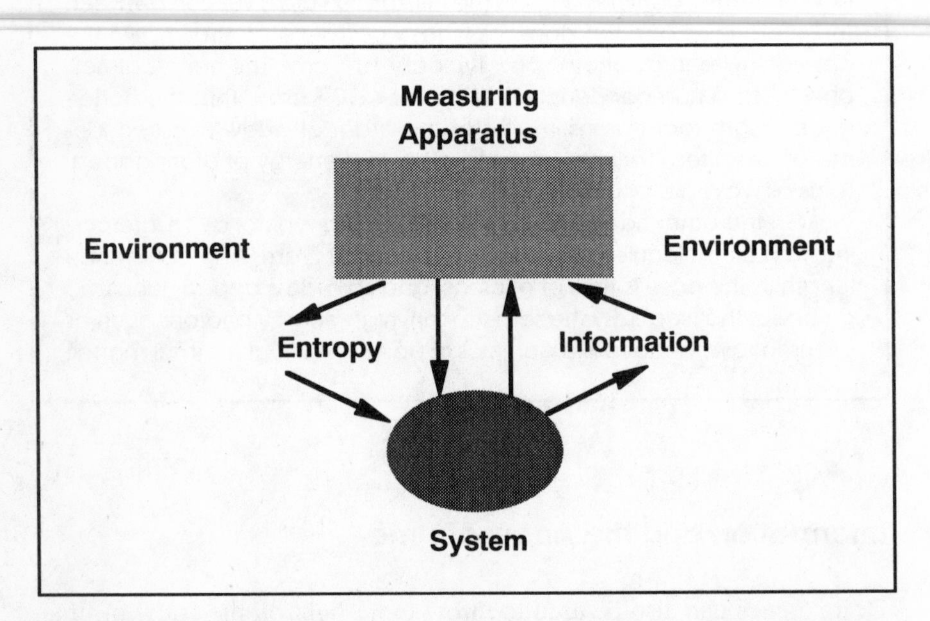

Fig. 6.6. During the act of measurement, the measuring apparatus gains information and thus loses entropy, becoming more ordered. The system becomes more disordered by interactions with both the apparatus and the environment.

ficult to retrieve, although some effects, such as increased temperature or pressure changes in the vicinity of the apparatus, can be measured.

Granularity

Let us now return to the picture illustrated in figure 6.4, where we have a large number of parallel screens with multiple slits, some that have detectors behind them and some that do not. We can take away the screens and imagine space divided up into various regions. And we can imagine each detector replaced by a body in the environment that scatters a photon at a point, changing the photon's phase in a random way so that it decoheres. The environmental body can be heavy, in which case the photon will be deflected from its original path, as when you have a screen with slits. Or the scattering body can be very light and cause no appreciable deflection, except for changing the phase sufficiently to produce decoherence.

In the classical view, we can divide space into regions as small as we want, with material always present at each point to produce decoherence. Note also that we can, in principle, make our measuring devices sufficiently small so that they do not interfere with the systems being observed.

Alternatively, we can think of classical space being completely filled by a continuous medium, an aether or some new type of field that scatters the photon. In either case, all paths of the photon are possible and we have what is called a completely fine-grained set of alternative histories.

So we see that the classical universe can be characterized by the absence of granularity. Quantum mechanics successfully describes the observed granularity of the universe. Classical mechanics, with its continuous fields, does not. Granularity results from *atomicity*—the key fact of nature that matter cannot be indefinitely subdivided. The uncertainty principle then arises because there is a limit to how small we can build detectors. At some point these detectors no longer play a passive role in the observation process but must be included as part of the system being analyzed.

In the decoherent histories view, we always possess a set of histories that is more coarse-grained than in classical physics. We do not have particles at every point, because the universe is not smooth and continuous but rather bumpy and broken. But where particles are present, a particle can scatter, decohere, and follow a path corresponding to an allowed history.

When performing experiments, we insert our detectors at particular points where they scramble the particle's phase.[35] This explains why reality appears different for different experimental arrangements. Placing a detector at a certain point where previously no material object was present allows that point to become part of an allowed history—a part of reality. This may appear to have a metaphysical ring to it, reality again being determined by the act of measurement. However, I think it is not saying much more than the obvious: measurements interact with the system being measured in a granular universe.

In the Copenhagen interpretation, this leads to a dual universe of quantum system and classical observer. However, now we see that the classical observer is simply an agent of decoherence, and its role can be performed by the environment that is part of the overall quantum system. After all, a detector is part of the environment too. Clearly a conscious observer is not necessary. The tree falls in the forest, even when no one is looking.

Notes

1. Albert Einstein, "Reply to Criticisms" in Schilpp 1949, p. 671.
2. Specifically, the relativistic four-vector potential A is used, where the three space-like components correspond to the classical vector potential **A** and the fourth component corresponds to the classical scalar potential ϕ. Upon quantization, this becomes the wave function for an ensemble of photons.
3. The electric field surrounding a charge distribution can be mapped at various points in space by measuring the force on a test charge placed at each point and dividing the result by the value of the test charge. The gravitational field is similarly mapped with a test mass. The magnetic field is mapped by putting a tiny dipole magnet, such as a compass needle, at the point. It can be argued that these fields are not "real" either, since they are replaced by particles, or field quanta, in quantum mechanics, but that is another issue. My present argument holds even in classical physics, where the potentials are recognized as arbitrary while the fields are taken to be real.
4. Aharonov 1959. An earlier discussion of the effect has been attributed to W. Ehrenberg and R. E. Siday.
5. Chambers 1960.
6. O'Raifeartaigh 1991. The authors do not discuss the two-slit experiment, but rather consider the energy shift of an electron circulating around the outside of the solenoid. However, their result applies equally well to the double slit experiment.
7. Aharonov 1993a,b, Anandan 1993.
8. Aharonov 1993b.
9. For a detailed critique of the Aharonov protection scheme, see Unruh 1994.
10. Anandan 1995. I appreciate the patience shown by Anandan and his colleague Harvey Brown in trying to help me understand the concept during the sabbatical I spent at Oxford in spring 1995.
11. Ghirardi 1986, Bell 1987, p. 201.
12. Scientists are often criticized for using concepts such as "an ensemble of universes," the big bang, and the evolution of human life that are not directly observed. But

this is part of the abstraction and generalization of the scientific method. Galileo could not drop objects in a vacuum, but he could hypothesize the situation. Scientific progress would grind to a halt without this kind of generalized modeling and idealization.

13. Here I am again speaking a bit loosely. Experts will recognize that the term *density matrix* is more appropriate, but an unnecessary complication for the current discussion.

14. Everett 1957. See also DeWitt 1973.

15. See, for example, Wolf 1991.

16. Rae 1986, p. 82.

17. See "Quantum-Mechanics Debate," *Physics Today* 24, April 1971, p. 36.

18. As reported in Tipler 1994, p. 170.

19. Murray Gell-Mann and James B. Hartle, "Quantum Mechanics in the Light of Quantum Cosmology" in Zurek 1990, pp. 425–58. See also Hartle, "The Quantum Mechanics of Cosmology" in Coleman 1991, pp. 65–157, and Gell-Mann 1994, pp. 135–65.

20. Zeh 1971; Zurek 1981, 1982; Griffiths 1984; Joos 1985; Omnès 1988, 1992, 1994. See these papers and books for further references.

21. Some authors, such as Tipler 1994, p. 170, muddy the waters by equating post-Everett quantum mechanics with the many worlds interpretation.

22. Griffiths 1984.

23. A not-too-highly technical discussion has been given by Zurek 1991. For a taste of the debate, see the response letters printed in *Physics Today* 82, April 1993, pp. 13–90. The new book by Omnès, 1994, argues that the modern interpretation is logically consistent, complete, and absent of any paradoxes. However, see Dowker 1994 for a dissent.

24. Feynman 1946, 1965.

25. Goldstein 1980.

26. Feynman 1965, p. 9.

27. Feynman 1965, p. 13.

28. Griffiths 1984.

29. In practice, both measurements would have finite width distributions, related by the uncertainty principle, but this need not concern us here.

30. Griffiths 1987.

31. Gell-Mann and Hartle in Zurek 1990, p. 425.

32. Stapp, 1993.

33. Zurek 1991, Omnès 1994 and references therein.

34. Joos 1985.

35. I do not mean to imply that every photon scattering produces complete decoherence, that is, a random change in phase. But the net effect of many scatterings will lead to a good approximation of complete decoherence.

7

Restoring Reality

So irrelevant is the philosophy of quantum mechanics to its use, that one begins to suspect that all the deep questions about the meaning of measurement are really empty, forced upon us by our language, a language that evolved in a world governed very nearly by classical physics.

—Steven Weinberg[1]

Subverting Common Sense

At the price of some repetition, let me bring together what we have found in our survey of the interpretations of quantum mechanics and their various implications.

The Copenhagen interpretation, promulgated by Bohr, Heisenberg, and their followers in the late 1920s but somewhat evolved since, is still presented in one form or another in most quantum textbooks. When they stand in front of a quantum mechanics class, most physics professors follow these texts and speak a Copenhagen-inspired language that unfortunately leads to considerable misinterpretation of the ontological meaning of quantum mechanics.

The words used are intended to explain and motivate the formalism. However, since the formalism is precise and unchallenged, while the words are vague and controversial, perhaps it would be better to skip the latter altogether. But then, you can't walk into a classroom the first day of a semester and simply start writing equations on the board.

In my own classes I have said things like, "The act of measurement causes the wave function of the electron to collapse," and, "The electron in an atom has no position until it is measured." Only in recent years have I begun to appreciate the caveats behind these words and the depth of the philosophical debate over the meaning of quantum mechanics that has been carried on from its inception.

As a student, I did not learn from any professor or textbook that alternate interpretations can be found that lead to the same formal theory and predict the same experimental results as Copenhagen. I became aware of this by virtue of my side research on pseudoscience and science mysticism, and not through any work I have done in three decades as a physics teacher and researcher.

Few physicists share my interest in the borderlands of science. Perhaps a handful are vaguely aware of other quantum languages such as hidden variables, many worlds, or decoherent histories, but they pay them little attention. And few will take notice until one language is shown to be superior either in terms of providing unique, testable predictions or making calculations easier and more powerful.

The Copenhagen interpretation is often associated with the philosophical school of positivism. Relativity and quantum mechanics underscored the critical role of the measuring apparatus in defining what is measured, which is a prime tenet of positivism. Time is defined by clocks, temperature by thermometers. But taken to its extreme, the dubious conclusion is drawn that measurements constitute the extent of reality. This view subverts the common-sense belief in an objective reality independent of and external to human consciousness. Surely the universe does not care about human existence! Common sense can be wrong, but proposed violations of common sense should be taken seriously only after they have been amply demonstrated.

Within the philosophy and sociology of science today there are four schools of thought: positivism, realism, relativism, and pragmatism.[2] Although scientists often write books and articles that utilize a positivist perspective, most probably think of themselves vaguely as realists and pragmatists. Certainly they believe that what they do as scientists bears some relation to objective reality, that it's not just all in their heads.

Few disagree that the best case to be made for science is that it provides benefits and practical applications that anyone plainly can see. However debatable its philosophical foundations and moral value, science works better than any other mode of thought we humans have been able to invent so far.

To the practicing scientist, this is sufficient justification for the scientific enterprise, no matter what philosophers, sociologists, and preachers say about it. Buried in the pleasurable details of their research, most scientists are at best dimly aware that this practical perspective on the nature of science is not automatically endorsed by everyone.

Perhaps the most annoying challenge to the contented view that scientists hold about their profession is the new relativism. As my Hawaii colleague and antirelativist philosopher Larry Laudan defines it, relativism is "the thesis that the natural world and such evidence as we have about that world do little or nothing to constrain our beliefs."[3]

The beliefs referred to here are not just those of churchmen and laymen, but those of scientific professionals as well. Science, in the relativist view, is akin to every other human activity—it is primarily a social phenomenon. In the relativist view, the so-called theories and laws of physics have less to do with the real world than with the traditions and interactions of the scientific community.

Those who write on modern metaphysics from perspectives outside mainstream science often propound elements of relativism, and no little New Age solipsism: Reality is what we make it to be, by our conscious acts of observation and measurement. Growing to a great extent out of the work of Thomas Kuhn,[4] who denies he is a relativist, relativism has now spread its wings beyond science; it has become the fashionable view among postmodernists and deconstructionists in the humanities.

However, I cannot think of a single working physical scientist who is a relativist. That handful who are even aware of Kuhn's work (most would not even recognize his name) scoff at the idea that scientific truth is an arbitrary social convention. They all can produce examples that belie the notion.

One of my favorite examples is the magnetic moment of the electron, mentioned in chapter 1, which can be first calculated and then measured to one part in ten billion, with the two results in perfect agreement. To characterize this spectacular achievement as nothing more than social convention is absurd. The magnetic moment of the electron (and thus the electron itself) is as objectively real as any concept that humans can bring to mind, including the chair you are sitting on.

Another example is the precession of the perihelion of Mercury, the slow rotation of the major axis of the planet's elliptical orbit. In 1859, French astronomer Joseph Le Verrier found that observations of the precession disagreed with Newtonian calculations by 43 arc seconds per century, an incredibly small but nonetheless measurable amount. In 1915, Einstein used his general theory of relativity to compute this anomalous precession and obtained exact agreement with the observations. After a decade of accurate, modern measurements terminating in 1976, the measured value was 43.11±0.21 compared to the best calculated value of 42.98.[5] When confronted with a success like this, it is again difficult to see how anyone can maintain that science is simply a social activity, like church bingo, with no deeper connection to reality (and perhaps less).

Positivism has not been so readily dismissed in scientific circles as has relativism. Obviously, our knowledge of the physical world depends on our sensory apparatus and how we design detection devices. Time can be given no unambiguous meaning without the concept of a clock. And my beloved electron magnetic moment depends on the operational definition of magnetism. This is common sense. Unfortunately, the application of extreme positivism to quantum mechanics has encouraged the nonsensical notion that human consciousness is the agent that creates physical reality.

David Bohm offered a possible route to (but not a workable specific model for) a realistic, causal subquantum theory that avoided the consciousness connection. He proposed that still-unknown forces act as the agents of quantum behavior, the way once-invisible atoms produce thermal behavior. At first glance, Bohm's theory appeared to provide an alternative that restores traditional deterministic ideas and practical common sense to quantum phenomena, answering ontological questions about their source. But John Bell, and the experiments that tested his theorem, showed that subquantum agents, if they exist at all, must be nonlocal and thus allow for superluminal connections that violate at least the spirit of Einstein's relativity.

This was not regarded as a fatal blow to Bohm's theory, although perhaps it should have been. The invisible agents were simply said to be nonlocal—holistic—as Bohm later said he knew all along. Although this conclusion appears to conflict with Einstein's speed limit, Bohm and other proponents of hidden variables have argued that no signals are transferred faster than the speed of light, and thus no violation of relativity is implied.

I regard this as a disingenuous position. The proof that no superluminal signals exist in Bohm's theory and other published alternate interpretations of quantum mechanics utilizes the formalism of conventional relativistic quantum field theory.[6] It does not invoke any unique features of hidden variables or other subquantum theories, and applies in general for any theory that gives the same results as conventional quantum mechanics. Since Bohm's theory gives the same results, it cannot allow superluminal communication despite its nonlocality.

One often hears the statement that Bohm's hidden variables are consistent with relativity. Technically this is true. However, as we have seen, superluminal motion is actually allowed by relativity, provided the particles that move superluminally always do so. That is, they must represent a new class of particles, called tachyons, that are not simply normal particles accelerated beyond the light barrier. Einstein's speed limit resulted from the additional hypothesis of causal precedence and is supported not so much by theoretical argument as experimental fact: no tachyons have yet been observed. Thus a Bohmian subquantum theory with tachyons would not be in conflict with relativity, but would require that the labels cause and effect depend on the choice of reference frame.

It is not a virtue but a vice for a theory to agree with all conceivable experiments. For then it cannot be tested, and no way exists for it to be shown to be right or wrong. Such a theory is useless, since it make no contribution to knowledge or technology. I cannot see how one can escape the conclusion that nonlocal hidden variables, to have any unique observable effects that make the notion meaningful and useful, must yield superluminal motion.

The way to demonstrate the validity of hidden variables theories is clear: measure superluminal effects. The primary characteristic of the hidden variables would be their superluminality. In the absence of observed superluminal effects, or some other comparably unique prediction of the theory, no evidence exists for hidden variables.

The holistic ideas that follow from nonlocal hidden variables, such as Bohm's holomovement and the implicate order, have great emotional appeal. They provide some people with a reason to believe that they are not simply insignificant specks of matter but instead are intertwined with the totality of existence. Unfortunately, this comforting construct has no rational basis. So far, nonlocal phenomena have not been observed and the nonlocality of objectively real quantities is not required to explain the results of any experiment.

As we have seen, the nonlocality that many believe is inescapable in quantum mechanics resides in subquantum theories or interpretations in which the wave function is assumed to be a real field associated with individual particles. Nonlocality in the behavior of objectively real objects, not just abstract mathematical functions, does not exist in conventional quantum mechanics, nor in any actual observations. And even if superluminal motion exists at the elementary level, it still may have no role to play in human life, which can be very well understood in familiar, local terms. That is, the quantum world could be superluminal with cause and effect relative while the macroscopic world remains exactly as demanded by common sense.

At present it is incorrect to state, as is often done, that the EPR experiments require nonlocality in the real world. These experiments, if correct, only require nonlocality in any theory that provides a quasi-Newtonian causal mechanism for the motion of individual particles, or retains some other classical notion of the role of physical theory, such as completeness or separability. I have shown how the simple procedure of viewing the Bohm-EPR experiment in a time-reversed frame of reference leads to a perfectly local picture, provided indeterminism is retained. Paradox appears only when one insists on imposing the time direction of common experience, which need not apply to quantum events.

While these issues remain the subjects of intense debate, the indeterministic nature of conventional quantum mechanics appears sufficient to retain pragmatic locality and still agree with EPR experiments. If it does not, perhaps an effort needs to be made to cast quantum mechanics in such a form, rather than taking for granted that the universe is nonlocal. At least this would put an end to mystical speculations about quantum mechanics demanding a holistic universe.

Two Steps Are Better Than One

The existence of a deterministic universe ruled by real, nonlocal hidden variables can probably never be totally ruled out. But, in the words of Pauli from another context, such a theory is "not even wrong" as long as it remains untestable.

To the extent that the models of Bohm and de Broglie are not testable, they should not be taken as seriously as they are in the popular and philosophical literature. Their main merit was to show that multiple interpretations of quantum mechanics were possible. This got us out of the metaphysical morass of the consciousness connection that pervades the common misinterpretation of Copenhagen. But hidden variables contribute a metaphysical morass of their own by requiring superluminal motion and its attendant implausibilities.

The many worlds interpretation provides a way of viewing the entire universe as a quantum system that is not possible in the Copenhagen picture, where a classical observation device resides outside the quantum system under study. Something different from Copenhagen is a necessity for quantum cosmology. Many worlds also avoids the notion of wave function collapse, although, as we have seen, collapse represents no philosophical problem for Copenhagen once we recognize that it does not require that the wave function be a real field. Many worlds has itself been interpreted as requiring a continual splitting of the universe into parallel branches, again a rather bizarre, nontestable notion that obscures the virtue of Everett's original work in developing a quantum formalism that incorporates the detection apparatus.

Out of the Metaphysical Morass

The more recent, post-Everett reinterpretation of many worlds in terms of alternate, consistent, or decoherent histories seems to offer the most promise of solving our conceptual problems. In this modern view, the wave function of the universe contains all the information needed to predict the probabilities for the alternate paths particles can take, but these paths are taken by chance.

The universe, by virtue of its atomicity, is fundamentally discrete, or coarse-grained. This was the first observation of Planck—what put the *quantum* in quantum mechanics. In the alternate, consistent, decoherent histories view, particle paths cannot be precisely defined unless they pass through spatial regions containing other matter from which they scatter and decohere. Quantum mechanics allows one to calculate the probability for each such path, but does not predict with certainty which one is actually taken except when that probability turns out to be close to unity.

The wave function of the universe not only describes the paths that are taken by all the particles of the universe, it also describes all the other possible paths that might have been taken consistent with the laws of physics. Note that the theory does not tell us "why" the universe takes a particular path. No further selection is made beyond a probabilistic one.

Although the analogy is not perfect, one might compare the alternate histories of quantum mechanics with the alternate histories of the theory of evolution. Evolutionary theories cannot predict the exact path along which species will evolve. The unknowns are too many and random effects play too large a role. At best, theory can predict which paths are more likely than others. Quantum mechanics may be very much like that.

Those, following Einstein, who desire a deterministic universe will be unsatisfied and will regard theories such as alternate histories (or evolution) as incomplete. Perhaps an underlying hidden determinism still exists; the issue is not settled. But after almost a century of quantum mechanics, the prospect is likely that an element of indeterminism, or randomness, will remain.

If indeterministic quantum mechanics is the complete story, then the choice of paths is done by a toss of the dice, despite Einstein's horror at the prospect. If, on the other hand, quantum mechanics is incomplete, then some underlying causal mechanism, yet undiscovered, may decide on the path taken (although an underlying indeterministic theory is still possible). Proposals for the underlying mechanism range from Bohm's quantum potential to the "mind" itself. With no evidence supporting any such mechanism, and with nonlocality as the consequence, an application of Occam's razor leaves as the most rational alternative a nonmechanistic, noncausal, random selection of particle paths.

The idea of consistent histories and decoherence along allowed paths may not turn out to be the final solution to the mysteries of quantum mechanics. Some think they are insufficient as they stand and need to be supplemented by additional axioms.[7] But, at the very least, these notions demonstrate that the metaphysics of either holistic hidden variables or many worlds is not currently required to move beyond Copenhagen to a modern, pragmatic interpretation of quantum mechanics.

Decoherence produced by the measuring apparatus or environment provides for wave function collapse without paradox or mysticism. Certainly the qualities and quantities of observation are objectively real. They do not exist solely in the mind of the observer. Further, the theory retains a rational notion of objective reality. The decoherent histories approach provides a clear concept of reality consistent with common usage. Sequences of real events in spacetime form sets of paths for which relative probabilities can be assigned.

The probability for a given path need not be either 0 or 100 percent, as in classical physics, where all motion is determined by previous motion. The actual paths that occur in the quantum universe are randomly selected from the allowed set of paths according to their probabilities.[8] If the probability for following branch A along a particular path, given by the square of the wave function of the universe at the point where the paths diverge, is 75 percent, and the probability for path B is 25 percent, then path A is three times as likely to be followed, though it is by no means guaranteed. One way to think of this example is to visualize four paths of which three follow branch A and one follows B, with all four equally likely.

The laws of physics are then associated with setting limits and likelihoods for events and restrictions on the possible paths. They do not determine that a given event will happen in an exact way at a given place and time.

Physical reality is associated with predictability, and predictability with probability, but not necessarily with certainty or determinism. We imagine a huge wave function that describes the whole universe (technically, an ensemble of similarly prepared universes), including observers and measuring devices. This wave function contains the information needed to compute the probabilities for all allowed histories, but only broadly controls the outcome of events, which are random within limits prescribed by conservation principles. These events are real, by definition. The contents of dreams and illusions have neither limits nor predictability, and so are not real by an application of common notions of reality.

Obviously we have not yet developed the cosmological theory that enables us to compute the full wave function of the universe. Such a theory may be beyond human capacity. But we do have partial theories that enable us to compute various pieces of that wave function. In this regard, we retain the practical notion of reductionism, which asserts that we can usefully analyze the parts of a system without knowing all the details about the whole. This, after all, is the best we can do. The popular alternative of holism is empty of utility, insisting as it does that we must know everything before we know something, that we must be God before we can become a scientist.

From the pieces of the wave function calculated reductively, we are able to compute the probabilities of alternative histories for the particular region of the universe under study. Now, the relative probabilities for two alternative paths of a particle can only be calculated when the wave functions for the two paths decohere. This decoherence can be produced by a particle detector, or by particles in the environment itself. Quantum mechanics gives the same results as classical mechanics when wave functions decohere, and so objectively real histories represent our familiar observations, which are conventionally described in classical, macroscopic terms.

The Source of Quantum Effects

Quantum effects are not limited to the microscopic realm. Authors usually point this out with examples such as superconductivity or superconducting quantum interference devices (SQUIDs). But light provides a more familiar example. The interference and diffraction of light, effects observable to the naked eye, are fundamentally quantum in nature. True, they can be "explained" by the classical wave theory of light, but once we accept the fact that light is composed of photon *particles*, we must acknowledge that light interference is a quantum phenomenon, just as is the interference of electrons. It simply was not originally recognized as such. Indeed, interference and diffraction occur at some level for all objects in the universe.

Quantum effects, microscopic or macroscopic, manifest themselves in observations that are described in classical terms because we and our devices are macroscopic. In the Copenhagen interpretation, a distinction seems to be made between the quantum system being observed and the classical apparatus doing the observing. This is unsatisfactory to many since the detection apparatus is fundamentally quantum—made up of atoms just like everything else in the universe.

In the language of post-Everett quantum mechanics, quantum effects result from the ultimate granularity or discreteness of the universe, the discreteness first elucidated by Planck when he observed that light is not continuous but exists in quanta. Classical effects result when this granularity is too small to be observable. In the classical world, matter, light, and force fields appear continuous.

The apparent continuity of the classical gravitational, electric, and magnetic fields is only approximate, as is the apparent continuity of matter on the macroscopic scale. Ultimately these fields are discrete. Although effects of quantum gravity are yet to be observed, the detection of photons testifies to the quantization of electromagnetism. Electromagnetism is fundamentally not a continuous field phenomenon, despite the great utility of describing it that way for many practical applications.

If light and matter were smooth and continuous on all scales, infinitely divisible as was believed by many in the nineteenth century (despite the already widely known atomic theory), then the alternate histories of post-Everett quantum mechanics would be completely fine-grained. That is, the universe would have no granularity at all; every path through space would be a possible, real, decohering path.

Furthermore, some paths would be allowed and others forbidden. The motion of bodies would be determined, predictable with unit probability. Deterministic classical physics implicitly assigns unit probability to one path, and zero for all the others. Indeterministic quantum mechanics leads to deterministic classical mechanics in the fine-grained limit, as it must if it is to agree with the great bulk of human observation.

Undoubtedly, the classical picture works to a good approximation for most phenomena of common human experience. And determinism naturally follows in this limit from the Feynman path integral picture. The wave function, in the classical limit, decoheres at each point because something material exists at that point. (Recall that the classical "vacuum" contains an all-pervading etheric medium for the propagation of light.)

Put another way, a completely fine-grained alternate histories version of quantum mechanics is indistinguishable from continuous, deterministic classical mechanics. Thus we do not have to make an arbitrary division between the classical and the quantum, between the observed and the observer. Everyday phenomena are fundamentally quantum, like everything else, but are largely well explained by classical mechanics because of the high level of decoherence of most of the macroscopic world.

With the important exception of light, we live in a world that is accurately described by the classical limit of quantum mechanics. No wonder light has been always regarded as a mystery, something more profound than mundane matter. It is the one quantum phenomenon that is part of everyday experience.

The limit to divisibility implied by the ideas of Planck and the other founders of quantum mechanics is interpreted, in post-Everett quantum mechanics, as a coarse-graining of the allowed alternative histories that exists as a fact of nature. This is what leads to quantum effects. When a detector or an environmental particle is present at a particular point in space, then an allowable path can pass through that point and the alternate histories then include that path. The objective reality of that path exists, independent of conscious observers.

For example, suppose the double slit experiment is conducted in a transparent medium in which the photons in the beam have a wavelength larger than the distance between slits. Since the medium is transparent, the photons negligibly decohere. With no detectors behind the slits, so that no decoherence occurs there, the coarse-graining of the experiment is such that specific photon paths through either slit are not predictable by standard quantum theory.

Now consider this experiment from the point of view of the photons incident on the slits. Suppose the photons are the same color or wavelength. The higher the photon energy E, the lower its wavelength $\lambda = hc/E$. When the energy is high enough so the wavelength is very small compared to the slit separation d, then the photons "see" both slits and each photon passes through one slit or the other in the normal, classical way. In this case, no observable interference pattern results.[9]

If, on the other hand, the energy is low enough so λ is comparable to or larger than d, then the photons do not resolve the two slits. From the point of view of someone using those photons to observe the screen containing the slits, two separated slits are undetectable. In that case, the photons are not known to pass through one or the other and the rules of quantum mechanics tell you to add their amplitudes rather than their probabilities. An interference pattern is then the result.

If we place a detector behind one slit we return to the classical picture because the experimenter now has sufficient information to resolve the slits, although we still cannot predict, before the fact, which slit the photon passes though. Alternatively, the environment can cause the paths to decohere, also leading us back to the classical picture.

Here is where much of the confusion about whether or not quantum mechanics is local can be cleared up. If we are to insist that the photon has a predictable path in all cases, then we must introduce some nonlocal influence such as the quantum potential that passes simultaneously through both slits. But if we simply admit that the slit through which the photons passes is unpredictable at the level of coarse-graining that exists in the region of space that encloses the experiment, then nonlocality need not be introduced.

While we cannot escape the role of experiment in defining the quantities observed, this role is not so exclusive as it is in Copenhagen physics or positivist philosophy. We can simply view the placement of a detector at a particular point as a change in the environment produced by humans, sort of like the ozone hole. We simply place a decoherer at a place where nature originally did not provide one. At other places, nature provides.

Decoherence is produced by the environment, which includes any human modifications. We are part of the environment too, as are the detectors we place at different points when we perform experiments. No artificial distinction should be made between observer and observed.

Even far from the atmosphere and oceans of earth, well beyond the scope of human intervention, macroscopic regions of totally empty space are impossible. Low-energy photons in the 2.7°K microwave background pervade all space, and so-called "virtual" particles are constantly being created and destroyed in the vacuum, according to modern quantum field theories. These will produce some decoherence at all levels, but especially on macroscopic scales. In fact, decoherence caused by the environment explains why the macroscopic world appears classical, why the moon does not appear in one place in the sky at one instant and some other place at another instant.

In the decoherent histories view, the coarse-graining of the universe can be of macroscopic dimensions when the environment of a particle is highly transparent to that particle. Such is the case with visible photons in air, water, glass, or the vacuum. An extreme example is provided by neutrinos, to which even the earth appears as empty space and the corresponding coarse-graining has astronomical dimensions.

Thus the granularity of the universe is not the same for all forms of matter, but depends on interactions between particles and their energies.

Our current theories of particles and forces enable us to calculate the granularity for photons and the other known particles as they move through known media. Thereby we can predict, with some uncertainty, the paths these particles will take.

The Time-Symmetric View

As we probe more deeply, we find that the various interpretations of quantum mechanics overlap more than they may appear to on the surface. The one characteristic that they all share, what seems to be a minimal requirement for any quantum theory, is *contextuality*. Explain contextuality and you explain quantum mechanics.

Contextuality refers to the fact that quantum theories must consider the entire experiment in making their predictions, not just one detector off in some corner. Changing a detector off in another corner changes the experiment.

Contextuality can be shown to be a natural feature that follows directly from a fundamental fact about quantum phenomena—the apparent symmetry of time. When no distinction is made between past and future, the future experimental arrangement of detection devices, such as polarizer orientations, must have as much effect on the system as any initial conditions. Final and initial conditions are conceptually equivalent.

How contextuality follows from time symmetry can be easily understood from the illustration in figure 7.1. There a system is seen to start from some initial state i. In classical mechanics, the final state f is completely determined. In quantum mechanics, a range of final states is possible for a given initial state. These are illustrated by f1, f2, and f3 in the figure.

When we then consider the time-reversed situation, these three final states becomes different initial states i1, i2, and i3. From this point of view, we have three separate experiments corresponding to the three different starting points. The only way we can maintain time symmetry is to insist that both the initial state and final state be specified in any given experiment. Quantum mechanics then calculates the probability for the transition i–f to take place.

Time symmetry also goes at least part of the way in satisfying our need for visualizations of quantum phenomena that correspond closely to common-sense experience, although this can never be fully achieved since quantum phenomena and time symmetry deviate from common-sense experience. Consider once more the double slit experiment. One of the "mysteries" of quantum mechanics is how the interference pattern can

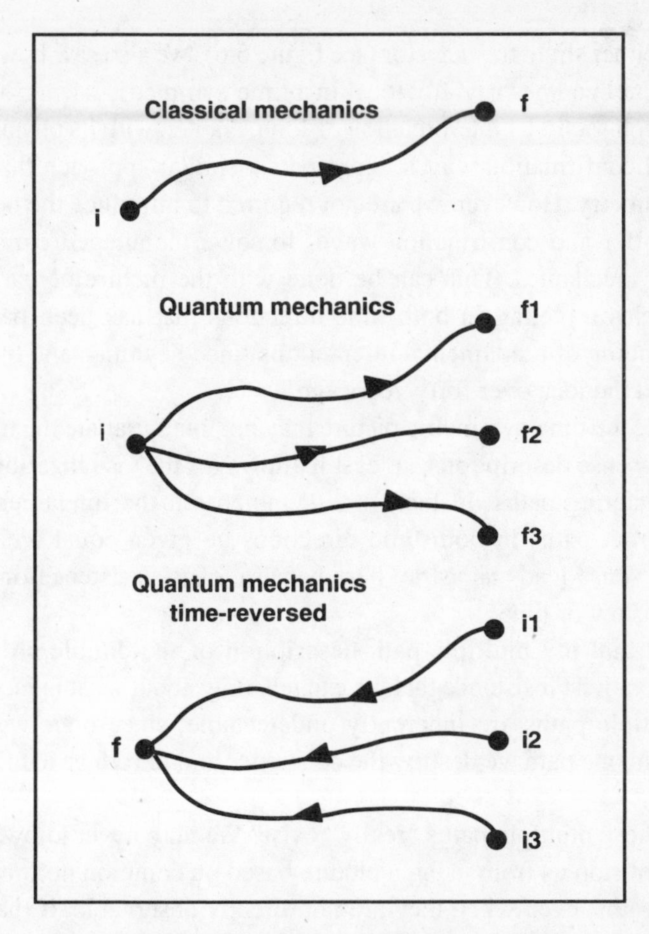

Fig. 7.1. How time symmetry implies contextuality. In classical mechanics, a particle in an initial state i will always move to a specific state f. In quantum mechanics several final states are possible. In the time-reversed situation, the three final states shown become different initial states, and thus different experiments. Thus both initial and final state must be specified for a given experiment.

depend on a choice of detector arrangement made after the particles are emitted from the source. Furthermore, the very existence of the interference pattern seems to defy any attempt to describe the process in terms of familiar particle paths. How can a photon or electron go through both slits simultaneously so that their corresponding wave functions can interfere?

I have shown in chapter 5 how the notion of multiple specific paths for a single particle can be retained in the double slit experiment by viewing the particle as going through one slit to the detector, then *back in time* to the source through the second slit and finally forward in time once more through either slit to the detector (see figure 5.6). We also saw how the EPR experiment becomes trivially local in a time-symmetric scenario.

The transactional interpretation of John Cramer, with its physically real offer and confirmation waves, represents a similar approach that utilizes time symmetry. However, we are not required to introduce the novel concept of offer and confirmation waves to solve the alleged paradoxes of quantum mechanics. This can be done with the picture of particle and antiparticles traveling in both time directions that has been part of our understanding of fundamental interactions since Feynman and others first suggested the idea over forty years ago.

While the time-symmetric picture may not totally satiate the hunger for common-sense descriptions, at least it allows for the visualization of multiple interfering paths. In fact, time symmetry at the fundamental level requires that paths in both time directions be given equal weight. Our common sense leads us astray here because of its insistence on a unidirectional flow of time.

Note that my multiple-path description of the double slit violates Bohr's positivist insistence that we can talk only about what is measurable. The multiple paths are inherently undetectable, since once we place a detector in one path we destroy the coherence and introduce a unique time direction.

Do these multiple paths "really" exist? We may never know. But this should not stop us from using a picture based on common notions such as particle paths, even when they are not directly observable, if that picture aids in our understanding and can be used to obtain results that agree with experiments that are conducted. In this case, the multiple-path picture leads to the observed interference pattern.

Finally, I cannot emphasize too strongly that speaking about time symmetry at the level of fundamental interactions implies nothing about the possibilities of human travel backward in time. The arrow of time remains present in macroscopic or many-body phenomena as the direction of most probable occurrences. At the scale of human experience, the random scatterings of particles in many-body systems that is responsible for decoherence and quantum phenomena appearing classical is also responsible for the generation of entropy that defines the direction of time. Going backward in time is technically possible, but statistically so improbable that we cannot expect it to happen for a single macroscopic body in the age of the universe. Humans are governed, for all practical purposes, by the emergent properties of many-body systems, including cause and effect and the second law of thermodynamics.

Uncertainty and Rationality

Consider again how classical mechanics is viewed from the perspective of decoherent histories, as the limit in which the universe has no granularity, where every point in space can decohere a particle moving through that point. In this case, all paths in spacetime are possible paths, since they all have definable probabilities. But note: Nothing in this statement requires that these probabilities be either zero or unity, that all allowed paths are necessarily completely certain. This occurs only as an approximation in the classical limit.

In the usual formulation of classical mechanics, the certainty of specific paths results from an additional assumption that given the initial position and momentum of a particle, the future path of that particle is completely determined by certain equations of motion, or, equivalently, by the principle of least action. As mentioned previously, the Feynman path integral formulation of quantum mechanics can be used to establish the classical principle of least action. This is the determinism of the Newtonian world machine. It results in a probability calculation yielding essentially 100 percent for one path and 0 percent for all others.

But classical calculations rarely make predictions with 100 percent certainty. Any application of the classical equations of motion to predict the path of a particle requires an exact knowledge of the initial conditions, which is impossible, since such measurements cannot be made with infinite accuracy.

Measurement errors are inevitable, even where we can imagine building a measuring device with unlimited precision. To represent the position of a particle in a continuous universe with perfect precision would require an infinitely long number. However, calculations must be performed with calculators or computers (or fingers) using numbers of finite length. This is not a practical problem for many of the traditional applications of classical mechanics, where sufficient accuracy for a very precise calculation can be made and compared with an equally precise measurement.

As we will see in a later chapter, certain physical systems, even if modeled by the completely deterministic equations found in Newtonian physics, can be so sensitive to initial conditions as to be unpredictable for all practical purposes. These systems are chaotic.

Physics Is Counting

The mathematical representation of completely continuous, fine-grained space, time, and matter requires the use of **irrational** numbers—numbers like π and $\sqrt{2}$ that are not integers or the ratio of integers, like ⅓, but calculate out to infinite, nonrepeating length. These are taken for granted in classical physics, and even the formulas of quantum mechanics. For example, when we solve Schrödinger's equation, we assume the resulting wave function Ψ is a function of continuous spatial coordinates x, y, and z, and a continuous time coordinate t. Infinitesimal calculus, which is a mathematics of irrational numbers, is used.

However, physics is counting. Physical observables are operationally defined by counting procedures and so in principle should always be represented by **rational** numbers—integers or ratios of integers. The observables of physics are only approximately represented by continuous, irrational numbers, despite the widespread belief of the opposite, that the irrationals are fundamental and the rationals are approximations.

Consider time. By international agreement, a time interval equal to one second is operationally defined as 9,192,631,770 vibrations of cesium-133, an integer. Suppose we count the number of cesium vibrations between two events. The result will be an integer N. Of course, the time interval can be less than one second, or less than one cesium vibration. We can easily accommodate smaller times by dividing the second into fractions and writing $t = N/M$. Since N and M are both integers, t is a rational number.

Now, the above definition is arbitrary and not the source of the ultimate discreteness of time. The minimum definable time is given by quantum mechanics and gravitational theory to be 10^{-43} second, the *Planck time,* which will be discussed in the next chapter. In principle, a time interval is the number of Planck times between two events—an integer.

Irrational numbers, as applied to physical quantities, occur only in a universe without granularity, or in the granular universe as a process of approximation that should not be taken as valid to unlimited precision. We normally assume that the time t can take on an infinite number of values in any time interval. That is, we take t to be part of a continuum. But ultimately time intervals can be no smaller than the Planck time.

Distance, the quantity of space between two points, is also defined to be proportional to the time it takes for light to go between the two points in a vacuum. If that time is one second, the distance is 299,792,458 meters, again an integer. We call this number c, the "speed of light" in a vacuum, but note that this is simply the definition of the meter; the Einstein principle that c is a constant (in a vacuum) is assumed. More generally, where time is in fractional units, distance will be a rational number, say cN/M. The smallest distance is c times the Planck time, 10^{-35} meter, called the *Planck length.*

In the second century B.C.E., Euclid, assuming a continuum of space, proved that the length of the diagonal of a unit square was not a rational number. This length is conventionally written as $\sqrt{2}$. However, if you measure the diagonal of a square by counting the number of spatial units from one end to the other, you will have an integer in those units. The diagonal becomes an irrational number only if space is a continuum.

Another familiar irrational number is π = 3.14159. . . . The surface area of a sphere of radius r is $4\pi r^2$. But our measurements are fundamentally rational numbers, such as the number of unit area squares that fit on the surface of a sphere. And in computations we approximate π by a rational number such as 3.14159 = 314,159 /100,000.

Discreteness, manifested by rational numbers, is built into the way in which we operationally define and perform actual numerical calculations on the quantities of physics. Continuity, manifested by irrational numbers and associated with mathematical tools such as calculus, is applied in physics only as a convenient approximation.

This is not to say that we should give up our "irrational" mathematical methods. They, along with the rest of the mathematics of real and complex numbers, which includes infinitesimal calculus and other powerful methods, have proven their immense value in the classical physics that applies, to an excellent approximation, for most of the macroscopic world. My point here is simply this: We should not regard as logical paradoxes any conclusion drawn from a mathematical analysis applying our system of irrational numbers, infinities, and infinitesimals to a universe that is fundamentally rational, finite, and not continuously divisible.

As we have seen over and over again, the problems and paradoxes of quantum physics are the direct result of thinking in continuum terms. That darn aether refuses to die. At the most esoteric levels of quantum field theories, infinities occur in calculations that utilize conventional continuum mathematics; to get finite results, a very elaborate trick called *renormalization* must be used. This trick often works, but not always. When it does not work, then discrete calculations must be employed. Unfortunately, these are far more laborious than the techniques of continuum calculus, though supercomputers are making the procedure increasingly viable.

At the more elementary level, we can show how the uncertainty principle arises naturally once we accept the fundamental granularity of space and time. Imagine a time sequence as a discrete series of unit steps. That is, if we start at $t = 0$, the next time is $t = 1$, the next $t = 2$, and so on. Let x give the position of a particle of mass m at a certain time and x' give the position at the next time step. Then the velocity of the particle is $(x'-x)/\Delta t$ and its momentum $p = m(x'-x)/\Delta t$, where $\Delta t = 1$. Since the position of the particle at two different times is needed to measure p, we conclude that position and momentum cannot be measured simultaneously.[10]

In other words, the indeterminism in quantum mechanics is not some added assumption in the theory but a fundamental consequence of the discreteness or granularity of the universe. Any deterministic subquantum theory will have to restore the spacetime continuum to physics.

One of the questions that has been raised about the decoherence mechanism is that it makes the interference terms only small, not identically zero. Thus Schrödinger's cat is not half alive and half dead, but 0.999999 probability alive and 0.000001 probability dead, which still presents us with a logical paradox. Discreteness rescues us from this dilemma by effectively rounding off the irrational numbers to the nearest rational number, in this case 1 and 0.

Full (But Subluminal) Speed Ahead

And so, the fundamental granularity of the universe seems an inescapable fact about nature, at least when we attempt to describe nature in terms of familiar concepts such as distance, time, mass, and energy. But continuity is so deeply embedded in our thinking that even when they developed the discrete quantum theory, physicists retained many of the notions of classical continuity, particularly those of a space-time continuum underlying all reality, with spatial coordinates and time intervals represented by the infinite set of irrational numbers.

The persistence of the notion of continuity within the formulation of quantum mechanics has not prevented it from being applied with great success. This can be attributed to the fact that the ultimate granularity of space and time occurs at the Planck scale (smallest distance 10^{-35} meter, smallest time 10^{-43} second), far beyond existing experimental reach. So, treating space and time as continuous is a perfectly justifiable approximation at current levels of observation, as long as we recognize it as an approximation and do not cry "Paradox!" when it leads to inconsistent results.

Where the discreteness of nature is manifest in observational data, such as with mass, energy, and angular momentum, it is built into theory, and has been since the 1920s. But right from the beginning, conceptual problems raised their heads when physicists tried to explain quantum effects in familiar terms. In the Bohr atom, for example, the electron jumps "instantaneously" from one orbit to another—at infinite speed, if you insist on space-time continuity in the picture you form of the event in your mind.

Note how the paradox disappears in the discrete view: the electron jumps from x at time 0 to x' at some time t that can be no less than one unit. Its speed is (x'-x)/t, which is finite.

Bohr's energy-level formula correctly reproduced the observed spectral lines of hydrogen that had no classical explanation, so his theory must have had something to do with reality. With the more powerful tools that were later developed, atomic theory was extended to the rest of the periodic table of the elements, providing an understanding of the fundamental basis of chemistry. Armed with that knowledge, twentieth-century chemists have been able to develop thousands of new and useful substances. The unique spectra of atoms are used to determine the chemical composition of materials, in the laboratory and in the cosmos billions of light-years distant. X-rays produced by atoms excited by high voltages are now as common a medical diagnostic tool as the stethoscope.

Studies of the nuclei of atoms have led to nuclear power, medical applications of nuclear radiation, and radioactive dating for the study of human and earth history. Although nuclear energy is regarded by most as a mixed blessing at best, its development also provides dramatic testimony that the world of the quantum is not a fantasy but a very real component of our universe. These results amply demonstrate that the universe is not a continuum in space, time, matter, or energy.

All these developments came about without the settling of the philosophical disputes over the interpretation of quantum mechanics. To the extent that these disputes rest on the assumption of a continuum and an arrow of time, they may be safely set aside while we get on with the business of further developing and using our new knowledge of the structure of the universe.

Notes

1. Weinberg 1992, p. 84.
2. See Laudan 1990.
3. Laudan, p. viii.
4. Kuhn 1970.
5. Will 1986, pp. 93–95.

6. Eberhard 1989.

7. Dowker 1994.

8. The word *random* technically means *equally likely.* This does not preclude the notion of unequal probabilities. For example, in the toss of two dice we assume all faces are equally likely to fall facing up, but a total of seven still has six times the probability of two ones, or "snake eyes." These probabilities are determined by adding the various ways that the desired result can happen assuming an equally likely $1/6$ chance for each face on each die. Snake eyes can happen only one way, both die falling with aces up ($1/6 \cdot 1/6 = 1/36$ of the time), while a seven can happen six different ways (for a probability of $6/36 = 1/6$).

9. Interference patterns are observable with λ small compared to d if the screen is placed sufficiently far away. I am considering here the case when $\lambda << d$ and the interference pattern collapses to a point in any practical experiment. This can be regarded as taking the classical limit, where the quantum wavepacket still exists, but is very narrow compared to the other relevant dimensions of the experiment.

10. For a discussion of discrete physics, see Kauffman 1994 and Noyes 1994.

8
Cosmythology

> The physicists are getting things down to the nitty-gritty, they've really just about pared things down to the ultimate details and the last thing they ever expected to happen is happening. God is showing through. They hate it, but they can't do anything about it. Facts are facts. . . . God, the Creator, Maker of Heaven and Earth. He made it, we can now see, in that first instant with such incredible precision that a Swiss watch is just a bunch of little rocks by comparison.
>
> —John Updike, *Roger's Version*[1]

Intelligent Design

The preconceptions of centuries are not easily altered by reasoned argument or objective evidence. We have discussed at length the attempts to use the strange properties of quantum phenomena to buttress the widely held doctrine of mind-matter dualism and a modern version of a holistic cosmic consciousness. This is supposed to confirm the ancient teachings of Hindu and Buddhist philosophy, and perhaps provide a scientific basis for Judeo-Christian beliefs as well.

In the meantime, a second mythology based on the remarkable discoveries of modern cosmology has taken its place alongside quantum mysticism in the continuing effort to put science to work as a defender of faith. This *cosmythology* purports to confirm the Judeo-Christian concept of a Creator God.

Astronomical evidence is now compelling that the observable universe is the expanding remnant of the **big bang** that happened about fifteen billion years ago.[2] This is interpreted as meaning that our universe came into being at one defined instant, $t = 0$. To the cosmythologists, this confirms the teachings of *creatio ex nihilo* in the Bible and other ancient scriptures.[3] Quoting astronomer Robert Jastrow, "The scientist . . . has scaled the mountains of ignorance . . . as he pulls himself over the final rock, he is greeted by a band of theologians who have been sitting there for centuries."[4]

In a 1951 address before the Pontifical Academy, Pope Pius XII argued that the big bang proves the existence of God:

> In fact, it seems that present-day science, with one sweeping step back across millions of centuries, has succeeded in bearing witness to that primordial "*Fiat lux*" (let there be light) uttered at the moment when, along with matter, there burst forth from nothing a sea of light and radiation, while the particles of the chemical elements split and formed into millions of galaxies. . . . Hence, creation took place in time, therefore, there is a Creator, therefore, God exists![5]

In California, science, traditional religion, and the New Age are blended together as cosmologist Joel Primack and his attorney wife Nancy Abrams present programs on the cultural repercussions of the revolution in cosmology. While Nancy sings about "The Handwriting of God," Joel lectures on how religious imagery illuminates cosmology, borrowing terms from the Kabbalah as "metaphors for some of the new cosmological theories."[6]

Most religions teach, in one way or another, that the universe and life is the result of divine plan, with humanity occupying a very special place in that plan. But since Copernicus and Galileo, science has continually undermined this comforting faith. Perhaps the most powerful blow was struck in the early nineteenth century by Charles Darwin, who showed how organisms as complex as the human body could have evolved from simpler forms, with no help from the outside. Evolution is a natural process, proceeding according to established physical laws and requiring no preexisting design.

Later discoveries in paleontology, genetics, biology, and many other disciplines, including physics, might have proven Darwin's theory false in any one of a thousand ways. Instead they have proved so beautifully consistent that only those whose personal beliefs are so powerful as to block out reality can question the fact of evolution. We can be as certain of evolution as we are that the earth is round.

As an example of how physics might have proven Darwin in error, consider the age of the earth. At the time Darwin's theory was announced, science had no knowledge of nuclear reactions. Known principles of gravity, chemistry, and thermodynamics predicted that the sun's lifespan would have been far too short to allow life to evolve on earth. Only the discovery of nuclear energy allowed sufficient time for evolution to take place. If it had been demonstrated instead that the energy source of the sun was gravitational or chemical, Darwin's theory would have been proved false.

I do not mean to imply that attempting to prove scientific theories false is the only valid way to test them. But, when any theory has passed risky test after risky test and failed to be falsified by a single one, we can reasonably proceed on the assumption that the theory represents a goodly portion of the reality it purports to represent. There have been many ways evolution could have been shown to be wrong by subsequent developments. It never has been.

Nevertheless, many rational people who accept that life on earth evolved naturally still find compelling the argument that the universe as a whole has some supernatural, externally imposed structure and purpose. Being so alien to the teachings by which they were raised, they find the alternative that the universe is somehow spontaneous or self-creating impossible to comprehend.

Having grown up in the tradition of Judeo-Christian mythology, many Western scientists who do not themselves take biblical fairy tales seriously will still use God as a metaphor for the order of the universe. Einstein often did, and has been equally often misinterpreted. Cardinal O'Connor of Boston criticized his general theory of relativity in 1921 as "a ghastly apparition of atheism." When Rabbi Herbert Goldstein of the International Synagogue in New York sent him a cablegram bluntly demanding, "Do you believe in God?" Einstein replied, "I believe in Spinoza's God, who reveals himself in the orderly harmony of what exists, not in a God who concerns himself with the fates and actions of human beings."[7] This was sufficient to get him out of trouble with the good rabbi.[8] I wonder if it would work in today's less liberal religious climate.

In his bestselling book *A Brief History of Time,* famed cosmologist Stephen Hawking concludes that the discovery of a "complete theory" of the universe would be "the ultimate triumph of human reason—for then we will know the mind of God."[9] Hawking, however, has insisted (elsewhere) that he is an atheist and his book argues that the universe does not require a creator. He proposes that space and time are boundless.

Spacetime, Hawking suggests, is like the surface of a sphere, imagined in four dimensions instead of just two. The axis of time and the three axes of space are analogous to the great circles of longitude on the surface of the earth, with neither a beginning nor an end. This comes about by means of a change in our normal representation of time. By considering time to be an imaginary instead of a real number, it can be treated the same as any of the three spatial coordinates.

> In relativity, time is taken as a fourth coordinate of four-dimensional spacetime, where spatial points are located with respect to three axes such as the Cartesian axes x, y, and z. If we let $x_1 = x$, $x_2 = y$, $x_3 = z$, and take $x_4 = ict$, where $i = \sqrt{-1}$, then the proper distance s from the origin of the coordinate system to a point in spacetime is given in Euclidean fashion by $s^2 = x_1^2 + x_2^2 + x_3^2 + x_4^2$. The four coordinates are then interchangeable, with the time coordinate represented by an imaginary number. A sphere in this spacetime is the locus of points with constant s.
>
> Points in three dimensions can be located by spherical coordinates r, θ, and ϕ, where r is the distance to the origin, θ is the polar angle or colatitude, and ϕ is the azimuth or longitude. A sphere has constant r and its surface two dimensions. We can visualize Hawking's picture in terms of the surface of a sphere, with a single spatial dimension represented by ϕ and the imaginary time represented by θ. The polar point $\theta = 0$ is arbitrary; no boundary exists for the θ axis.

Having no beginning of time, there is no place in Hawking's universe for a creator. And with no boundary conditions, no flexibility exists in the laws of nature; only one set of still-undiscovered laws is possible. As Hawking explains it, "So long as the universe had a beginning, we could suppose it had a creator. But if the universe is really self-contained, having no boundary or edge, it would have neither a beginning nor end: it would simply be. What place, then, for a creator."[10] And, as Carl Sagan says in the introduction to *A Brief History of Time*, Hawking's cosmology leaves "nothing for a Creator to do."[11]

Nevertheless, the last words of Hawking's main text, "the mind of God," have become a familiar title for magazine cover stories on science and God as the popular press promotes the notion that science and religion are finally coming together. The phenomenal success of Hawking's book is attributed to this religious implication, not the well-written expositions of black holes and elementary particles.

Physicist-author Paul Davies entitled his 1992 book *The Mind of God.*[12] He has received a million-dollar prize for his clever fogging of the boundary between science and religion. The recurring theme in this and his earlier works is that the ingenuity, economy, and beauty of the world have a "genuine transcendent reality."

God as the metaphor for cosmic order can be found in media coverage of the 1992 observations of anisotropies in the cosmic microwave background made by the Cosmic Background Explorer (COBE) satellite. Seeing "the face of God" was how the media described the results, which strongly supported the big bang model. When, at a meeting in France that summer, I personally asked mission scientist George Smoot about the source of this statement, he denied that anyone involved in the experiment had used the term "face of God." He recalled that he said, "If you're religious, this is like looking at God."[13]

Still, patterns are often in the eye of the beholder. At that same meeting, physicist Simon Swordy jokingly told me he saw the face of Elvis—the young Elvis of the recent U.S. postage stamp issue. When I looked at the pictures from COBE, however, all I saw was the face of chaos.[14]

The COBE data showed a deviation of only 1 part in 100,000 from what one would expect from complete smoothness. Certainly order exists in the universe, but it is tiny. We should not allow our personal proximity to a small pocket of enhanced order (the earth) to close our eyes to the far greater disorder of the whole.

Physicist-theologian Willem Drees has written an excellent book on the relationship between modern cosmology and theology entitled *Beyond the Big Bang: Quantum Cosmologies and God.* [15] Drees knows too much science to think that it can provide the solid basis for faith that so many other theologians desperately seek. He recognizes the futility of using the supernatural to fill the gaps of scientific knowledge, because science has a way of eventually filling its own gaps. Drees sides with "those who freely admit the impossibility of a doctrine of God based upon natural knowledge."[16]

In Drees's opinion, "Cosmology . . . does not support a design argument for the existence of God."[17] Drees remains a worshipper, but explains that what he worships is a hypothesis. In his book Drees hypothesizes a transcendent, immanent God who "shines through" in our desire for perfection and justice in the world. In the last chapter of this book I will explore the currently fashionable notion that a transcendent reality is shining through in the form of "evidence" that our mental capacities exceed those that can ever be achieved by strictly material systems.

But for now, let me concentrate on the claim made by others than Drees that modern cosmology implies a supernatural creation. Certainly no scientific consensus currently exists on the genesis of the universe. But, as I will show, we have no basis to think that it was supernatural.

The Free Lunch

To many people, common sense and intuition imply that the universe could not have just appeared by any process that lies within the boundaries of physical knowledge. That is, the creation had to be a miracle, or at least something beyond human understanding.

The arguments are basically twofold:[18] (1) The **no free lunch argument**: "How can you get something from nothing?" and (2) The **argument from design**: "How could all of this (gesturing to the world around us) have happened by chance?" The debate over these issues has of course gone on since long before the development of modern science, and it is not my intention here to review this history nor to provide a comprehensive philosophical analysis of these questions. In this book, I am focussing on the various ways twentieth-century science is currently being misused to provide a modern rationale for traditional beliefs that had previously been challenged by Newtonian physics and other scientific developments of the past few centuries.

Thus I will examine the two arguments stated above from a strictly scientific viewpoint. They can be directly translated into the precise language of physics. In fact, they represent vernacular statements concerning the apparent violation, at the beginning of the universe, of two of the most fundamental laws of physics.

The no free lunch argument claims a violation of the first law of thermodynamics in producing the universe. The argument from design claims that the second law of thermodynamics must likewise have been broken at the beginning of time.

The **first law of thermodynamics** is equivalent to the principle of **conservation of energy**. Conservation of energy describes commonly observed facts about our physical situation such as our need to inhale oxygen, eat regular meals, and pull our car into the filling station when the fuel gauge reads empty. In the world we see with our own eyes, and in the world seen with our most sensitive scientific instruments, energy does not appear out of nothing. Nor does it disappear into nothing. Or so it seems.

What holds for energy must also hold for matter. Modern physics, thanks to Einstein, makes no distinction between matter and energy. They are the same stuff, related by $E = mc^2$. Thus conservation of energy also implies conservation of matter.[19]

A physical system isolated from its environment, called a **closed system**, is characterized by a constant total matter-energy. If the system interacts with the environment, say by absorbing heat energy from some source, then that energy must go into increasing the internal energy, usually marked by a rise in temperature of the system, or by the act of doing work in moving a piston or some other external object.

When people say, "You can't get something for nothing," they are asserting the requirement of energy conservation. They are expressing the common-sense notion that to get something out, such as useful work or higher stored energy, you must put something in.

Applying the first law of thermodynamics to the origin of the universe, we are inclined to think that since the universe now contains matter, either matter was always present or it must have been miraculously created out of nothing at the beginning of time.

In fact, our best current estimate of the total energy of the universe, including the rest energy of matter, is essentially consistent with zero. Within observational accuracies, the rest and kinetic energies of the material bodies of the universe are almost exactly balanced by the negative potential energy of their gravitational interactions.

The visible matter in galaxies is not sufficient to do the job. While the issue is still controversial and unsettled, both theory and observations continue to indicate that an additional component of invisible *dark matter* exists throughout the universe in just the amount needed to give zero total energy.[20]

If the universe in fact has zero total energy, consistent with current observations, then no violation of energy conservation is implied at the start of the big bang. We needed zero energy then to get zero energy now.

In the big bang theory, the universe exploded from an initial tiny region of space about fifteen billion years ago, give or take five billion years. How big was that region? Extrapolating the standard cosmology based on Einstein's general relativity back to time zero, you get a region of space with zero volume—a **singularity**.[21]

However, general relativity, for all its majesty, is a classical theory. That is, quantum effects are not incorporated. While general relativity has never been empirically refuted, no one doubts that it has to break down at the quantum level.

Quantum mechanics tells us that mathematical points can only approximately represent physical systems. The smallest distance that can be considered, even in principle, is 10^{-35} meter, the **Planck length**. This tiny distance is twenty orders of magnitude smaller than a nucleus of an atom. At distances less than the Planck length, space and time cannot be operationally defined in their usual ways. All our fundamental physical theories utilize the notions of space and time. Without space and time, we can only speculate on the physics involved, if any, at the Planck scale.

By similar arguments, the smallest unit of time that can be operationally defined is 10^{-43} second (the **Planck time**). This is the time it takes light to travel the Planck length. If we extrapolate the big bang backward in time to the earliest moment that is definable even in principle, that moment is the Planck time. No singularity existed at that time, since the singularity is only present in general relativity and general relativity breaks down at the Planck time.[22]

At the Planck time, the universe did not have zero energy. Because of the uncertainty principle, no physical system has exactly zero energy; some minimum called the **zero point energy** will always be found. Thus, when the universe was Δt old it had a zero point energy given by the uncertainty principle to be something of the order of $\Delta E = h/\Delta t$. Currently Δt is so large that ΔE is negligible. However, the smaller Δt, the larger ΔE.

The uncertainty ΔE in the energy corresponding to a time interval Δt equal to the Planck time, 10^{-43} second, is of the order of 10^{28} electron-volts. While this may seem like a large number, that is only because of the microscopic units being used. In fact, this is the rest energy of 10^{19} protons, about the energy contained in the matter of a speck of dust. This is the zero point energy of the universe.

Assuming energy conservation has been maintained since the Planck time, the total energy of the universe remains today fixed to that original minuscule amount. By contrast, the current visible universe contains about 10^{79} hydrogen atoms, so a factor of 10^{60} increase in rest energy ($10^{79}/10^{19}$) was needed to produce the mass of these atoms.

Thus, since the Planck time, an enormous transfer of energy into the rest energy of matter and kinetic energy of radiation must have taken place. We can view that energy as being removed from the gravitational field of the universe, making its potential energy negative while keeping the total essentially zero.

> In simple algebraic terms, the total energy of the universe $E = T + V \approx 0$, where $T = K + Mc^2$, the sum of the total (positive) kinetic energy K and total rest energy Mc^2, and V is the (negative) gravitational potential energy. At the beginning, $T \approx |V| \approx 0$. Now both T and $|V|$ are large, with most of K in the microwave background and Mc^2 in the dark matter and the matter of galaxies. The algebraic sum, however, remains essentially zero.

Inflation

So no energy was required from the outside to "create" the universe. But why did it not remain a sphere of Planck dimensions? As I will show, such a sphere will necessarily expand exponentially.

In his general theory of relativity published in 1916, Einstein showed that a universe totally empty of matter and radiation can still contain gravitational energy. This energy is stored in the curvature of space, sort of like the potential energy of a bent bow. The energy density of curved space is proportional to its curvature.

The universe at the Planck time was very likely an empty, curved space, with the zero point energy discussed above stored in its spacetime curvature. In such a situation, Einstein's equations require that the volume of space will expand exponentially. In recent years, the exponential expansion of an empty universe has been dubbed *inflation.*[23]

For cosmological purposes, the curvature of spacetime can be represented by the quantity a/r, where r is the scale factor that multiplies distances as the universe expands and a is the acceleration of that quantity (the second time derivative of r). In the presence of matter and radiation (where radiation is simply matter moving at the speed of light), Einstein's general theory of relativity gives the following expression for the curvature, which corresponds to the equation of motion for r(t):

$$-a / r = 4\pi G (\rho + 3P) /3 - \Lambda/3 \qquad (8.1)$$

where ρ is the energy density of matter (rest energy), P is the radiation pressure, and Λ is the so-called **cosmological constant** introduced by Einstein to counterbalance gravitational attraction and produce a static universe. (I am using units with $c = 1$.) In the current expanding, "cold" universe, P is negligible compared to ρ. Λ is also apparently negligible, or at least very small. In this case, the above equation gives the Newtonian equation of motion for a mass within a uniformly dense sphere:

$$-a / r = 4\pi G \rho /3 \qquad (8.2)$$

In the early, radiation-dominated universe, when all matter moved near the speed of light, the pressure P obeyed the photon equation of state $P = \rho/3$. Then, with $\Lambda = 0$, we get the equation of motion for a photon in a uniformly dense sphere

$$-a / r = 8\pi G \rho /3 \qquad (8.3)$$

where we have the now-familiar factor of two that characterizes the gravitational effect for photons and other extreme-relativistic matter compared to nonrelativistic particles.

In the absence of matter or radiation, P and ρ are each zero. If Λ were zero, we would have nothing: 0 = 0. But, if Λ ≠ 0, we have the equation of motion of an *empty* universe—no matter or radiation—with spacetime curvature Λ/3:

$$a / r = \Lambda/3 \tag{8.4}$$

In this case, when Λ is a constant, the above equation has the simple exponential solution:

$$r(t) = r_o \exp\left(\sqrt{\frac{\Lambda}{3}}\, t\right) \tag{8.5}$$

That is, an empty spacetime with positive curvature will expand exponentially. Einstein made nothing of this, but today it is called the **inflationary universe**.

Note that if we put P = -ρ in Einstein's equation, 8.1, then the cosmological constant term is equivalent to a radiation (relativistic) energy density

$$\rho_v = 8\pi G\, \Lambda/3 \tag{8.6}$$

where the subscript signifies that this is the energy density of the vacuum, that is, the energy stored in the curvature of spacetime rather than in radiation and matter. Since Λ is positive for r(t) to grow with time, the equivalent pressure is negative.

As the empty universe expands, its own negative internal pressure does work on the universe itself, increasing the internal energy of the system the way the external force applied by the crankshaft on a piston will increase the internal energy of the combustion gas in a cylinder of an automobile engine. No violation of the first law of thermodynamics takes place in either case.

This scenario may seem bizarre, but only because of unfamiliarity. In fact, it follows directly from general relativity and the first law, by now well-established principles of physics. These principles allow, and indeed predict, that in a tiny fraction of a second, about fifteen Planck times, an empty sphere of Planck dimensions will inflate more than twenty orders of magnitude to the size of a proton with a total contained mass energy, removed from the spacetime curvature, sufficient to create all the visible matter in the universe. No miracle was required to create this matter since this process is fully consistent with all known laws of physics.

While the physics that is applied to the problem of the formation of the early universe—general relativity and quantum mechanics—may be baffling to the person on the street, these theories have been tested to high precision. How the universe evolved to its current form is still not understood in detail, but that is a matter for physicists and cosmologists to work out. They will undoubtedly do this without invoking any supernatural forces. And so, within the framework of modern particle theories, we are beginning to understand how the basic properties of matter spontaneously developed after the Planck time.

Order by Chance

Having explained how the presence of matter in the universe does not violate the first law of thermodynamics, let me now move to the second law. I need to explain how order could have arisen from all that chaos at the Planck time and still be consistent with the second law of thermodynamics.

We must answer the argument from design: "How could all of this (gesturing to the world around us) have happened by chance?" When people ask this question, they are voicing their inability to comprehend how order can be created without the intervention of some outside agent to perform the act of ordering. Does design not ultimately require a designer? Does an effect not ultimately require a cause? Not necessarily.[24]

My purpose in this book is not to provide a comprehensive review of the many philosophical and theological arguments made for and against the existence of a creator or other supernatural forces.[25] My focus is on what we can know from science, both experimental and theoretical. However, I feel compelled to comment on what strikes me (and many others) as an obvious flaw in an oft-heard assertion that goes something like this: *Nothing can exist without being created; thus a creator exists.*

Who created the creator? Presumably it is uncreated. We can see that introducing a creator does not solve the problem of creation; it just pushes it back one more level. Why make things more complicated than they have to be? This violates the law of parsimony. If it is logically possible that something can exist that was not created, then why can't that something be the universe itself? This is certainly a more economical hypothesis than the one of a supernatural creator for which no independent evidence exists.

Only if we can show that spontaneous, natural (that is, nonsupernatural) creation is impossible, that some law of physics was violated at the origin of the universe, can we legitimately entertain the added hypothesis of miraculous creation. First we must rule out any possibility that the universe was uncreated, thus requiring an uncreated creator. As we will see, this cannot be concluded from our current understanding of physics and cosmology.

Still, most people are rarely swayed by such arguments. Their everyday experience does not encompass order forming from chaos. Common sense seems to rule it out.

The common-sense belief that order requires an orderer translates into physics as the second law of thermodynamics. Like the first law, the second law also arose out of mundane observations. The first law by itself does not forbid a perfect heat engine from converting all its absorbed heat into work. It only requires that the work done by the system not exceed the heat that comes in from outside. The heat input could equal the work output, conserving energy and making possible a perfect heat engine.

However, nineteenth-century engineers found that some energy was always lost in a heat engine, which led to the development of the second law. The second law has survived intact throughout all the upheavals of twentieth-century physics. Even with today's technology, no one has built a perpetual motion machine that simply absorbs heat energy from the environment (or any other type, for that matter).

Similarly, the first law does not prevent the flow of heat from a colder body to a hotter body. Energy conservation alone allows the possibility of cooling a room with an air conditioner that is not plugged into the wall socket or some other source of energy. But common experience tells us that cooling does not come free.

Thus the first law is insufficient; the second law must be added if our observations are to agree with common facts. The second law determines that perfect heat engines are impossible and heat will only flow from a hotter body to a colder one, requiring the expenditure of energy to cool a room. Engineering and everyday experiences confirm that perfect air conditioners and perpetual motion machines cannot be built. These observations are codified in the second law.

The second law is often expressed in terms of a quantity called **entropy**, which is usually explained, in a somewhat oversimplified fashion, as a measure of the disorder of the system. According to the second law, the entropy (or disorder) of an isolated system must increase or at best remain equal with time. More precisely, we can define negative entropy, or **negentropy**, to be associated with the **information** content of a system, where in today's scientific lexicon the term *information* has a well-defined, quantitative meaning.

Viewing the second law as a fundamental principle, we say that heat does not flow from a lower-temperature body to a hotter one (when no work is done) because this would result in a net decrease in the entropy of the system of two bodies. When the flow is from hot to cold, as is always observed, the cold body increases in entropy while the hotter body decreases by a lesser amount.

However, the second law does not forbid the lowering of the entropy of a part of an isolated system, so long as another part of the system compensates by increasing at least the same amount. Since entropy is a measure of disorder, the subsystem that loses entropy becomes more orderly. Since in common experience most systems interact with their environments, they can lose entropy and become more orderly without some outside agent acting to do the ordering.

A common misconception of the second law of thermodynamics is that it forbids the natural, spontaneous formation of order, and thus that order requires an external orderer. Order occurs spontaneously each time a snowflake is formed from water vapor in the atmosphere, as the entropy of the water vapor/snowflake system is reduced at the expense of increased entropy of the environment. No conscious agent or intricate machine is needed.

The Failure of Common Sense

The first and second laws of thermodynamics have formed the basis of much of our understanding of physics for over a century, with nary a hint of violation in countless experiments and observations over that time. The violation of either could reasonably be defined as a supernatural event, if by supernatural we mean the transcending of natural law.

Three nineteenth-century concepts strongly implied that the origin of the universe must have violated natural law:

1. The universe was believed to be a firmament. The earth and planets moved about the sun, but the sun and other stars were assumed to be essentially fixed in space.

2. The second law of thermodynamics, as it was formulated in the nineteenth century, seemed to require that the universe had started out in a state of maximum order, or minimum entropy, and was evolving toward a final end of total chaos, called the *heat death.* Thus an original order, a grand design, must have existed at the beginning of the universe.

3. It was believed that matter cannot be created or destroyed by natural processes. Since matter now exists, it must therefore have been created supernaturally.

The revolutions in physics and astronomy that occurred in the early twentieth century turned these conclusions on their heads. There were three relevant developments:

1. From astronomical observations, the universe was found not to be a firmament. Rather, it is expanding as if it began in an explosion, the big bang, fifteen billion to twenty billion years ago. An expanding volume has continually increasing room for disorder, that is, entropy. So it becomes possible for local pockets of order to form at the expense of disorder elsewhere. This can happen even when the original state was one of maximum entropy—total disorder, devoid of design or plan.

2. $E = mc^2$. Einstein's famous equation showed that matter and energy were equivalent and that matter could thus be created from energy.[26] Further, Einstein's general theory of relativity allowed a region of space to be completely empty of either matter or radiation but still contain energy stored in its curvature. This curvature energy can appear as a quantum fluctuation "from nothing." Matter can then appear by the conversion of vacuum curvature energy to mass.

3. Using the interpretation we have shown is the most economical, not requiring nonlocality nor the intervention of consciousness, quantum mechanics demonstrated that certain events, such as atomic transitions and nuclear decays, happen spontaneously and unpredictably, and without visible cause. Accidents happen. Everything that occurs in the universe is not precisely predetermined by natural law. In fact, our so-called "laws" apply only on average to ensembles of physical systems, not to individual ones. The behavior of individual systems is left to the vagaries of chance.

Each of these three statements violates common preconceptions derived from everyday observations about the world. Thus it should come as no surprise that they make possible a universe whose origin and contents also fail to follow prejudices based on the limited experiences of human beings on earth.

The Planck Time and Beyond

Nothing can escape from deep inside a black hole. The gravitational attraction is such that even light rays are bent back down and swallowed up in the interior. If we try to look inside a black hole by sending in light, or some other probe, that light cannot get back out to tell us what is going on inside.[27] As a consequence, we have no way of obtaining information about the inside of a black hole. From our perspective then, a black hole is a state of maximum entropy.

Put another way, the entropy of a black hole is as large as it can possibly be for an object of that size. Perhaps something is happening inside a black hole, but we have no way of knowing since no information can escape. For all practical purposes, those of us on the outside of a black hole can assume that any internal activity is nonexistent.

If our universe was ever as small as a Planck length, it was at that instant necessarily indistinguishable from a black hole.[28] At that time, the Planck time, the entropy of the universe was maximum and a condition of total chaos existed. Thus we can conclude that at the Planck time the universe must have been without structure or design.[29]

Obviously we have structure in the universe today, although it is tiny by comparison with the chaos of the whole. That structure must have happened after the Planck time. Existing physical knowledge allows a surprisingly firm conclusion to be drawn: If the universe was ever as small as 10^{-35} meter, then at that time it must have been in total chaos. If the order that now exists in the universe was imposed from without, it must have been injected into the universe sometime after the universe had expanded beyond Planck dimensions.

I realize that this concept of the origin of the universe is difficult to grasp. Readers are still likely to ask, "What 'caused' it to happen? What existed 'before' the Planck time?" We can never know what went on before the Planck time. If there was a "before," if such a concept even makes sense, no information about this "before" remains in our universe.

As with our attempts to make sense of the quantum world, we cannot assume that the common notions of everyday experience apply far outside the realm of such experience. While it might be argued that the same goes for physics when it comes to understanding the early universe, high-energy colliding beam accelerators have already achieved energies that existed when the universe was a tiny fraction of a second old. The physics at these energies is well understood, with no unexplained anomalies as of this writing.[30] The view I present here is fully consistent with this and all other current knowledge. Obviously, future knowledge can change these conclusions, but this is the best we can do for now.

The need for causes simply reflects our everyday experience that things must always happen as the direct result of preceding things. In the quantum world, assuming no Bohmian hidden, nonlocal forces, things can simply happen. And if Bohmian forces exist, they simply happened. And if God exists, he simply happened. I have shown that directional causality, or *causal precedence,* is in fact a classical, macroscopic concept that does not apply at the fundamental level of elementary particle interactions, where fundamental interactions make no distinction between cause and effect.

At the Planck time, the universe was a sphere with a radius equal to the Planck length. Since the universe was a black hole at that time, its entropy was as large as it possibly could be. Now, the second law says that the entropy of a closed system cannot decrease with time. Such a system would not spontaneously become more orderly; ordering energy must be supplied from the outside. Thus it would seem that, barring any outside help, order cannot occur in the universe beyond the Planck time.

A closed system of constant volume, as the universe was assumed to be in nineteenth-century cosmology, has a fixed maximum entropy. On the other hand, a closed system of increasing volume, like the now-established expanding universe, will have an increasing maximum entropy. Thus, as the universe expands beyond the Planck time the maximum possible entropy grows faster than its actual entropy, leaving ever-increasing room for order to form.

Consider the situation in the early universe a few tenths of a second after the Planck time. At that time, the universe was an exploding gas of relativistic particles in the throes of the big bang. An expanding sphere of gas, with particles moving about randomly, is undoubtedly a system of high entropy. But its entropy is not as large as it possibly can be; it is not a black hole. It is disorderly, but it has more order than a black hole. Measurements on the particles can be made; the information contained in those measurements is negative entropy, subtracting from the whole.

The situation is analogous to a rubbish can sitting in the corner of a well-tended yard. The rubbish is disorderly within the can, but the yard is orderly because the rubbish is concentrated in the can. The can could be made more orderly by dumping its rubbish into the yard. However, the yard would then get more disorderly as it absorbed the entropy removed from the can.

So, soon after the Planck time, when the universe was a tiny sphere of expanding relativistic particles, the entropy of the universe was less than its maximum possible value. Being less than maximum, the entropy-generating processes of spontaneous formation of local order could then take place. The rubbish can could be dumped in an ever-expanding yard. The second law is upheld as the rest of the universe absorbs the entropy subtracted from the local systems that are being ordered.

Once the universe exploded in the big bang, it had increasingly more room for the formation of local order, despite the fact that it started out with no room at all. Today the entropy of the universe is many orders of magnitude below its maximum allowable value. Local structures can and do continue to form, without violation of the second law of thermodynamics. The decrease in entropy of the ordered system is compensated for by an increase in the entropy of the microwave background.

Reiterating, the entropy of the universe at the Planck time was as low as it could be. Yet it was at its maximum—total chaos. During a fraction of a second, in the inflationary and early big-bang epoch, when the matter of the universe was being created out of curvature energy, the entropy increased to essentially its current value. Thereafter, the maximum allowable entropy increased as the universe expanded. With the entropy of the universe well below maximum, order could then spontaneously form.

Cosmologist Roger Penrose has objected to this answer to the problem of entropy in the early universe. He asks what happens if the universe has more than the critical density it needs for gravity to eventually reverse the expansion and contract back again into a black hole. As the universe collapses, the entropy ceiling will get smaller and smaller, leading to a "gross conflict with the second law of thermodynamics."[31]

I disagree. We can imagine a universe with very high but not maximum entropy collapsing back on itself until its entropy, which continues to increase, hits the now-decreasing ceiling. The second law does not forbid the ceiling from decreasing, as Penrose seems to be saying. At that point, disorder is maximum and no further disordering needs to take place. If you wish, time stops at that point, but since insufficient order exists to even make a clock, much less an intelligent organism to read and interpret that clock, the point becomes moot.

Admittedly I have not demonstrated that the universe appeared spontaneously out of nothing.[32] I don't know how anyone could ever establish that, one way or the other. I do not even know what that means. The universe simply is, and our physical notions are undefined in the absence of that universe. One such physical notion is time, the number one reads off a clock. With no clocks, there can be no time and the very notion of a "before" to the universe makes no sense.[33]

The universe may have been created supernaturally, but I think I have shown that those who believe this cannot call upon the first and second laws of thermodynamics to bolster their belief. Supernatural creation is not suggested, much less required, by any basic physical principles.

Likewise, the existence of the big bang does not confirm that the universe appeared as a creative act at one point in time. On the contrary, the current inflationary big bang theory provides a framework by which a spontaneous quantum fluctuation could have grown from complete chaos to the slightly less chaotic collection of material particles that forms our universe today.

Tuned for Life

Recently another development in modern cosmology has been seized upon to provide a scientific basis for the notion of intelligent design to the universe. This newest concoction of the argument from design is constructed around the so-called *anthropic coincidences.*[34] Here the claim is made that the values of fundamental constants of nature, and various other parameters of our universe, are incredibly fine-tuned for the production of life, perhaps even human life. This fine tuning is said to be far too unlikely to have been accidental. The only reasonable conclusion is intelligent design, with human life as the intent.

No doubt the universe would look quite different with the tiniest variation of the basic constants of physics. A slight difference in the strength of gravity, the charge of the electron, or the mass of the neutron, and life as we know it would not exist. The human race could not have evolved in a universe with different constants. But it does not necessarily follow that no other race could have evolved. Nevertheless, those who promote the notion of intelligent design think they have found confirmation in the way the universe seems to be exquisitely balanced on the tip of a needle for the purpose, it seems, of producing us.

As yet no theory, including the currently highly successful Standard Model of elementary particles and forces, predicts the values of the fundamental constants of the universe. None is able to specify such basic facts about the universe as why the proton has the mass it does, or why the hydrogen atom has the size it does. In the Standard Model, the basic constants of the universe must still be put in by hand. No known first principle prevents any of these constants from taking on a random value from zero to infinity.

Several physical constants have values that one would not expect from naive arguments of symmetry or ideas about the unity of phenomena. Recent developments in particle physics suggest that all the fundamental forces of nature were unified as a single force in the extremely high energy of the early moments of the big bang. While the forces are no longer identical, the huge differences between force strengths that we now find are difficult to explain. For example, the ratio of strength of the electric and gravitational forces in an atom is 10^{39}. That is, gravity and the electromagnetic force differ by 39 orders of magnitude. For later purposes, I will call this large number N_1.

In the nineteenth century, the electric and magnetic forces were found to be different aspects of the same basic electromagnetic force. This unification occurred despite the fact that the magnetic force on a charged particle is normally much smaller than the electric force.

In the 1980s, electromagnetism and the weak nuclear force were found to be different aspects of the same basic *electroweak* force. This also came about in the face of the large difference between force strengths observed in the laboratory. In both of these examples, the differences in the observed strengths of unified forces is explained in a natural way. When and if gravity is unified with the other forces, its comparative weakness may be shown to be similarly natural.

Starting with Hermann Weyl in 1919,[35] many have speculated about the large size of the dimensionless number N_1 and its possible connection with other large numbers in cosmology and microphysics. For example, the ratio of a typical stellar lifetime to the time for light to traverse the radius of a proton, N_2, is another dimensionless number, 10^{39}, which is the same order of magnitude as N_1. This was the first of what are now called the anthropic coincidences.

Most physicists greeted the $N_1 = N_2$ "coincidence" with the same Bronx cheer, "pbzzzpht," that they give to new interpretations of quantum mechanics. It seems like nothing more than numerology. Look around at enough numbers and you are bound to find some that appear connected.[36]

However, in 1961, R. Dicke argued that N_1 is necessarily large in order that the lifetime of main sequence stars be sufficient to generate heavy chemical elements such as carbon.[37] Furthermore, N_1 must be of the same order of N_2 in any universe with heavy elements. If the gravitational attraction in stars were comparable in strength to the electric repulsion between protons, stars would collapse long before nuclear processes could build up the chemical elements from the original hydrogen and deuterium (heavy hydrogen).

The formation of chemical complexity is only possible in a universe of great age. Biological life needs time to evolve, a stable source of energy over that time, and raw material from which to build complex structures. That raw material includes carbon and other heavy elements to provide the diversity needed for building proficient organic systems. While hydrogen, helium, and lithium were readily synthesized in the first few minutes of the big bang, heavier nuclei did not appear until much later, after they were synthesized inside stars and released into space upon the explosive demise of these stars. The existence of elements heavier than lithium in our universe depends on what also appears to be highly unlikely coincidences.

Billions of years were needed for stars to form, to burn all their hydrogen fuel while manufacturing heavier elements, and finally to explode as supernovae, spraying their atoms into space. Once in space, these elements cooled and accumulated into planets. Billions of additional years were needed for at least one star to provide a stable output of energy so that one of its planets could develop life.

In a debate on the existence of God held at the University of Hawaii on April 13, 1994, Christian theologian William Lane Craig was asked from the audience how he could believe that human beings have a special place in a universe that is so enormous and so old compared to humankind. His answer was that the universe had to be very big and very old. Paraphrasing Craig's response, all the billions and billions of stars and galaxies that spread over billions of light years in billions of years were put there by God so that the chemistry needed for life and human beings had time to evolve.[38] My reaction: Why not cockroaches?

The element-synthesizing processes in stars depend sensitively on the properties and abundances of deuterium and helium produced in the early universe. Deuterium would not exist if the neutron-proton mass difference were just slightly different from its actual value. The relative abundances of hydrogen and helium also depend strongly on this parameter.

The hydrogen-helium abundances also require a delicate balance of the relative strengths of the gravitational and weak nuclear interactions. A slightly stronger weak force and the universe would be 100 percent hydrogen, since all the neutrons in the early universe would then have decayed. A slightly weaker weak force and few neutrons would decay before being bound up with protons in helium nuclei where energetics prevent their decay. All the protons would also be bound up, leading to a universe that was 100 percent helium. Neither of these extremes would have allowed for the existence of stars and life based on chemistry.

The electron also enters into the tightrope act needed to produce the heavier elements. Because the electron mass is less than the neutron-proton mass difference, a free neutron can decay into a proton, electron and neutrino. If this were not the case, the neutron would be stable and most of the protons and electrons in the early universe would have combined to form neutrons, leaving little hydrogen to act as the main component and fuel of stars. It is also rather convenient that the neutron is heavier than the proton, but not so much heavier that neutrons cannot be bound in nuclei.

The evolution of life on earth thus depends critically on the relative force strengths and mass differences. With the slightest change of these values, the variety and diversity of the chemical elements would not exist. In their tome *The Anthropic Cosmological Principle,* John Barrow and Frank Tipler have gone to great length in seeking many similar connections—some quite remarkable, others a bit strained—between the physical parameters of our universe and the formation of complex, low-energy material structures.[39]

Carbon appears to be the chemical element best suited to act as the building block for the type of complex molecular systems that develop life-like qualities. Even today, new materials assembled from carbon atoms exhibit remarkable, unexpected properties, from superconductivity to ferromagnetism. However, it is carbon chauvinism to assume that only carbon life is possible. We can imagine life based on silicon or other elements chemically similar to carbon, but these would still require cooking in stars. Hydrogen, helium, and lithium, which were synthesized in the big bang, are all chemically too simple to be assembled into diverse structures.

Furthermore, it seems like molecular chauvinism to rule out other forms of matter in the universe as building blocks of complex systems. While atomic nuclei, for example, do not exhibit the diversity and complexity seen in the way atoms assemble into molecular structures, perhaps they might be able to do so in a universe with different properties. Sufficient complexity and long life are probably the only ingredients needed for a universe to have life. Carbon may be unlikely, but, as I will show, long life and complexity are not.

Fingers and TOEs

The anthropic coincidences resonate with the mystical notions we have already discussed at length: that human existence is deeply connected to the very nature of the universe. However, from the time of Copernicus, cosmology has been based on the principle that the universe is indifferent to humanity and human concerns. Most physicists are not quite ready to give up on the Copernican principle. They believe it should be possible to derive the values of the fundamental constants of nature from a yet-undiscovered *Theory of Everything* (TOE) that arises from a set of principles that operate at the level of subnuclear particles, not biological cells.

It is very unlikely that a direct causal connection will ever be found between fundamental processes that apply at scales as small as the Planck scale and those that apply at the macroscopic scale of everyday life. I doubt if any TOE will tell us why we have five toes on each foot, or why the three-toed sloth has (I assume) three. Most of the properties of the macroscopic world were not predetermined by events at the Planck time, but emerged by the processes of chance and natural selection. I can conceive of the possibility that some of the constants of physics also took on the values they did by chance.

The chance that any initially random set of constants would correspond to the set of values they hold in our universe is very small. Cosmologist Roger Penrose has calculated that the probability of our universe is one part in $10^{10^{123}}$. In *The Emperor's New Mind,* Penrose has a cartoon of the Creator pointing a finger toward an "absurdly tiny volume in the phase space of possible universes" to produce the universe in which we live.[40] This has given comfort to believers. In the previously mentioned debate in Hawaii, theologian Craig argued that this unimaginably low probability illustrates the need for a Creator—our universe could not have happened by chance. The bulk of the audience applauded this assertion.

But claiming our universe is a miracle because of its unlikelihood, calculated after the fact, is reminiscent of the television advertisements for publisher sweepstakes that sing, "Miracles can happen, can happen to you" if you simply send in your entry. It may seem like a miracle to the person who wins $10 million, but the probability that someone would win was unity.

Every human being on earth is the product of a highly elaborate combination of genes that would be a very unlikely outcome of a random toss. Think of what an unlikely being you are, the product of so many chance encounters between your male and female ancestors. What if your great-great-great-grandmother had not survived that childhood illness? What if your grandfather had been killed by a stray bullet in the war, before he met your grandmother? Despite all the alternate possibilities, you still exist. If you ask today, after the fact, the probability for your particular set of genes existing, the answer is 100 percent—certainty!

Similarly, the probability for the universe we live in existing as it does, having the values of the fundamental constants that it has, is not one in $10^{10^{123}}$. It is 100 percent! Some universe happened, and it happened to be the one we have.

Still, it is argued that if a universe were created with random values for its physical constants, a universe with no life would have almost certainly been the result. Of course, no one would then be around to talk about it and the fact is we are here and talking about it. Unfortunately, we have no way of talking about it with strict rationality. We do not have enough information, examples of other universes, to use as data for drawing reasonable conclusions.

As with the many interpretations of quantum mechanics that are empirically indistinguishable, the safest response is to shut up until some testable consequence of all these ideas can be formulated. "A closed mouth gathers no feet." Still, people demand explanations that "feel right" and "make sense," even if such explanations do not yet meet all the tests necessary to qualify as legitimate scientific theory. The field cannot be abandoned to those who promote the mystical point of view, those who can cash in on public gullibility because there is no effective counter voice.

So I will respond to speculation with my own speculation, with the knowledge that my explanation is no more testable that those it seeks to counter, but also with the insistence that it is no less so. I claim that my explanation (not really original with me) is the simplest explanation and thus to be preferred by Occam's razor. The other side will undoubtedly disagree and say that nothing could be simpler than the notion of intelligent design. At the minimum, however, I hope to show that there are non-supernatural alternatives to the argument-from-design explanation for the anthropic coincidences.

An Infinity of Universes

One way to "sensibly" explain the anthropic coincidences within the framework of existing knowledge of physics and cosmology is to view our universe as just one of a very large number of mini-universes in an infinite super-universe.[41] Each mini-universe has a different set of constants and physical laws. Some might have life of different form than ours, others might have no life at all or something even more complex that we cannot even imagine. Obviously we are in one of those universes with life.

This multi-universe picture should not be confused with the many worlds interpretation of quantum mechanics discussed in earlier chapters. They are not at all related.

Several commentators have argued that a cosmology of many universes violates Occam's razor. I beg to differ. The entities that the law of parsimony forbids us from multiplying beyond necessity are theoretical hypotheses, not universes. Although the atomic theory multiplied the number of bodies we consider in solving a thermodynamic problem by 10^{24} or so per gram, it did not violate Occam's razor. It provided for a simpler, more powerful exposition of the rules obeyed by thermodynamic systems.

Similarly, the cosmology of many universes is more economical if it provides an explanation for the origin of our universe that does not require the highly nonparsimonious introduction of a supernatural element that has heretofore not been required to explain any observations.

An infinity of random universes is suggested by the modern inflationary model of the early universe described above.[42] Recall that a quantum fluctuation can produce a tiny, empty region of curved space that will exponentially expand, increasing its energy sufficiently in the process to produce energy equivalent to all the mass of the universe in a mere 10^{-42} second.

Cosmologist Andre Linde has proposed that a spacetime "foam" empty of matter and radiation will experience local quantum fluctuations in curvature, forming bubbles of "false vacuum" that individually inflate, as described above, into mini-universes with random characteristics.[43] In this view, our universe is one of those expanding bubbles, the product of a single monkey banging away at the keys of a single word processor.

I thought it might be fun (and instructive) to see what some of these universes might look like. From the values of just four fundamental constants, it is possible to estimate the physical properties of matter, from the dimensions of atoms to the length of the day and year. Two of these constants are the strengths of the electromagnetic and strong nuclear interactions. The other two are the masses of the electron and proton.

This is not, of course, the whole story. Many more constants are needed to fill in the details of our universe. And other universes might have different physical laws. I have no idea what those laws might be; all I know is this universe and its laws. Varying the constants that go into our familiar equations will still give many universes that do not look a bit like ours. The gross properties of our universe are determined by these four constants, and we can vary them to see what a universe might grossly look like with different values of these constants.

I have written a program, *MonkeyGod,* listed in the appendix to this chapter, that the reader is welcome to use. Try your hand at generating universes. Just choose different values of the four constants and see what happens. While these are really only "toy" universes, the exercise illustrates that there could be many ways to produce a universe old enough to have some form of life.

To illustrate this important point, figure 8.1 shows a scatter plot of N_2 vs. N_1 for a hundred universes in which the values of the four parameters were generated randomly from a range five orders of magnitude above and five orders of magnitude below their values in our universe, that is, over a total range of ten orders of magnitude. We see that, over this range of parameter variation, N_1 is at least 10^{30} and N_2 at least 10^{20} in all cases. Both are still very large numbers. Although many pairs do not lie exactly on the diagonal $N_1 = N_2$, the coincidence between these two quantities is not so rare.

The distribution of stellar lifetimes for these same hundred universes is shown in figure 8.2. While a few are low, most are clearly high enough to allow time for stellar evolution and heavy-element nucleosynthesis. I think it is safe to conclude that the conditions for the appearance of a universe with life are not so improbable as those authors enamored by the anthropic principle would have people think. Perhaps only one universe exists, ours. Life in that universe was not at all unlikely.

The Descent of the Universe

Lee Smolin of Syracuse University has suggested another speculative idea, a mechanism for the evolution of the universe by natural selection.[44] He proposes a multi-universe scenario in which each universe is the residue of an exploding black hole that was previously formed in another universe.

An individual universe is born with a certain set of physical parameters, its "genes." As it expands, new black holes are formed within it. When these black holes eventually collapse, the genes of the parent universe get slightly scrambled by fluctuations that are expected in the state of high entropy inside a black hole. So when the descendant black hole explodes, it produces a new universe with a different set of physical parameters, similar but not exactly the same as its parent universe. (To my knowledge, no one has yet developed a sexual model for universe reproduction.)

Smolin's black hole mechanism provides for both mutations and progeny. The rest is left to survival of the survivor. Universes with parameters near their "natural" values can easily be shown to produce a small number of black holes and so have few progeny to which they can pass their genes. Many will not even inflate into material universes, but quickly collapse back on themselves. Others will continue to inflate, producing nothing.

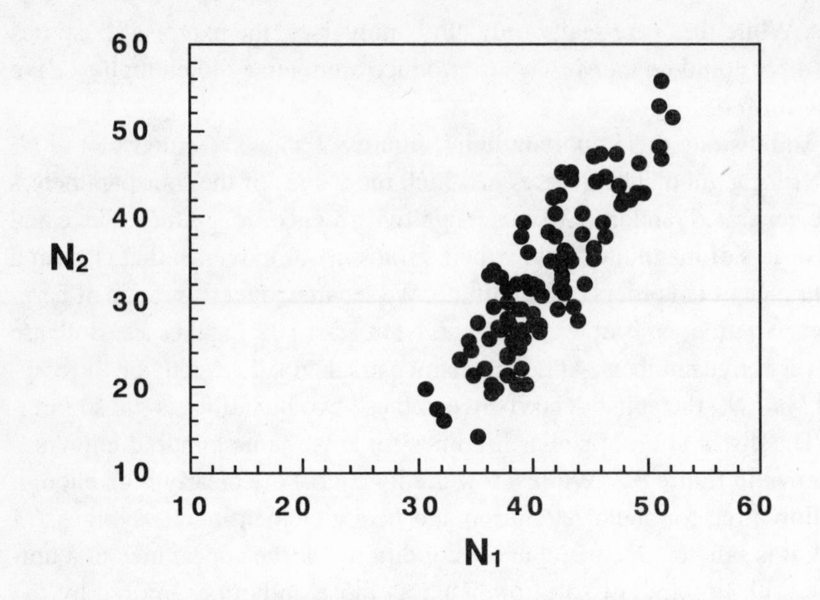

Fig. 8.1. Scatter plot of N_2 vs. N_1 for 100 universes in which the values of the four parameters were generated randomly from a range five orders of magnitude above and five orders of magnitude below their values in our universe. We see that, over this range of parameter variation, N_1 is at least 10^{30} and N_2 at least 10^{20} in almost all cases.

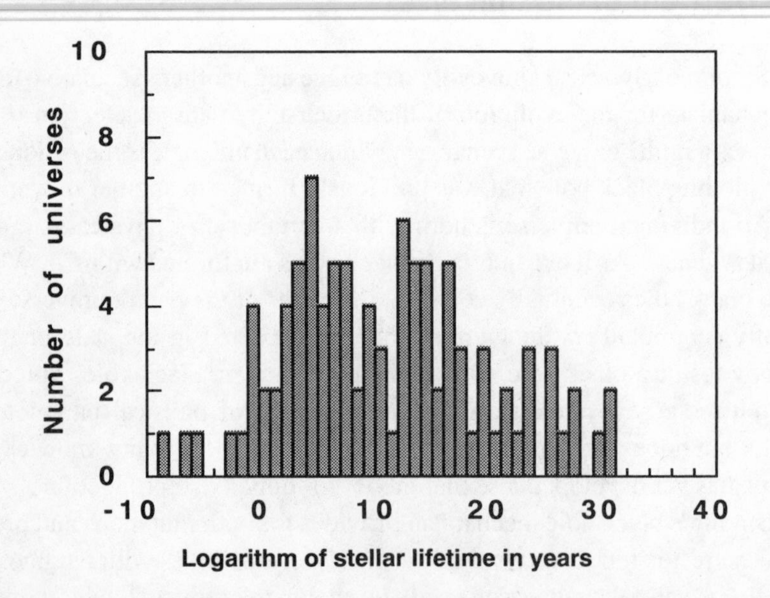

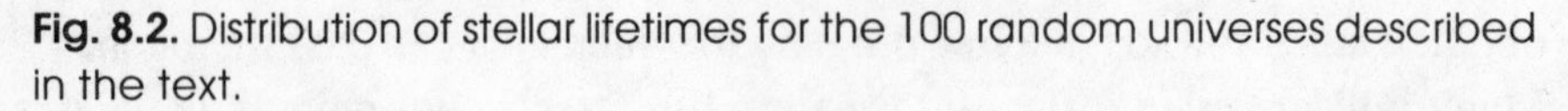

Fig. 8.2. Distribution of stellar lifetimes for the 100 random universes described in the text.

However, by chance some small fraction of universes will have parameters optimized for greater black hole production. These will quickly predominate, as their genes get passed from generation to generation.

Smolin's main task is to demonstrate how the survival capabilities of "species" of universes is enhanced by virtue of their containing the genes for producing complex structures. Like living organisms, universes need to survive long enough to produce offspring. But why tens of billions of years? Many low-mass black holes could form quickly after a universe appears, and these will have the shortest lifetimes, leading to a high rate of mutations. Some additional universe-species survival value must attach to long life and complexity for this model to work.

The evolution of universes by natural selection provides a mechanism for explaining the anthropic coincidences that appears far-out but may be worthy of further consideration. Smolin suggests several tests. In one, he predicts that the fluctuations in the cosmic microwave background should be near the value expected if the self-energy of the field responsible for inflation in the early universe is just below the critical value for inflation to occur.

It is no coincidence that the idea of the evolution of universes is akin to Darwin's theory of biological evolution. In both cases we are faced with explaining how unlikely, complex, nonequilibrium structures can form without invoking even less likely supernatural forces. Natural selection may offer a natural explanation.

Scientist Nitwit Atheist Proves Existence of God[45]

Finally, let us take a look at the most recent attempt to use cosmology to confirm deeply held beliefs. The book jacket to *The Physics of Immortality: Modern Cosmology and the Resurrection of the Dead* tells us that author Frank Tipler arrived at a "stunning" conclusion: "Using the most advanced and sophisticated methods of modern physics, relying solely on the rigorous procedures of logic that science demands, he had created a proof of the existence of God."[46]

Conservative radio newsman Paul Harvey read the jacket and exclaimed, "Professor Frank Tipler was a typical scientist nitwit and an atheist. As a physicist, he could not accept anything that he could not prove. But when he began to calculate the ultimate end of the universe—wow! He discovered God!"[47]

Tipler hedges no bets. On page one he assures the reader who may have lost a loved one, or is afraid of death, "Be comforted, you and they shall live again." Tipler claims his deductions follow straight from the laws of physics as we now understand them.

Tipler is professor of mathematical physics at Tulane University. I have already referred to his earlier book, *The Anthropic Cosmological Principle,* coauthored with John Barrow, which has become an authoritative source for the new generation of Christian apologists. I urge the apologists to read Tipler's latest effort very carefully before they get too excited.

Tipler's idea is not new, being the sort of thing cosmologists prattle about when they sit around drinking beer, but he has added a few interesting wrinkles. Tipler argues that the robots we should be able to build by the next century will ultimately spread themselves throughout the universe, each generation of robot producing ever-superior versions of itself. He estimates that robotic life will blanket the galaxy in a mere million years. In a hundred million years, it will spread to the Virgo cluster of galaxies. By then, *homo sapiens* will likely have long vanished from the universe.

Finally, after the passage of a billion billion years, give or take a hundred billion or so, the universe will be uniformly populated with an extremely advanced form of life that will be capable of feats far beyond anyone but Tipler's imagination.

At that point, Tipler assumes the universe will begin to contract toward what is called *the big crunch*, the reverse of the big bang. Now, it should be noted that most cosmologists currently do not expect that the big crunch will happen. As we have seen, the best guess based on current observation and theory is that the universe is open; that is, it will expand forever. Tipler, however, claims that his theory predicts that the universe is closed. It is a strange sort of scientific prediction, when a desired result far in the future is used to predict a current fact. But at least we have a testable claim: If someday cosmologists convincingly demonstrate that the universe is open, then Tipler will be refuted.

Tipler makes other predictions, such as the masses of the top quark and Higgs boson. But these are essentially based on the unrelated calculations of others and he is being a bit disingenuous to claim them as his own.

The big crunch is not sufficient for Tipler's predicted immortality. The crunch must happen in a highly specific way in order to maintain causal contact across the universe and provide sufficient energy for what life must then accomplish in order to avoid extinction. In other words, the collapse of the universe must be very carefully controlled.

Now, if Tipler believed in a supernatural cosmic mind controlling everything, he could simply say "anything is possible." But he does not escape to supernaturalism. He escapes instead to chaos. Tipler notes that the equations of general relativity imply that the collapse of the universe is chaotic, meaning that it is very sensitive to the conditions that exist at the start of the collapse. According to Tipler, the "butterfly effect" that characterizes chaos (see the next chapter) will be utilized to guide the collapse of the universe.

The advanced life form that evolves from our twenty-first-century robots must collapse the universe in a highly controlled way. Assuming it can manage this, life then converges on what the French Jesuit Pierre Teilhard de Chardin called the *Omega Point.*[48] Tipler associates the Omega Point, as did Teilhard, with God.

Being the ultimate form of power and knowledge, the Omega Point would also be the ultimate in Love. Loving us, it would proceed to resurrect all humans who ever lived (along with their favorite pets and popular endangered species). This is accomplished by means of a perfect computer simulation, what Tipler calls an *emulation.*

Since each of us is defined by our DNA, the Omega Point simply emulates all possible humans that could ever live, which, of course, includes you and me. Our memories have long dissolved into entropy, but Omega has us relive our lives in an instant, along with all the other possible lives we could have lived. Those that Omega God deems deserving will get to live even better lives, including lots of sex with the most desirable partners we can imagine. Even this Tipler places on a mathematical basis, computing the relative "psychological impact" of meeting the most beautiful women whose existence is logically possible compared to simply the most beautiful woman in the world. He finds this to be $[\log_{10} 10^{1,000,000}]/[\log_{10} 10^{9}] = 100,000$ (page 257).

Those deemed undeserving by Omega will be put through purgatories, but if they perform satisfactorily they may gain heaven. We all get the chance to correct our mistakes. I will live a life where I learn to hit a curveball. Hitler will live a life in which he is Jewish. Bill Clinton will be president over and over again until he finally gets it right.

Tipler claims that the Omega Point represents the God of Judeo-Christian religion. Omega is the God of the Jews who told Moses, in Hebrew, "*Ehyeh Asher Ehyeh,*" which Tipler translates as "I WILL BE WHAT I WILL BE" in place of the conventional "I AM WHAT I AM." Omega is the God of the early Christians who will reassemble the complete bodies of all humans on Judgment Day. Omega is the God of Islam, who continually destroys and recreates the universe from moment to moment and provides for his warriors a paradise of total pleasure.

Tipler finds parallels between Omega Point immortality and the views of rebirth in Taoism and early Hinduism. He finds them with Buddhism as well, interpreting nirvana as "heaven" despite its literal meaning of "extinction." Not one to be politically incorrect, Tipler also finds parallels in African and Native American religions.

Scientists have been less kind than media reporters in their evaluation of *The Physics of Immortality.* George Ellis starts his review in *Nature* magazine, "This has to be one of the most misleading books ever produced . . . a masterpiece of pseudoscience."[49] Other prominent scientists have called the book "awful" and accused Tipler of writing it for the money. Paul Harvey would undoubtedly call them "nitwits," so I will not mention their names.

However, Harvey and other believers should read *The Physics of Immortality* more carefully, despite the fact that much of the text is incomprehensible to nonphysicists. Tipler's Omega Point God is not the supernatural, spiritual being that they imagine. Tipler's Omega is completely material rather than spiritual, natural rather than supernatural. His resurrected humans do not have bodies or souls—they are bits in a computer. I doubt very much that this is what Harvey or the pope have in mind.

Is the Omega Point possible? Who can say what will happen in a billion billion years? Tipler, despite his claims, cannot reasonably predict that we will be resurrected at the Omega Point. And I cannot reasonably predict that we will not be. Maybe we are living a simulation right now!

Others have imagined computers and robots as a means for extending human survival. While a purely material immortality may be problematic, the chances that it is possible are surely better than those provided by supernatural fantasies. It's too bad Tipler makes his case so poorly, providing so many targets for ridicule. I am not sure he isn't pulling our legs.

Appendix

!MonkeyGod. A True Basic™ program by Victor J. Stenger

```
!OPEN #1: name"monkey.out", create newold
!ERASE #1
    CLEAR
    CLOSE #1
    OPEN #1: screen 0,1,0,1
    WINDOW #1
```

```
DIM p(4),ctrue(4),con$(4)
!Conversion factors. Basic units S.I.
LET ly=9.4e15
LET eV=1.6e-19
LET GeV=1e9*eV
LET MeV=1e6*eV
LET cm=0.01
LET km=1000
LET hr=3600
LET day=24*hr
LET yr=365*day

!Arbitrary Constants
LET c=2.99792458e8
LET h=6.626075e-34
LET hb=h/2/pi
LET G=6.67259e-11                        !Nm^2/kg^2 usual S.I. units

!Set parameter true values and names
LET con$(1)="1/alpha"
LET ctrue(1)=137
LET con$(2)="alpha_s"
LET ctrue(2)=0.2
LET con$(3)="mp (kg)"
LET ctrue(3)=1.67e-27
LET con$(4)="me (kg)"
LET ctrue(4)=9.11e-31
DO
   !Generate universe
   !Read in parameters. If zero, use true values.
   FOR i = 1 to 4
      INPUT prompt con$(i): par
      IF par <> 0 then
         LET p(i)=par
      ELSE
         LET p(i)=ctrue(i)
      END IF
   NEXT i

   LET alpha=1/p(1)                      !Fine structure constant
```

```
    LET alpha_s=p(2)                     !Strong interaction constant
    LET mp=p(3)                          !Mass of proton in kg
    LET me=p(4)                          !Mass of electron in kg

    PRINT #1: con$(1),con$(2),con$(3),con$(4)
    PRINT #1: 1/alpha,alpha-s,mp,me

    !Compute properties of universe
    LET rb=hb/(alpha*me*c)               !Bohr radius
    LET Eb=alpha^2*me*c^2/2              !Hydrogen binding energy
    LET rN=hb/(alpha_s*mp*c)             !Nucleon radius
    LET EN=alpha_s^2*mp*c^2/2            !Nucleon binding energy
    LET alpha_G=G*mp^2/hb/c              !Gravitational coupling
    !Lifetime and mass of main sequence star
    LET tstar=alpha^2/alpha_G*(mp/me)^2*hb/mp/c^2
    LET mstar=alpha_G^(-1.5)*mp
    !Radius and mass of typical planet
    LET Rplanet=sqr(0.1)*2*rb*(me/mp)^.25*(alpha/alpha_G)^.5
    LET Mplanet=(0.1)^(1.5)*2*mp*(me/mp)^.75*(alpha/alpha_G)^1.5
    !Length of day and year
    LET Tday=2*pi*2^1.5*rb/c*(mp/(me*alpha*alpha_G))^0.5
    LET Tyr=.2*rb/c*(mp/me)^2/alpha^6.5/alpha_G^.125
    !Large numbers
    LET N1=alpha/alpha_G*mp/me
    LET N2=c*tstar/rN

    PRINT #1: "Bohr radius = ",rb*100,"cm"
    PRINT #1: "Hydrogen binding energy = ",Eb/eV,"eV"
    PRINT #1: "Nucleon radius = ",rN*100,"cm"
    PRINT #1: "Nucleon binding energy = ",EN/MeV,"MeV"
    PRINT #1: "Minimum stellar lifetime = ",tstar/yr,"yr"
    PRINT #1: "Mass of star = ",mstar,"kg"
    PRINT #1: "Radius of planet = ",Rplanet/km,"km"
    PRINT #1: "Mass of planet = ",Mplanet,"kg"
    PRINT #1: "Length of day = ",Tday/hr,"hr"
    PRINT #1: "Length of year = ",Tyr/day,"days"
    PRINT #1: "N1 = ",N1
    PRINT #1: "N2 = ",N2
LOOP
END
```

MonkeyGod

The program computes the following quantities: the (Bohr) radius and binding energy of the hydrogen atom, the radius of a nucleon (proton or neutron) and its binding energy in a nucleus, the lifetime and mass of a typical star, and the radius, length of day, and length of year for a typical planet.[50] The program also computes the numbers N_1 and N_2 mentioned in the main chapter text.

The lifetime and mass of a typical main sequence star sets the scale for the age of a universe populated, in the vicinity of at least once such star, by complex material systems assembled from chemical elements produced in the stars themselves. Thus we can easily determine what a universe will look like if it possesses basic parameters that differ from our own.

The strength of the electromagnetic force is given by **alpha** (for greater familiarity, 1/alpha is printed out). The strength of the strong nuclear force is **alpha_s**. Both of these quantities are dimensionless (that is, they have no units). The electron mass is indicated by **me**, the proton mass by **mp**. Both are in kilograms. In the tables below I have rounded off most of the results since only orders of magnitude are really significant in a calculation of this type. The abbreviation for the units in the answers are standard in any physics text. Large and small numbers are shown in computer notation. For example, 1.05e-13 = 1.05 x 10^{-13}.

First we have the universe we know and love, in which 1/alpha = 137, alpha_s =.2 mp (in kilograms) = 1.67e-27, and me (in kilograms) = 9.11e-31

Bohr radius = 5.29e-9 cm
Hydrogen binding energy = 13.6 eV
Nucleon radius = 1.05e-13 cm
Nucleon binding energy = 18.76 MeV
Minimum stellar lifetime = 6.77e+8 yr
Star mass= 3.69e+30 kg
Planet radius = 5700. km
Planet mass = 5.0e+23 kg
Day length = 6 hr
Year length = 6 days
N1 = 2.2e+39
N2 = 6.0e+39

The fact that the day is shown as six hours and the year as six days should not worry the reader. Only orders of magnitude should be considered. Thus a day on a typical planet is of the order of ten hours and a year is of the order of ten days. Our planet earth is a bit atypical, with a year of the order of a hundred days, but that's only an order of magnitude higher, which is pretty good for these calculations.

The next example keeps all the constants the same but sets the proton mass equal to the Planck mass: 1/alpha = 137, alpha_s = .2, mp (kg) = 2.e-8, and me (kg) = 9.11e-31. This generates:

Bohr radius = 5.29e-9 cm
Hydrogen binding energy = 13.6 eV
Nucleon radius = 8.79e-33 cm
Nucleon binding energy = 2.2e+20 MeV
Minimum stellar lifetime = 6e-1 yr
Star mass = 2.6e-8kg
Planet radius = 8.e-21 km
Planet mass = 1.7e-29 kg
Day length = 1.6e-9 hr
Year length = 1.5e+34 days
N1 = 1.8e+20
N2 = 6.0e+39

Note how the age of a main sequence star is a fraction of a second, obviously far too small to allow time for the cooking of the heavy elements needed for life. This illustrates that a huge difference between the proton mass and the Planck mass is needed for a long-lived universe.

A final example makes all the parameters differ greatly from their values in our universe. A viable, if strange, universe results: 1/alpha = 1,000,000, alpha_s = .001, mp (kg) = 1.e-30, and me (kg) = 1.e-35. This generates:

Bohr radius = 3.5 cm
Hydrogen binding energy = 2.8e-12 eV
Nucleon radius = 3.5e-8 cm
Nucleon binding energy = 2.8e-7 MeV
Minimum stellar lifetime = 2e+14 yr
Star mass = 1e+37 kg
Planet radius = 3e+13 km
Planet mass = 1e+23 kg
Day length = 4e+15 hr
Year length = 1e+39 days
N1 = 5e+43
N2 = 5e+39

This universe has atoms with a diameter of 7 cm, days of 10^{15} hours, and years of 10^{39} of our days. Yet its stars live for 10^{14} of our years, which should be long enough to produce the materials of life.

Notes

1. Updike 1986, pp. 9–10.

2. The age of the universe is still uncertain by billions of years and the media continually report new astronomical results that seem to call into question the whole notion of a big bang. Furthermore, the big bang theory of the universe is not totally unchallenged among astronomers. I will not attempt to address those issues and challenges here. Suffice it to say that no serious alternate theory exists which comes even close to the big bang theory in quantitatively explaining a host of observations, from the microwave background to the abundances of light elements. Alternate theories either make no predictions (making them useless) or fail in their predictions. See Lerner 1991 for his case against the big bang and my review in Stenger 1992. For popular-level discussions of big bang cosmology, see Lederman 1989 and Riordan 1991. For a complete survey of early universe cosmology see Kolb 1990.

3. For a survey of the history of ideas of cosmic origins, see McMullin 1993.

4. Jastrow 1980, p. 125. For similar sentiments, see Jaki 1978, p. 278.

5. Pius XII 1972.

6. News item from *Insights: The Magazine of the Chicago Center for Religion and Science* 6.1, October 1994, p. 4.

7. Clark 1971, p. 502.

8. It didn't keep Spinoza out of trouble. When the synagogue of Amsterdam excommunicated him in 1656, the edict read, "Let him be accursed by day and accursed by night; accursed in his lying down and rising up, in going out and in coming in. May the Lord never more pardon or acknowledge him! May the wrath and displeasure of the Lord burn against this man henceforth, load him with all the curses written in the book of the law, and raze out his name from under the sky." Quoted in Johnson 1987, p. 290.

9. Hawking 1988, p. 175. For my review, see Stenger 1988/89.

10. Hawking 1988, pp. 140–41.

11. Carl Sagan in Hawking 1988, p. x.

12. Davies 1992, p. 214.

13. "What Does Science Tell Us about God?" *Time,* December 28, 1992, p. 39.

14. Stenger 1992/93. See also the comments by Pecker, Rothman, and Grünbaum in the same *Free Inquiry* forum, which was entitled "Does the Big Bang Prove the Existence of God?"

15. Drees 1990. For my review, see Stenger 1992a.

16. Drees 1990, pp. 5–6.

17. Drees 1990, p. 11.

18. For more details on the material in this and following sections, see Stenger 1988 and 1990a.

19. The m in $E = mc^2$ refers to the total relativistic or inertial mass of a body. It should not be confused with the rest mass m_o, which need not be conserved, as for example in nuclear reactions. A body's rest energy is $E_o = m_oc^2$.

20. Kolb 1990, Riordan 1991.

21. See Hawking 1988.

22. Hawking finds it quite amusing that he is given (well-deserved) credit for showing both that general relativity requires that the universe started out as a singularity and that quantum mechanics requires that the singularity did not occur. See Hawking 1988.

23. Kazanas 1980; Guth 1981, 1982; Linde 1982, 1987, 1990, 1994.

24. For a classical refutation of the arguments from design that are used to dispute the theory of evolution, see Dawkins 1987.

25. For a debate on this and other arguments for the existence of God, see Moreland 1993.

26. Cosmologists often make an artificial distinction between matter and radiation, which is sometimes confused in the popular mind as a distinction between matter and energy. Hydrogen atoms, for example, are matter and light is radiation, or "pure energy" in this imprecise view. But modern physics has shown that light is composed of material particles called photons, and the distinction becomes one between material bodies moving at speeds much less than light ("nonrelativistic" or "cold"), and those moving at or near the speed of light ("relativistic" or "hot"). In the first few seconds of the big bang, all matter was radiation, since particles moved at very near the speed of light. Today, the matter that constitutes familiar bodies is nonrelativistic, while light remains relativistic.

27. A black hole can radiate photons, but this happens at its surface. It also can disintegrate, with a mean lifetime proportional to its mass. In neither case does the resulting radiation contain any information about the interior of the black hole.

28. Experts will quibble that the volume of all space cannot constitute a black hole, but I believe the analogy is valid in the context used here.

29. For more details on this point, see Stenger 1988, 1990a.

30. For a review of the Standard Model, see Stenger 1990, pp. 33–52.

31. Penrose 1989, p. 329.

32. E. P. Tryon is generally regarded as the first modern scientist to suggest that the universe might have been initiated by an energy fluctuation in the vacuum. See Tryon 1973. One possible mechanism is quantum tunneling. For a recent discussion at not-too-high a technical level, see Atkatz 1994.

33. For a discussion of the "pseudo-problem of creation," see Grünbaum 1989.

34. The phrase *anthropic principle* was introduced in Carter 1974. See also Carter 1983. For a very complete, up-to-date discussion, see Barrow 1986. For popular-level surveys, see Davies 1982 and Gribbon 1989.

35. Weyl 1919.

36. For a hilarious satire on numerology, see Gardner 1967.

37. Dicke 1961.

38. For Craig's views on cosmology and theology, see Craig 1990, 1992.

39. Barrow 1986.

40. Penrose 1989, p. 343.

41. For a recent discussion of this idea, see Linde 1994.

42. I use the word "infinity" to mean a number much larger than any of the other numbers being used in the discussion.

43. Linde 1982, 1987, 1990, 1994. See also Atkatz 1994.

44. Smolin 1992.

45. This section is based on my review in Stenger 1995.
46. Tipler 1994.
47. *Conservative Chronicle,* October 26, 1994.
48. Teilhard de Chardin 1975.
49. *Nature* 371, September 8, 1994, p. 115.
50. The astronomical quantities were calculated using the formulas of W. H. Press and A. P. Lightman in Press 1983.

9

The Edge of Chaos

> We're often told that certain wholes are "more than the sum of their parts." We hear this expressed with reverent words like "holistic" and "gestalt," whose academic tones suggest that they refer to clear and definite ideas. But I suspect that the actual function of such terms is to anesthetize a sense of ignorance.
>
> —Marvin Minsky[1]

The New Science of Wholeness

Quantum nonlocality is not the sole basis for the holistic metaphysics that seems so soothing to the human ego. Quantum theory has been joined by *chaos theory* in providing examples of how modern physics has supposedly cast out the mechanical, impersonal universe of classical Western science and replaced it with what Maui astrologer Harriet Witt-Miller has dubbed the *You-niverse.* As Witt-Miller puts it,

> Chaos theory is continuing a process begun by quantum physics: shattering our illusion of the universe as a machine and revealing a profound natural order obscured by our mechanistic cosmology. Now that this obstacle is removed, we are gaining a more dynamic perspective on many issues, including the one at the core of our problem of how a complex society might benefit from simple, prescientific cosmovision.[2]

Symbols from chaos theory can now be found in many books with New Age and occult themes. In 1991, a familiar pattern from chaos theory, the Mandelbrot set, was discovered in a grain field near Cambridge in England. At that time, a large community had assembled to study the mysterious "crop circles" that had been appearing with ever-increasing frequency and complexity in England and elsewhere. Many believed that the formations were the product of extraterrestrial intelligence, so this example was taken as a confirmation of this belief. The fact that Cambridge University was nearby did not lead the true believers to accept the most economical explanation, that they were being tricked by a group of playful students.[3]

The fact that quantum and chaos theories are pure Western science is conveniently swept under the rug by Witt-Miller and other New Agers who promote similar views. She is closer to the mark when she characterizes her cosmovision as "prescientific." Yes, indeed it is prescientific, like fairies and wood sprites.

This latest candidate for a holistic paradigm to replace reductionist atomism has been cultivated from the collective gleanings of recent discoveries in several related classical, macroscopic sciences, in particular, the grossly misnamed chaos theory.

Chaos represents an area within the trendy new sciences of complexity that have emerged from the development of new techniques in computer science, information theory, and statistical mechanics. These have made it possible to analyze the behavior of complex, highly interactive systems that do not lend themselves to traditional techniques.[4]

Important insights have been gained on the nature of these systems, particularly with regard to the ways such systems can spontaneously organize themselves to produce lifelike and even mindlike behavior. These exciting developments suggest that traditional reductionist physics should be supplemented by new principles and laws that operate for large numbers of particles at the macroscopic scale.

This is not in itself unreasonable. What is unreasonable is the rush to read great mystical meanings into these developments. Such readings are no more justified than the mystical interpretations of quantum mechanics we have found to be so baseless.

As we have seen in preceding chapters, the physical principles that govern the quantum world do not incorporate several concepts that we have come to regard as fundamental from everyday experience. Specifically, elementary processes make no distinction between cause and effect and select no unique direction for the flow of time. I have argued that the apparently paradoxical nature of quantum phenomena can be at least partially traced to our stubborn, unjustified insistence that such principles must still apply at the elementary level.

Once we accept that physical principles need not apply at all levels, we can readily conceive the possibility that properties not present in the world of quarks and electrons can arise out of the complex behavior of many-body systems. These are called **emergent properties**.

Some authors have gone so far as to suggest that the new "science of wholeness" is the third major scientific revolution of the twentieth century.[5] Parapsychologist Stanley Krippner thinks chaos may hold the key to understanding ESP. Krippner suggests, "Chaos theory offers a basis for seriously questioning the universal applicability of linear, rationalistic, predictive methodologies in science."[6] In this Krippner ignores the fact that chaos theory results from a rationalistic, predictive methodology that is simply nonlinear rather than linear. The evidence for ESP looks just as bad from the perspective of nonlinear science as it does from that of linear science.

Such discussions often muddle together quantum nonlocality and chaos, as Witt-Miller does in the quotation above. In fact, they have little in common and may even be incompatible. Chaos theory and the other sciences of complexity are based on classical, Newtonian physics extended into portions of the nonlinear realm that previously have gone unexplored only because of their mathematical difficulty. Even here we must correct an impression: nonlinearity is very much present in classical mechanics, most notably in Newton's theory of gravity.

Quantum mechanics, as we have seen, remains reductionistic, indeterministic, linear, and fully rational. Chaos theory is reductionistic, deterministic, nonlinear, and fully rational. So, while complexity and deterministic chaos have been important developments, there is less to them than meets the eye; and, as we will see, more as well.

Like the quantum holists we have met on earlier pages, the new chaos holists have it all backward. They are the ones who promote a mechanistic cosmovision, a return to the concept of a universe in which everything that happens is already written in the stars. In fact, quantum mechanics and the new sciences of complexity represent the natural evolution of Western science, building on and expanding, but not discarding, the Newtonian physics and methods of the past. In the case of chaos, it makes no sense whatsoever to assign magical new meanings to discoveries that lie well within the frame of physical theories that have existed for centuries. Quite simply, chaos is interesting and important but not a revolution.

The recent developments in complexity physics have not resulted from any fresh anomalous observations as revolutionary as photoelectricity or the cosmic microwave background. Nor have they arisen from a dramatic theoretical breakthrough as deep as relativity or quantum mechanics. Instead they are simply the consequence of normal, gradual scientific progress, assisted by, and feeding back into, the technology of modern computers.

With new mathematical insights and powerful new computational tools, a whole host of classical, macroscopic problems that are very simple in principle but cannot be solved in practice with traditional mathematical means can now be explored. These are not, for the most part, new problems. They are, for the most part, Newtonian in nature and were simply previously unsolvable mathematically.

The solutions now made possible by the new methods have provided insights and surprises, but they hardly represent a paradigm shift. They constitute an important step in the progression of scientific knowledge, but no more than a step.

Methods of Classical Physics

Despite the immense applicability and profound philosophical implications of quantum mechanics, we still live in a macroscopic world that can largely be described by classical physics. While quantum mechanics is needed to understand atoms, Newton's laws still suffice for describing the motions of the objects formed from large numbers of these atoms. Indeed, we have seen that classical physics follows as an approximation to quantum physics whenever the inherent granularity of the spacetime region under consideration is too small to be detected. Among the components of the classical world are localized bodies, from planets to viruses to the neurons in the brain.

For centuries physicists have possessed the basic equations that describe the classical motion of bodies interacting with each other by means of common forces such as gravity or magnetism. However, until recently we have had to rely solely on mathematcis, which lacks sufficient tools to solve these equations exactly for systems of more than two bodies.[7] The classical equations of motion for a system with as few as three bodies are in principle unsolvable by direct mathematical analysis.

For example, the orbits of the earth and sun about their common center of mass are exactly calculable by standard mathematical techniques if one neglects all the other bodies in the solar system. However, this calculation only approximates reality, since the moon and other planets also contribute to the net gravitational pull on the earth. Adding only a third body, say the moon, results in a new mathematical problem that cannot be solved in the same complete manner.

Fortunately, the gravitational forces on the earth from the smaller or more distant bodies in the solar system are sufficiently small that they can be treated as *perturbations* that cause the earth to deviate only slightly from the paths determined in the two-body calculation. Techniques developed in the last century made it possible to calculate with fair accuracy the earth's orbit when the other planets and the moon were taken into account in this approximate way.

Other mathematical and statistical methods developed in the nineteenth century made it possible to approximate the behavior of elastic solids, fluids, and gases, all of which are composed of so many atoms that any attempt to compute individual particle motion is doomed at the outset. The field concept, which, we have seen, is still associated with the holistic aether, is an example of a technique developed to handle large numbers of particles that behave approximately as a continuous medium. Thus physics was able to progress as a useful activity capable of making many practical, quantitative predictions on a wide range of macroscopic, many-particle systems.

Some macroscopic many-particle systems do not lend themselves to these approximation techniques, however. Examples include some of the most important systems affecting human life: the earth's atmosphere, oceans, and geological structures; most biological systems; human and animal populations; and even human and animal social structures. Until recently, these had to be studied largely descriptively, by gathering and classifying data and looking for patterns in repeatable behavior.

Since it was impossible to derive most patterns from first principles, the scientists working on these problems needed to study little if anything about the basic entities with which the objects of their interest are ultimately composed. I suspect the word "quark" has hardly ever appeared in the literature of botany or zoology, yet all plants and animals are made of quarks.

With the development of computers, it has become possible to analyze the motions of many-body systems that evaded traditional mathematical methods. Instead of trying to solve for these motions by solving the equations of motion, they are simulated on a computer. By numerical means, the classical equations of motion of any number of particles can be programmed to predict their movements. Step by step, the net force on each particle is recalculated as the sum of the forces produced by all the other bodies in the system and the equations of motion are then solved numerically to give the position and velocity of each particle for the next step. In principle, the only limitation is one of computer power. Still, simplifications must be made in most interesting cases since the number of particles in a macroscopic systems is so huge. The three-body problem can easily be solved on a computer. The million-body problem still must be solved approximately.

Nevertheless, computers have made it possible to analyze various models that approximate the interactions between elements in complex, many-body systems. These models also are unsolvable by conventional mathematical means, not because the equations are particularly complicated but rather because their form does not lead to a mathematically soluable problem.

These numerical studies have led to a number of results that seem to have a surprising universality to them. Remarkable, orderly behavior has been discovered buried within complex phenomena that appear to be very disorderly or even random at first glance. Because of this apparent disorder, the phenomenon has been mislabeled "chaos." But unlike the chaos we associate with randomness, the chaos of a classical dynamical system is completely determined by its equations of motion, calculable, in principle if not in practice, with unlimited precision. Thus a more accurate label for this line of inquiry is *deterministic chaos.*

Deterministic Chaos

Many books and articles have been written about deterministic chaos,[8] so I will only briefly summarize the basic ideas needed for our critique of modern metaphysics. This is best done by illustration, using the common example called the **logistic map** (also known as the **quadratic map**). This example is very easy to describe yet exhibits many of the properties that appear in more complicated models. As a bonus, the logistic map models a problem of great interest: the growth of populations subject to limits on that growth.

In populations (animal, vegetable, or mineral) whose members reproduce themselves without limits, the number of new members in a given time interval is linearly proportional to the number of members in that time interval, provided that the interval is small enough that the population can be approximated by a constant during that interval. This leads to the exponential "population explosion" or Malthusian curve.

However, no physical population that depends on external resources can grow exponentially forever. Usually it needs energy and material for growth and these are always limited. The result is that the limits on growth feed back to lower the incremental population change as that population approaches these limits.

The logistic map is a simple mathematical model that illustrates this behavior of a limited population. It is obtained by modifying Malthusian exponential growth with a factor to account for the limits on growth. This destroys the linearity of the population system. Nonlinearity, which can result from feedback in complex systems, is a prime ingredient of systems that exhibit deterministic chaos.

Let x_n be the population at a particular time t_n. If we take steps of equal, unit time intervals, the population at time t_{n+1} is given by the logistic map as

$$x_{n+1} = r\, x_n (1 - x_n) \tag{9.1}$$

where r is a parameter that essentially represents the replacement rate (births minus deaths) of the population. Limited growth is thus modeled by the additional factor $(1 - x_n)$, which lowers the incremental population as the population increases. Note the nonlinear, quadratic relationship.

As the result of some insightful work by physicist Mitchell Feigenbaum in the 1970s,[9] we now know that any system of this general form will exhibit very remarkable features that have come to be denoted by the term chaos. A simple study of the logistic map on a computer shows that limited populations will grow in wildly different ways, depending on their replacement rates and initial conditions.

At a low replacement rate, the population will grow for a while and then die off. However, as the replacement rate is increased above a certain threshold, r = 1, the population will grow until it reaches a stable point, called an **attractor**. Thereafter, the population will remain constant.

But this comfortable stability holds only for the range of replacement rates r = 1 to 3. Above a second threshold at r = 3, the system oscillates back and forth between two quasistable populations in what is called a **limit cycle**. At that threshold, the system is said to undergo a **bifurcation**. Increasing the replacement rate results in further bifurcations. At each bifurcation we have **period doubling**, where the system will oscillate back and forth between twice as many quasistable populations as existed below the bifurcation threshold. Thus periods occur in the sequence 2, 4, 8, 16, 32, and so on. For example, at period 8, the population returns to the same value every eight steps.

Period doubling occurs with increasing frequency as we raise the replacement rate r, rapidly converging at r = 3.57 to the very complicated variation of populations that we call **chaos**, where the population changes wildly in each time step. However, although the population variation is chaotic, it is not totally random. It will tend to settle into a limited, specific range of values called a **basin of attraction**.

One might expect intuitively that chaos will now remain as we further increase r. But once again our intuition is wrong. As we continue to increase r, chaos dies down and a new range of values of the replacement rate r occurs for which the more orderly oscillation between quasistable populations, with period doubling, again occurs, for example at r = 3.63, 3.74, and 3.83, until chaos reappears.

Requirements for Chaos

A defining characteristic of the chaotic system is its very strong dependence on initial conditions. The slightest change in the initial population will result in a totally different sequence of populations when the replacement rate r is in a chaotic region. This sensitivity to initial conditions in nonlinear systems is called the **butterfly effect**. Meteorologist Edward Lorenz stumbled upon the butterfly effect while studying a simple weather model on a computer in 1961.[10] The term suggests that an extra flap of a butterfly's wings can completely change the weather a week or two later. While that may not be literally true, accurate weather prediction may be fundamentally impossible more than a few weeks ahead because of the weather's sensitivity to initial conditions that are probably impossible to measure with sufficient accuracy to predict the outcome of what is, for all practical purposes, a completely classical, deterministic system.

Another interesting, unexpected characteristic is **fractal** behavior. Exploring the region around a bifurcation in detail by expanding the scale reveals the same pattern of period doubling and the transition to chaos as on the larger scale. Fractals are patterns that repeat themselves on various scales. The result is often a pattern of great beauty, as with the Mandelbrot set, which is closely related to the logistic map. Fractals are used to produce very realistic-looking computer images of earthly and unearthly landscapes, but it should not be concluded from this that the real world is a fractal. The differences between behavior at the microscopic and macroscopic scales belie such a simplistic conclusion.

The characteristics of chaos illustrated with the logistic map—bifurcation, period doubling, sensitivity to initial conditions, fractal behavior—

are not limited to this particular simple model. Feigenbaum was able to show that these properties are universal—independent of the specific form, and true for all functions of the same general shape. That is, many of the general conclusions we draw, including the formation of chaos, will be true for any nonlinear system with certain minimal properties.

One particularly simple physical example is the damped, driven pendulum.[11] In this case, nonlinearity is provided by the fact that the restoring force is not directly proportional to the displacement from equilibrium when the initial displacement is large. But this is not sufficient for chaos. Damping, which is normally present in a realistic pendulum oscillating in air, is necessary to dissipate energy. But then, an external driving force is needed to keep the pendulum from just coming to a halt as it loses all its energy. While the pendulum might not seem to have much in common with the logistic map, it has a bifurcation pattern that looks very similar.

Many simple physical systems can be used to illustrate chaos.[12] And other mathematical mappings, such as the circle map and the horseshoe map, have been used to explore its detailed characteristics.

The three requirements for chaos are: (1) nonlinearity, which can be the result of a large initial displacement away from equilibrium or feedback in the system; (2) energy dissipation; and (3) an outside driving force. These apply to many dynamical systems and set chaotic systems apart from the linear systems studied in physics by traditional mathematical means.

The equations of motion for linear systems, such as the small-angle pendulum, can be solved by standard methods of linear differential equations even with damping and driving forces. Isolated systems, which do not exchange energy with the environment, also can be studied by the conventional methods of equilibrium mechanics and thermodynamics. Certain nonisolated systems, such as those in contact with a heat reservoir (a large source of external energy) that continually provides energy to maintain a constant temperature, also have simple behavior that has been well-understood for a century.

However, many physical, chemical, biological, and even social systems too complex to be calculated in great detail exhibit the three characteristics of chaotic systems: nonlinearity, damping or dissipation of energy, and external driving forces. The universal characteristics of mathematical models of chaos like the logistic map, or simple physical systems like the pendulum, can be applied to better understand the behaviors of these more complex systems and even to make useful predictions about how they will respond to various stimuli. This has found ready application in many fields, from fluid dynamics and the study of turbulence to heart arrhythmia.[13]

In each application we have a highly nonlinear system that exchanges energy with its environment. It can exist in a very stable state when the driving parameters, analogous to the factor r in the logistic map, are not too high. Quasistable states may occur for higher driving parameters. When the system is in one of these states, changes in its parameters may cause it to jump to a new state, or initiate chaos. By means of computer simulations, or laboratory experiments in which system parameters are varied to determine where the bifurcations occur, it becomes possible to predict and in some cases control behavior such as the onset of chaos. Chaos can even be deliberately introduced to provide better control.

Antichaos and Self-Organization

If order can lead to chaos in a classical nonlinear system, chaos can also lead to order. We have seen that starting with an orderly nonlinear system and increasing the driving parameter produces chaos. Since deterministic classical mechanics is reversible,[14] it follows that we can start with chaos and, by decreasing the driving parameter (or increasing it, in some cases), produce order.

The study of self-organizing systems has become an important activity within the new sciences of complexity. It is being pursued along a number of lines. In **cellular automata**, complex and beautiful structures are generated on a computer screen by the application of a few simple rules.[15] Other computer studies involve unprogrammed programs called **genetic algorithms**, in which initially random solutions to problems are allowed to evolve by natural selection to produce a workable solution.[16] Similar computer simulations in which highly interactive computer networks are allowed to evolve on their own without outside intervention have been labeled **Artificial Life (ALife)**. The resulting spontaneous behavior is very suggestive of living, and even thinking, organisms.[17]

Biochemist Stuart Kauffmann thinks that a good model for biological systems can be found in parallel computer networks. Living organisms are essentially self-regulating networks of genes. The complete set of genes (the **genome**) of a human being carries the information for the production of about 100,000 proteins. According to Kauffmann, the genome acts like a complex parallel-processing computer or network. The observations he and others have made of the behavior of this type of network has led Kauffmann to claim that spontaneous self-organization occurs in complex systems. He calls this **antichaos**.[18] The systems fall into ordered states even when the pressures of natural selection are not programmed in. Kauffman does not suggest that natural selection is unimportant in the evolution of life, however.

Certain types of computer networks, called **boolean networks,** are composed of connected elements of simple switches that give a "yes" or "no" output depending on some logical combination of inputs. These networks can exhibit chaotic behavior, although chaotic behavior is not limited to such networks. Christopher Langton of Los Alamos National Laboratory has suggested an analogy that relates solids to ordered networks and gases to chaotic networks, with liquids represented by networks in between. The driving parameter used above in our discussion of chaos corresponds here to a certain "bias" that is assumed in the network switching rules. In the description below, "high bias" will correspond to "low-driving parameter" and vice versa.

Networks with low connectivity between elements and high bias will freeze into ordered "solids." If you increase the connectivity, or decrease the bias, the solid melts to a "liquid" and finally becomes a chaotic "gas."

Inanimate objects, such as rocks, are stable, many-particle systems. The turbulent air in a hurricane is chaotic. But living organisms, in the view of Langton and Kauffmann, are in between. They have a certain quasistability, like the limit cycle. But by being not too far from chaos, they are very sensitive to environmental fluctuations and thus highly susceptible to the mutations needed to change their characteristics during evolution by natural selection. Life, they say, is then poised on the "edge of chaos."

These ideas, developed from the observation of the results of computer simulations, have led Langton, Kauffmann, and other ALife researchers, to suggest that life is far from the unlikely accident that some would have us think. Not all biologists have come around to this new view, which is rather alien from the ideas they learned in their conventional university curriculum.

From my standpoint, however, this proposal offers a plausible, albeit tentative, solution to a number of the metaphysical issues concerning the development of life and mind. I have argued here and in other writings that the universe is fundamentally about as simple as it could possibly be, with no evidence for design at its origin and structure arising spontaneously as the natural course of events.[19] Life and mind are usually raised as counterexamples of phenomena surely too mysterious and complicated to have occurred naturally and spontaneously. Not necessarily.

Emergent Properties

In general, ALife attempts to simulate lifelike behavior in synthetic systems composed of many elements that interact *locally* with one another according to simple rules. The systems contain no global, holistic rules, and the complex structures that evolve illustrate how emergent properties of the whole can develop out of the local interactions of the parts. This process is highly reminiscent of embryological development, in which higher-order structures grow and compete with one another for support among the low-level elements.[20] ALife views an organism as a large population of simple machines, and works upwards from there, constructing large aggregates of objects that interact with one another nonlinearly according to simple rules.[21]

In computer simulations of ALife, the basic elements can be simple switches, as in the boolean networks described above, or more involved logical units or programs that are coming to be known as **objects**. These may perform a rather sophisticated calculation in determining what output to provide for a set of inputs, but are usually small, simple units compared to a modern computer program.

The essential features of computer ALife models are as follows:[22]

1. They are composed of populations of simple objects.
2. No single object directs other objects.
3. Each object details the way it alone reacts to local situations in its environment, including encounters with other objects.
4. No rules in the system dictate global behavior.
5. Behavior at higher levels than individual objects is therefore emergent.

Again it is important to note the reductionist, nonholistic nature of the enterprise. Structures emerge as the result of elementary, local interactions, just as they do in a snowflake. Once again, as with the other forms of science mysticism we have discussed in this book, the mystics have it backwards when they claim that a new holistic paradigm must replace reductionism. Complex structure follows from a joining of parts. Structure is more than the sum of the parts, but the parts do not disappear from the universe when a structure is formed.

One early example of ALife in action was called *BrainWorks*.[23] This was a computer system that allowed the user to construct a nervous system of a simple "animal." Sensors such as eyes and touch bumpers could be added that responded to food and obstacles. The simulated animal moved around in a space of obstacles. A version was created that supported mutation and reproduction, with survival dependent on food-gathering performance.

In an amusing development, animals evolved to exploit a mistake in the program. Energy consumption was computed, in the program, as a multiple of the number of steps taken by the animal. However, a backward step was wrongly being counted negative and the animal gained energy instead of losing it when it stepped backward. The animal then proceeded to evolve a behavior pattern so that it took backward steps.

This type of result is characteristic of ALife studies. Not only do the systems develop emergent qualities, similar to metabolism, that are common to even most of the simplest forms of life, they develop at least a primitive version of the "intelligent" behavior that we associate with the highest forms. If we associate intelligence with information processing, ALife provides a proven example that intelligence can spontaneously result in a system on the edge of chaos. Thus a kind of "artificial mind" seems to go hand in hand with the properties of artificial life.

Artificial Intelligence and Artificial Mind

We should not associate the ALife form of machine intelligence too closely with the term artificial intelligence (AI) that dates back to the earliest attempts to build computers that "think." AI was conceived by Allen Newall, Cliff Shaw, and Herbert Simon at Carnegie Tech in 1955. For their early purposes, they defined intelligence as the ability to solve problems, and developed a program called *Logic Theorist* that attempted to do so by a heuristically guided search for problem solutions. Computer intelligence was measured as the degree of improvement over a random search for the solution, the latter being impractical for either computers or the brain because of the "combinatorial explosion" that occurs when all possibilities are considered with equal weight.[24]

Another early attempt at AI was based on networks crudely patterned after the neural network of the brain. In 1959, Frank Rosenblatt and colleagues at Cornell University built a device they called the "Perceptron," which was composed of 400 photocells and 512 electronic neurons called accumulators, a paltry number of logical units compared to modern electronic networks or the human brain.[25]

The low level of this early technology, and theoretical objections raised by Marvin Minsky and others, directed AI investigators to follow an approach based on the architecture of the computers then available to them. As a consequence, AI studies have tended to examine how existing computers, not humans, might go about doing pattern recognition and similar problems. Until artificial neural network computers were reintroduced in more recent years, when the technology had advanced sufficiently to make them feasible, it remained problematic whether the solutions obtained by AI had much at all to do with human thinking. Even so, computer intelligence is in itself a justifiable line of inquiry, if for no other reason than its many applications, such as in medical diagnosing systems.

Partly as the result of over-promotion (very common in big science because of the competition for funding) and excessive promises that were not quickly fulfilled, AI has gotten bad press. Despite the caveat, not always explicitly stated, that AI was not attempting to duplicate the mind in a machine, many have had their prejudices confirmed that the brain cannot simply "be a computer" and that mind must reach beyond the world of matter.

This metaphysical notion that "there must be something more" to human thinking was strongly reinforced by cosmologist Roger Penrose's popular 1989 book, *The Emperor's New Mind,* which was highly critical of artificial intelligence.[26] He has recently updated his arguments in *Shadows of the Mind.*[27] We will be discussing Penrose's ideas in some detail in the following chapter, where we will see that they have implications going far beyond artificial intelligence.

As described above, ALife studies strongly suggest that a complex network of processing units, crudely analogous to the brain's neurons, will naturally evolve the ability to make intelligent decisions. These decisions are not made by solving mathematical equations that determine the optimum sequence of steps to achieve a goal, the traditional algorithmic method. The algorithmic method is inflexible, unable to adapt to changes; the human mind does not work algorithmically. Survival is greatly aided by adaptability. In ALife, and perhaps the human brain, many solutions are tried and a kind of Darwinian natural selection sets in to choose a satisfactory, though not necessarily optimal, solution.[28] However, since neural networks can be simulated by serial machines, it would be naive to conclude that they can perform something more than computation.

Chaos and Wholeness

As previously noted, some people suggest that a new holistic paradigm is implied by developments in the sciences of complexity. They claim that there is a kind of design implicit in the structure of the whole that is not derivable from, and indeed transcends, or even controls, microphysics. Certainly the properties of chaotic systems, only touched on above, are remarkable and indeed result when a nonlinear system composed of many particles is considered as a whole. And we have seen that antichaos and artificial life are based very much on the upward development of structure from the local, elementary level. But none of this provides any basis for claiming that emerging properties represent new holistic axioms that must be added to those of the theory of elementary particles and forces.

From the time of Euclid, we have learned to distinguish between axioms and the theorems and corollaries that we derive by applying logical deduction starting with these axioms. In the development of scientific theories, the same basic procedure is followed. Axioms or, more appropriate in the scientific context, testable hypotheses, are initially assumed since any logical process must start with some assumptions. These are tentatively regarded as fundamental principles. Then, by deduction, secondary principles are derived. These secondary principles are usually closer to experiment and their derivations are motivated by the need to test theory in the real world.

Nowhere in our discussion of chaos and antichaos have we been forced to introduce new axioms or hypotheses to explain observations. The basic physics is essentially that of Newton—three centuries old. True, the results of computer simulations indicate certain universal features, such as bifurcation and the transition to chaos, that were not predicted ahead of time, and apparently cannot be deduced from Newtonian physics by any conventional mathematical derivation. Nevertheless, these features were always present within Newtonian dynamics and simply had to await the invention of the necessary calculational tools, in this case computers.

Perhaps this method of solving a physics problem is unfamiliar because of its lack of traditional application, but that is just an accident of history. If Newton had a computer, he probably would not have bothered to invent calculus. He would not have needed it since he would have had a tool that enabled him to numerically integrate his equations of motion and gravity.

No grand new holistic paradigm is revealed in these recent, admittedly interesting and important, developments in macrophysics. In particular, no basis exists for the claim made by chemist Ilya Prigogine and his disciples that time irreversibility, as manifested on the macroscopic, many-particle scale, is a new fundamental principle of nature that extends down to the microscopic scale.[29]

Prigogine's prize-winning work on far-from-equilibrium thermodynamics has merit, but his later ideas have become a major component of New Age, holistic physics literature.[30] A look at Prigogine's claims reveals little more basic science than was already present in the work of Ludwig Boltzmann of a century ago, cast in the new language of nonlinear systems and chaos. Nothing in this latest convolution of classical physics contradicts Boltzmann's idea that time's arrow is simply a statistical definition.

The arrow of time, as Boltzmann explained, is the direction of most probable occurrences. Prigogine accurately calls time's arrow a "broken symmetry," but this deserves no more special significance than the broken mirror symmetry (the symmetry is broken, not the mirror) that exists for most complex material structures. The fact that your face looks different in a mirror than it does straight on is not the result of any basic causal law of nature. It is an accident.

Prigogine is given considerable undeserved credit in new holistic literature for pointing out that many systems, such as living organisms, are not isolated but interact strongly with their environment. He labels them "dissipative" systems. A more general designation is *open* system (a *closed* system is one that is totally isolated from its environment). The false impression is that physicists previously limited their attention to closed systems that cannot self-organize because of the second law of thermodynamics. This impression is simply incorrect.

From its very inception in the last century, thermodynamics has been concerned with open systems and their environments. Since only the universe is a completely closed system, it would have been absurd for physics to ignore open systems. For example, thermodynamicists treated engines and refrigerators as open systems that operate between high and low temperature reservoirs. The earliest statement of the second law of thermodynamics asserted that engines and refrigerators must have both types of reservoirs. The fact that the entropy of an open, "dissipative" system can decrease, and thus become more organized, was known from the time the word "entropy" was introduced by Rudolf Clausius in the nineteenth century.

As we saw in chapter 7, the quantum-to-classical transition can now be understood as a consequence of the mechanism of decoherence. Pure quantum states remain pure only in closed systems. Even in outer space, bodies are not found inside closed systems. The microwave background, if nothing else, acts to produce decoherence and make quantum systems look classical, to make the moon appear just one place in the sky.

So again we find that those who claim revolutionary insights in the new physics have it wrong. If anything, chaos and the new sciences of complexity have helped us to better understand why the behavior of macroscopic systems such as the human body can be described so well by classical mechanics. Given that the fundamental processes of the universe are quantum, why does classical mechanics work at all? It works because the macroscopic world has so many interacting particles that the quantum states decohere and quantum mechanics becomes classical.

The basic physics that is found in the current studies of complex systems was known before the twentieth century saw its first sunrise. A small intellectual step, chaos theory, assisted by a huge technological leap, the computer, got us from there to where we are today in our knowledge of complex macroscopic systems.

Again I do not wish to belittle or demean the new discoveries of the sciences of complexity. I am as excited about them as the next person, but perhaps for a different reason. To me, they provide a clear basis for disputing the need for a holistic metaphysics, old or new, with or without a major role for quantum mechanics, in explaining the existence of life and mind in our universe. The sciences of complexity demonstrate that we are not required to introduce anything that goes beyond classical physics for understanding life and mind, or for seeing how they naturally came about in a universe without prior design.

Notes

1. Minsky 1985.
2. Witt-Miller 1994.
3. See Schnabel 1993 for the story of the crop circle craze, which was all the result of a hoax by two senior citizens from Southampton.
4. Pagels 1989, Waldrop 1992.
5. See, for example, Briggs 1989.
6. Krippner 1994.
7. "Exact" in the classical context means to arbitrary precision, limited only by the accuracy of our measuring devices and computational equipment.
8. See, for example, Gleik 1987.
9. See Gleik 1987, pp. 155–87.
10. See Gleik 1987, pp. 11–31.
11. For a detailed study of the damped, driven pendulum, see Baker 1990.
12. See, for example, Tufillaro 1992. The authors examine the bouncing ball on an oscillating table and provide computer programs that illustrate the properties we have discussed.
13. For interesting applications of these ideas to a wide range of physical and social systems, see Laszlo 1987.
14. The irreversible engines of thermodynamics are in principle reversible, when viewed at the atomic level. They are just highly unlikely to work backward. Recall our discussion of the statistical nature of the second law.
15. Perhaps the most remarkable example of this is Conway's game *Life,* whose

unexpected features are described in Poundstone 1985. The most beautiful example is found in the pictorial presentations of the Mandelbrot set, which have been reproduced in many books. See, for example, Gleick 1987.

16. Holland 1992.
17. Langton 1989, 1992.
18. Kauffmann 1991, 1993.
19. Stenger 1988, 1990a.
20. Langton 1989, p. xxii.
21. Langton 1989, p. 2.
22. Langton 1989, p. 3.
23. Michael Travers in Langton 1987, p. 421.
24. Haugeland 1985.
25. Johnson 1986, p. 45.
26. Penrose 1989.
27. Penrose 1994.
28. Holland 1992.
29. Prigogine 1984. After reading this book, be sure to read the scathing review in Pagels 1985.
30. See, for example, Briggs 1989, pp. 134–52.

10

Shining Through?

> The Astonishing Hypothesis is that "You," your joys and your sorrows, your memories and your ambitions, your sense of personal identity and free will, are in fact no more than the behavior of a vast assembly of nerve cells and their associated molecules.
>
> —Francis Crick[1]

The Burden of Disproof

As I write this, the twentieth century is drawing rapidly to a close. The media soon will be saturated with articles and documentaries looking back at the many events that have marked this turbulent period in history. We can easily imagine the stories about the Beatles and Elvis, Charlie Chaplin and Marilyn Monroe, Babe Ruth and Jackie Robinson. Future historians, with greater hindsight, will focus more attention on Gandhi, Lenin, Hitler, and perhaps Einstein. They will undoubtedly point to the world wars, the end of empires, the failed experiment of communism, the population explosion, the environmental crisis, space travel, and other largely political events that marked the century as a watershed.

Surely these historians will not fail to note that the unprecedented force of these dramatic events was a direct consequence of the rapid development of science and technology, continuing the trends of the centuries since Galileo and Newton. With modern technology has come the means for both mass human expansion and mass human destruction.

Science has proved itself to be the mightiest activity ever undertaken by the human race. What other enterprise has had within its reach sufficient power to destroy all life on this planet? No doubt, science has great potential for disastrous misapplication. But on balance, most people would rather live with it than without it. After all, who would want a tooth pulled by a shaman when a dentist, using the novocaine developed by much maligned Western science, can do it painlessly?

Technology has resulted in unprecedented, if unevenly distributed, wealth. It has given its beneficiaries longer, more fulfilling lives that are largely free of physical pain and suffering. Undeniably, many billions have not shared in this largesse, but the billions who have enjoyed the benefits of science constitute, by themselves, a greater number of human beings than have lived on this planet in the entire period prior to the twentieth century.

As a result, many people have come to look to science to solve all their problems. Worried about nuclear missiles? Let science build a shield. Fretting about running out of oil? Science will find us an endless source of energy, perhaps cold nuclear fusion. Too little food? Science will grow more. Too many people on earth? Science will launch them into space. Too much pollution? Science will find a way to clean it up. Sick? Science will heal you. Feeling depressed because you are going to die someday? Science will find a way for you to live forever, if not by medical means, then perhaps by confirming your deeply felt belief that your selfhood is intimately connected to the very fabric of reality.

How wonderful that science makes our lives so comfortable. And how wonderful that science has confirmed our long-held belief that human consciousness is the driving force behind the universe itself.

Quantum mechanics is arguably the greatest scientific theory ever invented. It has provided us with many of the tools of modern technology while describing matter at its most fundamental level. Some believe that quantum mechanics has done even more, demonstrating that an act of human consciousness at one point in space can instantaneously cause a material system to change its behavior, indeed its very nature, at a distant point in space—even across the whole universe. And not just instantaneously. Human consciousness, it is said, can cause changes at other points in space even *before* thoughts occur. After all, thoughts are part of the unbroken wholeness of all existence. The mind exists throughout all space and time. It always existed, and always will exist.

This is the profound implication that many believe to be confirmed by the experimental violations of Bell's inequality. The popular literature abounds with this theme as paranormalists of every stripe, from psychics to astrologers to physicists and cosmologists, proclaim the oneness of human mind and the fabric of the cosmos.

While articles in scientific magazines and journals often blazon the wonders of quantum nonlocality, they rarely carry the story through to its bizarre conclusions. That is left to the authors of metaphysical books. But scientists and the science media cannot continue to ignore their responsibilities, if they are to remain reputable. By their avoidance of the sticky issues raised by nonlocality and the quantum consciousness, they effectively help promote a pseudoscientific interpretation of quantum mechanics that is not demanded by data or logic.

The notion of a holistic universe, with everything instantaneously connected to everything else, occurs in a number of interpretations of quantum mechanics. If still-undetected forces operate on particles to determine their quantum mechanical motion, these forces must necessarily be nonlocal, according to the implications of Bell's theorem. It would appear inescapable—the universe is one and we are one with it.

But as we have seen, no empirical or theoretical basis can be found to support this assumption. Nonlocality is not required by the data; superluminal motion has never been observed.[2] Experiments that have tested Bell's inequality are fully consistent with six-decade-old conventional quantum mechanics, reductively interpreted without the metaphysical baggage of collapsing wave functions, nonlocal hidden variables, or parallel universes. Nonlocality only enters via interpretations that assign ontological meaning to mathematical objects like the wave function, which play only an epistemological role in the physical theory of quantum mechanics, or demand the existence of deterministic subquantum forces.

Similarly, while "unconscious" instruments undoubtedly record quantum effects, "consciousness," human or otherwise, however defined, is not required to play the active role in quantum mechanics that some have suggested, namely to provide the mechanism for wave function collapse.

It is true that it has been difficult to gain a consensus on the interpretation of quantum mechanics, or even on the need to interpret it at all, so long as its mathematics works. Many interpretations have been proposed that lead to the same empirical results, and so are indistinguishable except by their formal structures. Without experiment to provide its traditional, (usually) unambiguous adjudication of scientific disputes, the interpretation one prefers becomes somewhat a matter of taste.

Not all interpretations of quantum mechanics are equally economical, or equally useful, however. For example, the claim that human consciousness plays a role in determining the outcome of experiments is not parsimonious since it is not required by a scrap of reliable data. Likewise, nonlocal hidden variables theories are uneconomical, proposing as they do invisible holistic entities having superluminal connections for which no direct or indirect evidence exists. The many worlds interpretation is not parsimonious because, well, it has many worlds, at least when all the parallel universes are viewed as "equally real." The transactional interpretation, on the other hand, is somewhat parsimonious, although it requires the uncomfortable idea that a detector continually sends out "confirmation" waves backward in time.

More generally, we have seen that the paradoxes of quantum mechanics disappear once we recognize that elementary processes do not distinguish between past and future or cause and effect. While this violates our common intuitions, those intuitions are based on our experiences in a world of many particles in which phenomena that are fundamentally statistical behave very predictably because their probability distributions are highly peaked around the most probable outcome. Thus our notions of cause and effect and the direction of time are principles invented to describe the macroscopic world of our experiences, not the quantum world.

We have seen that the alternate (or consistent or decoherent) histories interpretation of quantum mechanics, which grew out of many worlds and the earlier path integral formulation of Feynman, offers a local, reductionist, nondeterministic, and economical answer to the apparent paradoxes of quantum mechanics.

Some will dispute the statement that the alternate histories interpretation is local. I admit it is still arguable (if one insists on a fundamental time direction). However, I reiterate my view that until superluminal motion or communication is empirically demonstrated, the dispute is one over theory, not data. Rather than claiming nonlocality in nature we should be seeking a local interpretation of quantum mechanics. Even if this requires making some change in its fundamental structure, we would still be doing less violence to fundamental physics by pursuing this end than by discarding the essential ingredient of modern physics, namely, absence of speeds greater than the speed of light.

Without regarding all parallel universes as equally real, alternate histories gives us a simple prescription for calculating the relative probabilities for the various paths a quantum system may take through space and time. Furthermore, it represents a modern, albeit tentative, explication of what really goes on when physicists do quantum mechanics. Within the alternate histories framework, measuring devices and the environment generate the decoherence necessary to produce the classical paths we observe in our own macroscopic world, where quantum effects are minimal.

However, the issue is not settled. Questions remain about how the choice between histories, and equivalent sets of histories, are made. Some insist that a mechanism that makes this choice remains to be discovered. Currently, that mechanism is blind chance, but what is the nature of that mechanism? While the alternate histories interpretation may not remove all the mystery from quantum mechanics, at least it provides a counter example that disproves the contention that quantum mechanics *requires* either a conscious or a holistic universe or both.

We have also seen that the phenomena of life and mind likewise provide no basis for the introduction of a nonmaterial component to the universe. Computer studies suggest that complex matter can spontaneously self-organize into systems that exhibit the properties we associate with life and mind. It seems reasonable that high-level capacities were bound to happen somewhere in the universe in its over ten-billion-year existence. While the specific form that life took on earth was largely accidental, some form of life was almost sure to arise under the given environmental conditions. Calculations that purport to demonstrate the great unlikelihood of life forming by accident assume, anthropocentrically, that only one form of life, namely ours, is possible.

I do not mean to imply that life is likely to evolve on a large fraction of the planets in the universe, or that intelligence is likely to evolve in a large fraction of the planets that develop life. The evidence from our own solar system, and the absence of signals or probes from extraterrestrial civilizations (UFOs notwithstanding), indicates that life and intelligence require conditions that are quite rare in the universe. And even if life is somewhat common, complexity and intelligence are less likely to be so. Noted paleontologist Stephen Gould has argued that life is not necessarily progressive, and the formation of complex life forms may be more of a fluke than the inevitable result of evolution.[3] But the universe is a vast place, and rare conditions have ample opportunity to occur by chance. Computer simulations indicate that complexity can develop by chance.

Mutually interacting, nonlinear systems poised at the boundary between order and chaos are highly susceptible to the Darwinian selection processes that so powerfully direct evolution toward adaptive behavior. The systems being simulated are purely classical, implying that quantum mechanics may play a minor role at best in these processes.

In all of this I may have oversold my case. Undoubtedly some of the views I have presented will be modified by myself and others as time progresses. Someday, some specific example or argument may be shown to be wrong. However, I have confidence that my basic conclusions will stand the test of time. These conclusions do not rise or fall on every statement made in this book.

I maintain that I do not have the burden of disproof. Nonmaterial, superphysical explanations for any given phenomenon cannot reasonably be introduced until all more economical explanations for that phenomenon are ruled out to the highest degree. I merely have to show, as I believe I have, that plausible explanations of all current data can be found within conventional, reductionistic, local, materialistic science, including quantum mechanics. I do not feel compelled to validate these explanations, for example, by generating life in the laboratory, or performing an experiment that rules out all but a local, unconscious interpretation of quantum mechanics.

Platonic Truths

But have I covered all the metaphysical bases? Perhaps not. In a pair of recent books, cosmologist Roger Penrose has argued forcefully, and controversially, that the human mind is capable of reaching into a realm of reality that lies beyond time and space, a Platonic world of timeless mathematical truth that many physicists still seem to think exists.

Penrose first introduced this view in a widely read 1989 book, *The Emperor's New Mind.*[4] Although filled with many unnecessary and irrelevant technical details that surely daunted the average reader, *Emperor* briefly made bestseller lists and caused quite a stir.

Penrose has now followed up with *Shadows of the Mind.*[5] While *Shadows* goes over much of the same ground as *Emperor,* the exposition is much clearer and to the point, with the author making a serious attempt to address each specific criticism of *Emperor.*

In both books, Penrose bases his argument on the issue of whether a computer can ever duplicate all the thinking processes of human beings. In *Shadows* he summarizes four different viewpoints:

A. All thinking is computation; in particular, feelings of conscious awareness are evoked merely by carrying out the appropriate computations.

B. Awareness is a feature of the brain's physical action; and whereas any physical action can be simulated computationally, computational simulation cannot by itself evoke awareness.

C. Appropriate physical action of the brain evokes awareness, but this physical action cannot even be properly simulated computationally.

D. Awareness cannot be explained by physical, computational, or any other scientific terms.[6]

View A is the view of *strong artificial intelligence* or what is termed *functionalism.* In view B, awareness is still a physical process that can be simulated computationally, but this simulation is not in fact awareness. Penrose does not say how one could distinguish a simulation from true awareness, but this is not his position anyway. Rather he holds to view C, in which awareness cannot even be simulated computationally.

In *Shadows,* Penrose is careful to distinguish his position from view D, in which awareness is not amenable to scientific study and thus must be mystical or supernatural. He says, "I reject mysticism in its negation of scientific criteria for the furtherance of knowledge." I take this to mean that if awareness is something that can be understood scientifically, then it might still be possible for it to be simulated. It just cannot be simulated, in Penrose's view, *computationally.* Some kind of noncomputational machine, made of matter and operating purely in the physical domain, would have to be devised to simulate awareness.

Clearly view D is the common religious viewpoint of a dual universe of matter and mind, or body and soul. In my previous book *Physics and Psychics* I argue that no empirical or theoretical basis exists for this view.[7] Penrose agrees, asking why, if view D is correct, "our minds seem to be so intimately associated with elaborately constructed physical objects, namely our brains."[8] Thus Penrose tries to distance himself from what was beginning to be referred to as *Penrose mysticism.* Whether he will succeed remains to be seen.

If awareness is a physical phenomenon that is not computable, that is, a property that a computer (though not necessarily some other physical system) can never simulate, then some change in our physical world view is required to encompass a new, noncomputable physics. That is, some new kind of physics is poking its head through the thoughts in our own heads.

In the second part of *Shadows*, Penrose speculates about the direction that new physics will take. He repeats, in modified form, the notion he promoted in *Emperor* that new physics will be found in what he calls the **R-process**, that is, the process of **wave function reduction**, which throughout this book I have called *wave function collapse*. Penrose labels the new process **OR** for **objective reduction**, making clear that the process takes place independent of human involvement. He rejects the notion that human consciousness is the agent of wave function collapse, and, by implication, the seat of the new physics.

Penrose believes the key to the new physics lies in quantum gravity, which somehow disentangles spatially separated, coherent quantum states. Actually, this sounds very much like the decoherence by the environment we have been talking about, where for Penrose the environment is the basic spacetime structure of the universe. However, Penrose does not prove that this mechanism is noncomputational, and he only speculates on what it can possibly have to do with human thinking. I personally find it incomprehensible that quantum gravity, which only comes into play at the Planck scale, can have a profound role on the scale of biological processes. I also find it rather anthropocentric to think that the next great revolution in physics will be found in the human body.

Penrose insists that the evidence for the OR process is to be found in human consciousness, even if consciousness is not its source. Of course, the thesis that the brain is not simply a computer is one that the average person will grasp with open arms. Few can imagine, or want to imagine, how a computer can ever have "feelings" and "spiritual experiences."

Wisely, Penrose makes only a cursory attempt to justify his view on the basis of prevailing popular opinion. Neither does he rely heavily on examples of the admitted stupidity of today's computers, despite their power in some areas, and the limited success of the artificial intelligence program. Still, he can't resist giving an example of how the chess program *Deep Thought* failed to make a move any neophyte, grade-school player would have immediately seen. He uses this as an example of how computers simply do not seem capable of "understanding."

Penrose understands that computers are in their infancy and that attempting to predict tomorrow's technology is a fool's game. So he rests his case, in both books, on the argument that mathematics contains truths that are not computable. In obtaining these truths, Penrose claims, mathematicians peer into the Platonic realm of true forms to gain insights that no computer can ever duplicate. Although a mathematician may go through a complicated chain of reasoning that is for the most part computable, or *algorithmic,* that chain inevitably contains elements of "obvious knowledge" that are assumed to be valid, that are self-evident truths. Furthermore, these mathematical insights are not limited just to mathematicians, but occur naturally to human children as they instinctively learn the meaning of numbers.

Penrose seems to agree with philosopher John Searle, who has argued that computer "intelligence" is fundamentally misguided because computation is mere syntax without semantics, the manipulation of symbols that have no reference to the real world.[9] Neuroscientist Gerald Edelman also views the computational model of the mental process as inappropriate, though he has no doubt that it is biological in nature.[10] All three men receive moral support from a famous physicist of an earlier era, Sir Arthur Eddington, who said, "We can only reason from data, and the ultimate data must be given to us by a non-reasoning process—a self-knowledge of that which is in our consciousness."[11]

Unprovable and Undecidable

The primary focus for Penrose's discussion of noncomputability lies with **Gödel's theorem**, which says that unprovable truths can exist within any formal mathematical system at least as complicated as arithmetic.[12] Gödel's theorem, in Penrose's view, demonstrates that "the mental procedures whereby mathematicians arrive at their judgements of truth are not simply rooted in the procedures of some specific formal system."[13] That is, mathematicians are able to develop true propositions by means other than the strict logic of mathematical procedures.

A related theorem was uncovered by the English mathematician Alan Turing in the late 1930s. Turing showed that any mathematical computation can be carried out on a simple idealized computer now called a **Turing machine**.[14] By treating every formal mathematical statement as a number, Turing concluded that all deductive mathematical procedures can be reduced to such a sequence of mechanical operations on numbers, what are called *algorithms.*

Independently and at about the same time, American logician Alonzo Church derived a theorem proving that a logical statement is valid independently of the meaning given to the symbols used in its expression.[15]

Both Turing and Church had in mind trying to solve a problem posed by the German mathematician David Hilbert in 1900. He called it the *entscheidungsproblem* (the "decision problem"). Hilbert had asked whether a general procedure exists for resolving mathematical questions, and indeed he thought there was. Turing and Church, in their differing ways, showed it was impossible. Specifically, Turing showed that no procedure exists for determining whether or not a Turing machine will ever stop and thus solve, in a finite time, a particular problem posed to it. Thus, not only do undecidable propositions exist, as shown by Gödel, but no procedure exists for deciding if a proposition is decidable![16]

While all this threw the world of mathematics into turmoil, you might ask what it has to do with the real world, since mathematics is a human invention anyway, a game played by arbitrary rules. If someone proved a theorem showing that it was impossible to prove that certain sequences of chess steps were impossible, no one would suggest that some deep, metaphysical principle had been uncovered. And proving that it is impossible to prove whether a team can win a baseball game without scoring a run (barring forfeit) would hardly warrant a headline on the sports page.

But Penrose argues, "Once it is shown that certain types of mathematical understanding must elude computational description, then it is established that we can do *something* non-computational with our minds."[17] And, if we are to assume that the phenomenon of mind is still part of the physical world, then we are forced to relate mathematics to that world.

Furthermore, mathematics may not be the arbitrary game it seems. It appears to be deeply connected to the real world. All our physical theories are expressed mathematically, and surely the basic mathematical operation of counting is fundamental to any concept we use in organizing the data of our senses and instruments.[18] So if we are able to peer into a Platonic realm of mathematical truth, that realm surely relates in some profound way to the physical universe.

Penrose adds, "There is something absolute and 'God-given' about mathematical truth." He admits he is very much a Platonist. "In my own mind, the absoluteness of mathematical truth and the Platonic existence of mathematical concepts are essentially the same thing."[19] In other words, mathematical truths are the reality beyond the appearances. As I have mentioned, this neo-Platonic view has come to be called *Penrose mysticism,* although the author vigorously objects to this characterization, firmly insisting that the noncomputational remains amenable to scientific study.

Mystical Matters and Minds

In his book with the catchy title *The Mind of God*, physicist-author Paul Davies has used Penrose's ideas in discussing the possible connection between mathematics and the traditional notions of mystical truths.[20] Mystics have universally claimed direct communication with deeper reality, called variously The One, The Good, God, the Cosmos, Being, and many other names. The mystical experience is supposed to open the mind to instantaneous flashes of insight about a realm beyond the senses. Distinguished scientists such as David Bohm and Brian Josephson claim to have found mysticism useful in developing their scientific ideas, and many of the founders of modern physics speculated about the mystical.

Ken Wilber has edited a collection of such writings in *Quantum Questions: Mystical Writings of the World's Great Physicists.*[21] Included are essays by Heisenberg, Schrödinger, Einstein, de Broglie, Sir James Jeans, Planck, Pauli, and Eddington. Wilber interprets the essays as showing that each author was in fact a mystic. However, he admits that "these theorists are virtually unanimous in declaring that modern physics offers no positive support whatsoever for mysticism or transcendentalism of any variety."[22] So if these giants of physics were mystics, which is highly debatable, their mysticism was not derived from their physics.

In her unique book *A History of God,* former Catholic nun Karen Armstrong explains how all the great religions of the world, with the exception of Western Christianity (Roman Catholicism and Protestantism), long ago gave up on using rational thinking as a means for obtaining knowledge about the transcendent, if they ever considered it at all.[23] Buddhists, Hindus, Moslems, Jews, and Eastern Orthodox Christians all developed various techniques to rid their minds of rational thought so that they could make contact with what they believed was the transcendent power beyond the material world of the senses.

In *Physics and Psychics* I argued that the feeling of oneness experienced by the mystic is almost certainly a delusion. One can find no independent evidence that the claimed insights obtained in a mystical state have anything to do with objective reality. No one can point to a previously unknown discovery made in a mystical state that was later confirmed by scientific observation. On the contrary, virtually every claimed mystical, nontrivial revelation about the nature of the universe and humanity's place within it has proved to be grossly wrong.[24]

In Western Christianity, mystics have been largely discouraged or embraced only hesitantly. Such was the case with Francis of Assisi. More frequently, the most influential Church leaders have been theologians, such as Augustine, Abelard, and Aquinas (just to mention the A's), who attempted to place their faith on a rational foundation and thus make it more palatable in the Greek-influenced West. But with no empirical facts with which to cement their hypotheses, the result has been a dogmatic insistence on a personal God amenable to human understanding. Armstrong claims that this has led to many of the excesses associated with Western Christianity, as each sect formed its own rationalized image of God and declared all others to be heretical.

Where other religions have left it largely up to the individual to develop his or her own inner contact with the envisaged transcendent, Western Christianity has insisted that its priests or the Bible act as intermediary, that individual human mental processes are not to be trusted. Thus, by relying on reason, the Western churches became the most dogmatic. Reason has its dangers, too!

So where do Penrose's ideas fit into the mystical perspectives of religion? Certainly, he attempts to be completely rational in applying the theorems of Gödel, Turing, and Church to demonstrate that we cannot determine all that is true by computational means alone. On the other hand, he asserts that the human mind nonetheless can formulate these truths, and that they have a Platonic reality to them. Is mathematics, despite Penrose's disclaimer, really then a mystical path to truth? Is it not then more like revelation than science, as it goes beyond sensory data and their numerical manipulations? Is the existence of the Ultimate "shining through," despite the complete lack of any physical evidence or any compelling need to introduce metaphysical elements into our most fundamental theories of physics and cosmology?

Most workers in artificial intelligence and neuroscience remain unconvinced by Penrose's assertion that the human mind cannot be simulated by a machine. After the publication of *Emperor,* a goodly portion of one issue of the journal *The Behavioral and Brain Sciences* was devoted to a complete discussion of Penrose's ideas. Penrose's own eleven-page *précis* of the book was followed by thirty-seven peer commentaries and the author's response to each.[25] At this writing, a similar critique of *Shadows* is being assembled on the Internet by the journal *Psyche.* Virtually every commentary disagrees with most or all of Penrose's conclusions. In his responses, Penrose has studiously attempted to answer each criticism, one by one, but I believe it is fair to say that he has not achieved a consensus for his views in any of a number of communities, from artificial intelligence to quantum computation and neurobiology.

Daryl McCullough has said the following, which I think nicely summarizes the existing consensus:

> Penrose's arguments that our reasoning can't be formalized is in some sense correct. There is no way to formalize our own reasoning and be absolutely certain that the resulting theory is sound and consistent. However, this turns out not to be a limitation on what computers or formal systems can accomplish relative to humans. Instead, it is an intrinsic limitation in our abilities to reason about our own reasoning process. To the extent that we understand our own reasoning, we can't be certain that it

> is sound, and to the extent that we know we are sound, we don't understand reasoning well enough to formalize it. This limitation is not due to any lack of intelligence on our part, but is inherent in any reasoning system that is capable of reasoning about itself.[26]

In other words, if computers have limits, then so do people.

Undoubtedly, the issue will continue to be hotly debated and I will not settle it here. For my purposes, however, the following tentative conclusion can be drawn: Even if the human brain is not a computer, this does not imply that the "mind," which is the name we give to what the brain does, has a mystical or metaphysical component. The view promoted by Penrose is one in which the brain still does the thinking by means of some physical process that remains to be determined. Whether or not he is correct on the need for new physics, he sees no need to transcend physics—just move it to a new level. But, as we have seen, no scientific observation demands such an interpretation.

Network Thinking

In the previous chapter, I briefly discussed those computer simulation studies on artificial life that indicate how primitive intelligent behavior develops spontaneously when highly interconnected parallel networks are poised at the boundary between order and chaos. Let us now pursue this further, to see how such networks, operating only on numbers, might become capable of at least a good deal of what we label "thought."

Although the generation of language and mathematics may be noncomputational at their very sources, and not represented in exactly this manner in the brain, all verbal and mathematical statements can be expressed as sequences of the binary numbers, or bits. The words in this book, for example, exist as bits on a computer disk. The book can be completely reproduced from that disk, though some would argue that only a human mind is capable of extracting its "meaning."

Scientific knowledge is not all that can be expressed numerically. Even a great painting can be reduced to bits and reproduced in its entirety. Since a poem is a string of words, it can also be expressed in terms of numbers. Parenthetically, computers can write quite acceptable poems and indeed have won poetry contests when pitted, without the judges' knowledge, against human contestants. Perhaps a computer can be programmed to recognize computer poetry better than a human judge!

Suppose that a statement, verbal or mathematical, is determined by a sequence of steps that follow some logical rule. For example, if two statements in a row are each true, then the output "true" is generated; otherwise we generate a "false." This is the logical operation **and**. The logical operation **or** would give true if either input is true, and false otherwise (more precisely, this is an *inclusive* or, what in the vernacular we often write as "and/or"; the *exclusive* operation **xor** gives true only if the inputs are different—true or true = true, but true xor true = false). The negations of *or* and *and* are called **nor** and **nand**. Groups of ands, ors, nors, and nands can be strung together to produce any logical operation. Actually, it can be done with just nors or (that's an *inclusive* or) nands.

The above rules expressed in terms of statements that are true or false can be represented on a normal, binary computer as operations on the bits 1 and 0. Computations can be reduced to these operations. Conversely, we can think of true/false statements as computational. More complex logical statements can be reduced to true/false questions.

Starting with some initial number, or set of logical statements, we thereby determine by our algorithm some final number or final set of logical statements. For example, the initial number may describe a set of primary propositions, such as Euclid's axioms of geometry. The final number may then describe a theorem derived from these axioms, such as the one saying that the sum of the angles of a triangle equals two right angles.

Turing showed that the number representing a theorem can always be calculated by a computer (a **universal Turing machine**), starting with the original number representing the axiom. Where do the axioms come from? Penrose would argue that at least some are noncomputable numbers, numbers that cannot be generated by a Turing machine. Those who disagree with Penrose would say they come from some higher-level algorithm.

Deterministic events in nature are analogous to the logical operation of a computer, starting with one number and generating, by a series of programmed steps, another number.[27] Seen in this way, the fact that mathematics applies so successfully to physics loses much of its mystery. The mathematical procedures used in most scientific applications are simply logical, algorithmic operations on numbers, and physical measurements are expressed as numbers. How else but with mathematics can we describe operations on numbers? That is mathematics, by definition.

A serial set of steps is not the sole way to proceed from one number to another. A number can be presented as a sequence of 0 and 1 bits at the input of a parallel processor or what is called a **neural network**, a computer that works as an interconnected network of switches in loose analogy to the human brain. The general folklore holds that such networks can be simulated by serial Turing machines. However, in a recent paper, Hava T. Siegelmann has claimed that a highly chaotic dynamical network called the *analog shift map* has computational power beyond the Turing limit.[28]

Whatever their Turing status, neural networks still have interesting properties when viewed as a complex net of interacting units. The input bit pattern of a neural network can be transformed into other bit patterns based on a programmed response to the inputs to each switch, or "object," where a decision is made on what bit pattern to transmit forward. While each switch may perform mechanically, the studies with neural networks show that network processing, as clearly exists in the brain, often offers a more efficient way to solve many problems than a linear sequence of steps. But if a neural network can be simulated by a Turing machine, it remains capable of solely numerical (computational) operations.

One possibility that is just beginning to be studied is the use of tristate or even more sophisticated forms of logic. Recall the discussion in chapter 3 of James Hartle's scheme for representing the quantum state of an individual system in terms of the three possible answers to the question of whether or not the system possesses a certain property: yes, no, or indefinite. I mentioned that this might be programmed on a tristate computer.[29]

In what is called **soft computing** or **fuzzy logic**, hardware and software are developed that are tolerant of imprecision, uncertainty, and partial truth.[30] In a simple example, the binary switches of normal computer circuitry can be replaced by switches that return a yes or no according to some probability function. In this manner, the uncertainty that usually accompanies much of decision making can be simulated. The technique of fuzzy logic is being increasingly used in so-called expert systems.

Fuzzy theorist Bart Kosko has argued that fuzzy logic represents a grand revolution in Western thinking that harkens back to Eastern philosophy.[31] This claim, like similar ones we have heard made for quantum mechanics and chaos theory, is somewhat overblown. Nevertheless, it seems reasonable to surmise that the brain operates more in a fuzzy way, rather than in the uncompromising true or false way of binary switches. But again, since both tristate and fuzzy logic can be simulated on a Turing machine, such systems remain computable.

Can a serial or parallel processor, operating completely computationally with binary, tristate, or fuzzy logic, generate the possible noncomputable number associated with an axiom? Here it is important to distinguish between the words *computable* and *deterministic.* Classical physics is both computable and deterministic. However, in the conventional view, quantum mechanics is computable while being nondeterministic. Certainly we can imagine systems (the brain?) that may be neither deterministic nor computable, but still physical.

Perhaps we can regard an axiom as a nondetermined number that is still computable. Does this finally open a place for quantum mechanics in the processes by which the brain produces axiom numbers? Possibly in the neural network of the brain a quantum fluctuation of one bit is all that is necessary to produce new knowledge, thus leading to a previously unknown, underivable idea.

Is the Brain a Quantum Device?

As we have now seen in great detail, consciousness is not needed to explain quantum mechanics. The remaining issue we need to explore here is whether quantum mechanics is needed to explain consciousness.

Many authors have speculated that quantum mechanics plays a part in the functioning of the brain. Their arguments vary from the thoughtful to the specious. Neuroscientist Sir John Eccles has presented a dualistic model in which mind exists as an entity separate from matter, initiating wave function collapse that releases neurotransmitters at neural junctions.[32] Penrose and his collaborator Stuart Hameroff have more recently proposed the "orchestrated objective reduction" of quantum coherence in the microtubules of the neurons of the brain.[33]

As we have seen in the preceding chapters, the decoherence mechanism of the alternate histories version of quantum mechanics removes the need to introduce nonphysical elements to account for wave function collapse. However, quantum mechanics may still play a nontrivial role in brain processes.

Henry Stapp thinks it does: "Brain processes involve chemical processes, and hence must, in principle, be treated quantum mechanically."[34] Following the logic of this argument, we cannot use Newtonian mechanics to calculate the trajectory of a rock tossed in the air, because the rock is made of chemical elements.

Stapp notes that the transmission of signals in the brain is triggered by small numbers of calcium ions at synaptic junctions. As Nick Herbert colorfully describes it, a neural synapse fires when it emits "little bags of drugs" called synaptic vesicles into the synaptic gap. Drifting across the gap, the transmitter molecules contained in the vesicles either stimulate or inhibit the adjacent nerve cell. Vesicle release is initiated by calcium ions that enter the synapse from the surrounding fluid, crossing through channels that are opened in response to electrical signals.[35]

Herbert and others have made order-of-magnitude arguments that they claim demonstrate a likely role for quantum mechanics in synaptic signals.[36] All such estimates essentially come down to an application of the uncertainty principle.

A simple way one can test whether a system needs to be described quantum mechanically is to ask whether the product of the system's typical mass m, speed v, and distance d is of the order of Planck's constant h. If mvd is much greater than h, than the system probably can be treated classically, barring very special long-range correlation effects such as in EPR experiments.

In the case of synaptic transmitters, m is of the order of 10^{-22} kilogram. Assume v is the average speed an object of this mass would have as the result of thermal motions at body temperature, about 10 meters per second. The gap distance d is of the order of ten atomic diameters or 10^{-9} meter. Then mvd $\approx$ 10^{-30} Joule-second. This is to be compared with the value of Planck's constant, $h = 6.63 \times 10^{-34}$ Joule-sec. That is, mvd = 1700h, which seems to indicate that transmitter motion can be treated classically to a fair degree of approximation. Not all objects that must be viewed with a microscope need to be described by quantum mechanics.

Note that the fact that the brain sits at body temperature results in much greater particle speeds than occurs in superconductors, the favorite example of macroscopic quantum systems. Even the so-called "high-temperature superconductors" must be cooled far below 98.6 degrees Fahrenheit to function. The high temperature of the brain makes the decoherence mechanism much more efficient and quantum coherent effects very unlikely.

To make a case for quantum effects in the brain, proponents have to squeeze mvd lower by about three orders of magnitude. This can be done with a certain amount of hand-waving. The mass m may be an order of magnitude lower; the speeds have a statistical distribution and lower values will occasionally happen; the relevant distance may be as low as an atomic diameter in the case of charge-driven, electromagnetic synaptic processes.

On the other hand, the process of neural stimulation is not conducted by a single transmitter molecule. Hundreds of vesicles with thousands of transmitter molecules are required and fluctuations are smoothed out by the homeostatic mechanism of feedback that characterizes most biological entities. Again, the large number of particles involved implies strong decoherence.

Penrose and Hameroff have proposed the new idea that the seat of quantum effects in the brain lies in **microtubules**, hollow fibers that form part of the cytoskeletons of most of the cells of animal and human bodies, not just brain cells. They suggest that these may be the cell's own "nervous system."[37] The diameter of a microtubule is even greater than the synaptic gap, and they are much longer, so they are certainly "macroscopic" objects by the above standard. Penrose suggests that microtubules act in a coherent way, but has no hard evidence to back up this notion. Behavioral neuroscientist Michael Lilliquist has asked why the microtubules in neurons should alone show quantum effects and not those of other cells, say, in the liver.

I conclude that, while quantum effects are not obviously required to describe the general motion of synaptic vesicles or other components of the brain's nervous system, room does exist for the possibility that some low-probability quantum fluctuation occasionally generates a bit error in the signal transmitted across a synaptic gap. Perhaps this generates the noncomputable number Penrose calls "understanding." However, the existence of long-range, EPR-type correlations in the brain seems far-fetched.

The fact that the probability for bit errors in the brain is small is fortuitous. Otherwise, our brains would be unable to function as even semireliable information processors. On the other hand, if the probability for bit fluctuations is too low, then this mechanism for the generation of nonalgorithmic knowledge will not work.

Of course, most of the noncomputable numbers generated by random fluctuations will be of no value. Only the rare fluctuations will reproduce the axioms of geometry or some other original thought that cannot be computed by a sequence of mathematical steps. A great many noncomputable numbers would have to be generated and then some selection process, which itself could be perfectly mechanical or algorithmic, would have to act to choose those noncomputable numbers that lead to useful ideas.

It may be difficult to imagine such a selection process being very efficient, given the fact that the useless bit strings will far outnumber the useful ones. This might strike the reader as being as unlikely as the monkey at the keyboard writing *Hamlet* ("To be or not to be, that is the gzetrnamfj"). However, Darwinian selection, as embodied in both natural and artificial life, comprises a highly efficient mechanism for rejecting useless mutations and accepting useful ones. This does not mean that the brain selection process is necessarily Darwinian in detail, just that we already have at least one example where the selecting out of useful mutations from a sea of useless ones does in fact exist.

While quantum fluctuations may be the source of noncomputable numbers, they are by no means the only source. The fluctuations that produce noncomputable numbers could result from cosmic rays or other interactions between the brain and its environment.

Whatever the source of the fluctuations that generate noncomputable signals in the brain, if any, we may reasonably view the brain as a neural network on the edge of chaos. A fluctuation can give the network just the nudge it needs to move from one metastable state to another. By operating at the edge of chaos, the network can move between different states and rely on self-organization to provide the selection mechanism to move through what Stuart Kauffmann calls the *fitness landscape,* eventually settling down to a configuration of neurons that contains new ideas to be tested against sensory input from the real world. Quantum mechanics need not be invoked at all.

The Force of Consciousness

Those who promote the myth of quantum consciousness are not content with the trivial possibility that quantum fluctuations may be responsible for introducing a certain indeterminism into the processes of the brain, thus allowing for some non-Newtonian behavior. They refuse to believe that that's all there is to it, that the "mind" does not play a greater role in choosing between the alternative paths that can be taken as the brain moves between metastable states.

This belief is not based on any external evidence. Rather the claim is that our "inner experiences" of consciousness, wholeness, and self-awareness require something more—a controlling agent capable of dealing with complex wholes. Stapp argues that such control is a logical impossibility "within a framework in which everything is asserted to be nothing but an aggregation of simple parts." Claiming kinship with William James, Stapp adds, "In order to accommodate an intrinsically unified thought, as distinct from an aggregation that is *interpreted* as an entity by something else, one must employ a logical framework that is not strictly reductionistic: a framework that has among its logical components some entity or operation that forms wholes."[38] Stapp alleges that quantum nonlocality provides him with that framework.

Herbert thinks that "consciousness will turn out to be something grand—grander than our most extravagant dreams." He proposes "a kind of 'quantum animism' in which mind permeates the world at every level" with consciousness "a fundamental force that enters into necessary cooperation with matter to bring about the fine details of our everyday world."[39]

However, Herbert does not tell us what makes humans different from rocks, which, after all, is the goal of the discussion.

In the Copenhagen interpretation, instantaneous wave function collapse occurs during the act of measurement. In the emerging modern post-Everett interpretations, decoherence provides a mechanism for wave function collapse. In an economical view that remains to be refuted, the event that happens is randomly chosen from the possibilities, according to probabilities calculated from the wave function.

The quantum mystics persist in their belief that human consciousness must act as an agent to bring about the specific choice among alternate paths. The conscious force, in the view of Stapp and Herbert, acts to "actualize" the event, changing a possibility into a happening. To physicist Euan Squires, consciousness interacts with the world in determining the choices between paths.[40] As Squires describes his model, the mind acts as the "selector" among alternate worlds, the way a television viewer chooses which channel to watch.[41] Philosopher David Albert has a similar model that he calls the theory of *many minds.*[42]

Penrose also argues for "some kind of active role for consciousness, and indeed a powerful one, with a strong selective advantage" to avoid blind randomness.[43] However, he disagrees with the quantum dualists in an important way. In the dualistic view, consciousness is some kind of extraphysical force that acts to cause events to happen, to collapse wave functions or actualize particular histories. In the dualistic view, mind controls the universe.

Penrose, on the other hand, proposes that some new physics is involved in consciousness—but it remains physics. In his view, the universe still controls the mind. Nevertheless, in claiming that new physics can be found in the operation of human consciousness, Penrose joins Stapp, Herbert, Squires, and other authors I have quoted throughout this book in assigning a very special role in the universe to what may be in fact a simple accident of evolution.

For, after all, what is "consciousness"? No consensus exists on its definition. When I have mentioned consciousness, it has always been in the context of what other people have said about it. I am very suspicious of words that are difficult to define. When the concepts supposedly represented are so unclear, I wonder if they even need to be defined.

At the same time, however, these arguments do not require the nonexistence of a purely local, physical phenomenon called consciousness. Such a phenomenon may exist, but the evidence of decades of experiments in the neurosciences now indicates that it is not so profound as we have been led, and have led ourselves, to believe.

Since a detailed discussion of the nature of consciousness would take me far from the focus of this book, I will refer the reader to recent books by Daniel Dennett,[44] Gerald Edelman[45] and Francis Crick.[46] Adam Carley has given a concise summary of an emerging neuroscientific view, expressed in varying forms in these books and others, that consciousness is in fact simply a trick of the brain.[47] In this view, our experience of "consciously" willing an act, such as moving a hand, has little to do with the actual process by which the brain makes the decision to perform that act. Rather, it seems to be an inaccurate recollection that is formed *after* the actual decision is made.

Over a decade of experiments have confirmed that there is a delay of several tenths of a second between a physical action performed by a subject and that subject's report of the time that he became aware of his "free will" intention to perform the act. This may mean that the decision-making process takes place unconsciously and is only, a fraction of a second later, imperfectly perceived as an act of will.

This is undoubtedly a revolutionary concept that will be intensely resisted by those who find such a minor role for consciousness impossible to accept. The debate will be long and heated, and I surely will not settle it here. My purpose is not simply to add to the already huge volume of words on the nature of consciousness, but rather to address those claims that modern physics has placed the human mind back at the center of the universe, where human ego continues to insist it be located.

My perspective is admittedly that of a physicist, one who insists on overwhelming empirical evidence before accepting any extraordinary claim. As we have seen, the evidence that quantum mechanics either requires the action of human consciousness or plays a role in mental processes is nonexistent. Certainly quantum mechanics is needed to understand the atoms in the brain. But it is also needed to explain the atoms in a rock, and this implies nothing about rock consciousness.

Perhaps quantum fluctuations cause random bit errors that the brain is able to organize into new operations, but this role is neither necessary nor compelling. The environment can produce the needed fluctuations. The self-organizing capabilities of the brain's nonlinear neural network, operating at the edge of chaos, may be capable of doing all the work of selection of the best path among all possibilities, with no further help from quantum mechanics. In fact, the human brain and body probably evolved with the dimensions it has in order to *avoid* quantum effects.

It seems to be nothing more than primitive, wishful thinking to view consciousness as some supernatural, or at least supermaterial, psychic force that provides basic control over the choices the universe makes between allowed alternative paths, either inside or outside the brain. Such a theory is verifiable. It should lead to phenomena such as ESP and psychokinesis that violate the laws that constrain matter. But, as described in chapter 1, psychic phenomena have failed to be verified after 150 years of attempts involving thousands of independent experiments. After all this time, we can safely assume they do not exist.

The Platonic Realm

We have seen that a deterministic universe is only possible within a continuum of space and time. The motion of a body can be determined only when the body has a well-defined position and velocity at every instant. However, two instants are required to specify a velocity. In classical mechanics, a fundamentally nonmeasurable *instantaneous velocity* is defined by a purely mathematical process in which the time interval is imagined to go to zero. This abstract technique, the infinitesimal calculus independently invented by Newton and Leibnitz, has proven to be the most important implement in the scientist's theoretical tool box.

The laws of classical mechanics and many of the principles of quantum mechanics are composed in the form of *differential equations* that contain the abstract entities of calculus. These include the various forms of the generalized equations of motion, Maxwell's equations of electromagnetism, and Schrödinger's equation. These differential equations are so simple, elegant, and powerful that most physicists regard them as the ultimate description of reality. They are the Platonic truths that lie beyond the grubby data gathered by scientific instruments.

However, I would like to suggest the opposite interpretation: The continuum differential equations of physics, including those of quantum physics, are only an approximate description of reality. The universe is fundamentally discrete, and only when that discreteness is too small to be noticed are we justified in using the continuum equations. Because most of our scientific interests lie on such scales, the continuity approximation leads to many useful results that agree beautifully with experiments. And so, on these scales, the universe appears orderly and predictable.

Even quantum mechanics is interpreted as operating in a spacetime continuum, and that is what leads us to think it is paradoxical. As we have seen, nonlocality such as seems to occur in quantum jumps occurs only when we assume a continuum.

On the ultimate quantum scale of Planck lengths and times, however, things are different. Space and time are discrete and motion is less predictable. At this scale, no evidence for a deeper Platonic realm that determines events is "shining through." Quite the contrary. Our differential equations and other idealized mathematical expressions appear simply as handy tools for making calculations that would otherwise require extensive numerical computation.

In chapter 8 I mentioned that physicist-theologian Willem Drees agrees with me that neither the data nor the theories of modern physics and cosmology provide any rational basis for a transcendent world beyond matter.[48] Nonetheless he chooses to believe in such a world, admitting that his is a freely chosen belief based on what he perceives as the human desire for justice and perfection. While I cannot bring myself to worship a hypothesis, I have no wish to disparage those who do. I simply ask that they not assume that science, in its current state, provides any buttress for their belief in a Platonic reality, whether it be mathematical forms or a more traditional deity.

The Me Decades

A decade ago, Fritjof Capra, Marilyn Ferguson, Gary Zukav, and others had predicted that the 1980s would be a revolutionary time "because the whole structure of our society does not correspond with the world-view of emerging scientific thought."[49] They blamed classical physics for all the ills of society and saw the new physics, especially quantum mechanics, as a savior.

In her 1990 book, *The Quantum Self,* Danah Zohar asserts that, "Cartesian philosophy wrenched human beings from their familiar social and religious context and thrust us headlong into . . . our I-centered culture, a culture dominated by egocentricity."[50] The new holistic physics was supposed to teach people to be less selfish, to recognize that they are part of a greater whole and to work cooperatively for the benefit of everyone.

As I write in 1995, however, I can perceive no great holistic revolution actually having taken place in the decade past. The facts indicate the contrary. Even more than the 1970s, which originally had the name, the 1980s have been characterized, in America anyway, as the "Me Decade." Far from recognizing that we are each an inseparable part of the whole, and everyone pitching in to make the world a better place for its inhabitants, life in the 1980s was characterized by an unprecedented level of individual self-absorption. And the 1990s so far show no sign of a change in this focus on self, as every element of our society is geared to provide maximal short-term self-gratification for its members, while those who fail to be gratified view themselves as victims.

Now some will argue that the ever-increasing fixation with self only reinforces the need for a holistic philosophy like that of Capra, Ferguson, and Zohar. They will say that the problem is simply that holistic philosophy has not yet taken hold. I disagree. In fact, no small portion of the blame for the current excessive self-absorption lies at the feet of the new mysticism. Anyone listening to New Age gurus and modern Christian preachers cannot miss the emphasis on the individual finding easy gratification, rather than sacrificing and selflessly laboring for a better world. Holistic philosophy is the perfect self-delusion for the spoiled brat of any age, all decked out in the latest fashion, who loves to talk about solving the problems of the world but has no intention of sweating a drop in achieving this noble goal.

Reductionist classical physics does not make people egoists. People were egoists long before classical physics. In fact, classical physics has nothing to say about humans except that they are material objects like rocks and trees, made of nothing more than the same atoms—just more cleverly organized by the impersonal forces of self-organization and evolution. This is hardly a philosophical basis for narcissism.

The new quantum holism, on the other hand, feeds our delusions of personal importance. It tells us that we are part of an immortal cosmic mind with the power to perform miracles and, as actress Shirley MacLaine has said, to make our own reality. Who needs God when we, ourselves, are God? Thoughts of our participation in cosmic consciousness inflate our egos to the point where we can ignore our shortcomings and even forget our mortality.

The modern versions of traditional religions also feed on this desire. Where once Christian preachers shouted hell-fire and brimstone from the pulpit, their successors in the very same sects now present the soothing message that we are all perfect, worthy, and destined for infinite happiness. The only sacrifice required is a regular check. Then Jesus will provide all.

As with traditional religions, New Age Christianity and other modern religions mainly provide a means for escaping reality. In the United States, where self-gratification has reached heights never dreamed of in ancient Rome, where self-esteem is more important than being able to read, and where self-help requires no more effort than popping in a cassette, the myth of quantum consciousness is just what the therapist ordered.

Quantum consciousness is a grossly misapplied version of ancient Hindu and Buddhist philosophy, which were based on the notion that only by the complete rejection of self can one find inner peace in this world of suffering and hopelessness. Far from rejecting the self, New Age holism puts it on a pedestal and proposes that it be worshiped. Capra and his colleagues say they are putting a modern face on ancient Eastern philosophy, but they are rather covering a noble edifice with graffiti. Where they see similarities between the new and the old mysticisms, I see only contrasts. Where they promote the new mythology as an antidote for self-absorption, I assert that they are manufacturing a drug that induces it. And while they blame rational science for the ills of the world, I hold rational science as a source of genuine hope for reducing their severity, if only we and our successors have the wisdom to use it properly.

Notes

1. Crick 1994, p. 3.
2. Reports of apparent superluminal effects are often reported in the literature, in astronomical observations and in certain quantum experiments. However, these are all currently explained within conventional theory and cannot be used as evidence for the nonlocality expected from hidden variables theories.
3. Gould 1989, 1994.
4. Penrose 1989.
5. Penrose 1994.
6. Penrose 1994, p. 12.
7. Stenger 1990.
8. Penrose 1994, p. 202.
9. Searle 1980, 1992.
10. Edelman 1992, pp. 228–52.
11. Eddington 1958, pp. 333–34.
12. Gödel 1931.
13. Penrose 1989, p. 110.
14. Turing 1937.
15. Church 1941.
16. For an excellent discussion, see Pagels 1989, pp. 294–96.
17. Penrose 1994, p. 51.
18. For a readable and thorough history of mathematics that goes from primitive counting to modern-day disputes, see Barrow 1992.
19. Penrose 1989, p. 112.
20. Davies 1992, p. 226.
21. Wilber 1984.
22. Wilber 1984, p. 5.
23. Armstrong 1993.

24. Stenger 1990.

25. *Behavioral and Brain Sciences* 1990, 13, pp. 643–705.

26. McCullough, Daryl, 1995. Electronically disseminated article from *Psyche.*

27. While it is possible to write down mathematically deterministic equations that do not have computationally tractable solutions, these have no value in science unless they produce useful numerical results. As usual, I am speaking here in pragmatic, scientific terms that may not satisfy the more purist mathematician or philosopher.

28. Siegelmann 1995.

29. The possibilities of quantum computers is discussed in Deutsch 1985 and Feynman 1986b.

30. Zadeh 1965, Kosko 1994, Aminzadeh 1994.

31. Kosko 1994.

32. Popper and Eccles 1977, Eccles 1986, 1990.

33. Hameroff 1994, 1996.

34. Stapp 1993, p. 42.

35. Herbert 1993, p. 253.

36. Herbert 1993, p. 254; Squires 1990, p. 222.

37. Hameroff 1994, 1996; Penrose 1994, pp. 357–77.

38. Stapp 1993, p. 25.

39. Herbert 1993, p. 5.

40. Squires 1990, p. 229.

41. Squires 1990, p. 201.

42. Albert 1992.

43. Penrose 1989, p. 446.

44. Dennett 1991.

45. Edelman 1992.

46. Crick 1994.

47. Carley 1994.

48. Drees 1990.

49. As quoted in Ferguson 1980, p. 145.

50. Zohar 1990, p. 18.

Bibliography

Aharonov, Y., and D. Bohm. 1959. *Phys. Rev.* 115, p. 485.

Aharonov, Yakir, J. Anandan, and Lev Vaidman. 1993a. *Phys. Rev.* A 47, p. 4616.

Aharonov, Yakir, and Lev Vaidman. 1993b. "The Schrödinger Wave Is Observable After All!" In *Quantum Control and Measurement*, H. Ezawa and Y. Murayama, eds., North Holland.

Albert, David Z. 1992. *Quantum Mechanics and Experience.* Cambridge, Mass.: Harvard University Press.

———. 1994. "Bohm's Alternative to Quantum Mechanics." *Scientific American*, May, pp. 58–67.

Alcock, James. 1990. *Science and Supernature: A Critical Appraisal of Parapsychology.* Amherst, N.Y.: Prometheus Books.

Alexander, Col. John. 1989. *New Realities,* March/April, p. 10.

Aminzadeh, Fred, and Mohammed Jamshidi, eds. 1994. *Soft Computing: Fuzzy Logic, Neural Networks, and Distributed Artificial Intelligence.* Englewood Cliffs, N.J: Prentice Hall.

Anandan, J. 1993. "Protective Measurement and Quantum Reality." *Foundations of Physics Letters* 6, p. 503.

———. 1995. "Reality and Geometry of States and Observables in Quantum Theory." To be published in the *Proceedings of the International Conference on Non-Accelerator Particle Physics*, Bangalore, Jan. 2–9, 1995, R. Cowsik, ed. Singapore: World Scientific.

Armstrong, Karen. 1993. *A History of God. From Abraham to the Present: The 4000-Year Quest for God.* London: Heinemann.

Aspect, Alain, Phillipe Grangier, and Roger Gerard. 1982. "Experimental Realization of the Einstein-Podolsky-Rosen *Gedankenexperiment*: A New Violation of Bell's Inequalities." *Physical Review Letters* 49, p. 91; "Experimental Tests of Bell's Inequalities Using Time-Varying Analyzers," p. 1804.

Atkatz, David. 1994. "Quantum Cosmology for Pedestrians." *Am. J. Phys.* 62(7), p. 619.

Baker, G. L., and J. P. Gollub. 1990. *Chaotic Dynamics: An Introduction.* Cambridge: Cambridge University Press.

Ballentine, L. E. 1970. "The Statistical Interpretation of Quantum Mechanics." *Rev. Mod. Phys.* 42(4), pp. 358–81.

———. *Quantum Mechanics.* 1986. Englewood Cliffs, N.J.: Prentice Hall.

———. 1988. "What Do We Learn about Quantum Mechanics from the Theory of Measurement?" *Int. J. Theor. Phys.* 27(2), pp. 221–28.

Barrow, John D. 1992. *Pi in the Sky: Counting, Thinking, and Being.* New York: Little Brown.

Barrow, John D., and Frank J. Tipler. 1986. *The Anthropic Cosmological Principle.* Oxford: Oxford University Press.

Bell, J. S. 1964. *Physics* 1, p. 195.

———. 1966. *Reviews of Modern Physics* 38, pp. 447–52.

———. 1971. "Introduction to the Hidden-Variables Question." In *Foundations of Quantum Mechanics,* B. d'Espagnet, ed. New York: Academic, pp. 171–81.

———. 1982. "On the Impossible Pilot Wave." *Foundations of Physics* 12, pp. 989–99.

———. 1987. *Speakable and Unspeakable in Quantum Mechanics.* Cambridge: Cambridge University Press.

Bem, D. J., and C. Honorton. 1994. "Does Psi Exist? Replicable Evidence for an Anomalous Process of Information Transfer." *Psychological Bulletin* 115(1), pp. 4–18.

Blackmore, Susan. 1994. "Psi in Psychology." *Skeptical Inquirer* 18, pp. 351–55.

Bohm, David. 1951. *Quantum Theory.* Englewood Cliffs, N.J.: Prentice-Hall.

———. 1952. "A Suggested Interpretation of Quantum Theory in Terms of 'Hidden Variables,' I and II." *Physical Review* 85, p. 166.

———. 1980. *Wholeness and the Implicate Order.* London: Routledge and Kegan Paul.

Bohm D., and B. J. Hiley. 1993. *The Undivided Universe: An Ontological Interpretation of Quantum Mechanics.* London: Routledge.

Bohr, N. 1928. *Atti del Congresso Internationale dei Fisici Como.* Bologna: Zanchelli, vol. 2, pp. 565–88.

———. 1935. *Phys. Rev.* 48, p. 696.

Born, M. 1926. *Z. Phys.* 38, p. 803.

———. 1949. *Natural Philosophy of Cause and Chance.* Oxford: Clarendon Press.

———, ed. 1971. *The Born-Einstein Letters.* London: Macmillan.

Briggs, John, and F. David Peat. 1989. *Turbulent Mirror: An Illustrated Guide to Chaos Theory and the Science of Wholeness.* New York: Harper & Row.

Butler, Kurt. 1992. *A Consumer's Guide to Alternative Medicine: A Close Look at Homeopathy, Acupuncture, Faith-Healing, and Other Unconventional Treatments.* Amherst, N.Y.: Prometheus Books.

Cabello, Adán. 1994. "A Simple Proof of the Kochen-Specker Theorem." *Eur. J. Phys.* 15, pp. 179–83.

Capra, Fritjof. 1975. *The Tao of Physics.* Boulder, Colo.: Shambhala.

———. 1982. *The Turning Point.* New York: Simon and Schuster.

Carley, Adam L. 1994. *Free Inquiry* 14(4), pp. 26–30.

Carnap, Rudolph. 1931. "Überwindung der Metaphysik durch logische Analyse der Sprache," *Erkenntnis,* vol. 2.

Carter, Brandon. 1974. "Large Number Coincidences and the Anthropic Principle in Cosmology." In M. S. Longair, ed., *Confrontation of Cosmological Theory with Astronomical Data.* Dordrecht: Reidel, pp. 291–98, reprinted in Leslie 1990.

———. 1983. *Phil. Trans. Roy. Soc. London* A 310, p. 347.

Chambers, R. G. 1960. *Phys. Rev. Lett.* 5, p. 3.

Chiao, Raymond Y., Paul G. Kwiat, and Aephraim M. Steinberg. 1993. "Faster than Light?" *Scientific American*, August, p. 52.

Chopra, Deepak. 1989. *Quantum Healing: Exploring the Frontiers of Mind/Body Medicine.* New York: Bantam.

———. 1993. *Ageless Body, Timeless Mind: The Quantum Alternative to Growing Old.* New York: Random House.

Church, A. 1941. *The Calculi of Lambda-Conversion.* Annals of Mathematical Studies, no. 6. Princeton: Princeton University Press.

Clark, Ronald W. 1971. *Einstein: The Life and Times.* New York: Avon Books.

Clauser, J. F., and M. A. Horne. 1974. *Physical Review* D10, p. 526.

Clauser, John F., and Abner Shimony. 1978. *Rep. Prog. Phys.* 41, pp. 1881–1927.

Coleman, S., J. B. Hartle, T. Piran, and S. Weinberg, eds. 1991. *Quantum Cosmology and Baby Universes,* vol. 7. Singapore: World Scientific.

Costa de Beauregard, O. 1953. *Comptes Rendus* 236, pp. 1632–34.

———. 1987. *Time. The Physical Magnitude.* Boston Series in the Philosophy of Science, vol. 99, Robert S. Cohen and Mark W. Wartofsky, eds. Boston: Reidel.

Craig, William Lane. 1990. " 'What Place, Then, for a Creator?': Hawking on God and Creation." *Brit. J. Phil. Sci.* 41, pp. 473–91.

———. 1992. "The Origin and Creation of the Universe: A Reply to Adolf Grünbaum." *Brit. J. Phil. Sci.* 43, pp. 233–40.

Cramer, John G. 1986. "The Transactional Interpretation of Quantum Mechanics." *Rev. Mod. Phys.* 58, p. 647.

———. 1988. *International Journal of Theoretical Physics* 27, p. 227.

Crick, Francis. 1994. *The Astonishing Hypothesis: The Scientific Search for the Soul.* New York: Charles Scribner's Sons.

Cushing, James T. 1993. "Bohm's Theory: Common Sense Dismissed." *Studies in the History and Philosophy of Science* 24, no. 5.

Cushing, J., and E. McMullin, eds. 1989. *Philosophical Consequences of the Quantum Theory: Reflections on Bell's Theorem.* Notre Dame, Ind.: University of Notre Dame Press.

Datta, A., D. Home, and A. Raychaudhuri. 1987. *Phys. Lett.* A 123, p. 4.

Davies, P. C. W. 1974. *The Physics of Time Asymmetry.* Los Angeles: University of California Press.

———. 1982. *The Accidental Universe.* Cambridge: Cambridge University Press.

———. 1992. *The Mind of God: The Scientific Basis for a Rational World.* New York: Simon and Schuster.

Dawkins, Richard. 1987. *The Blind Watchmaker: Why the Evidence of Evolution Reveals a Universe without Design.* New York: W. W. Norton.

De Baere, W. 1986. *Advances in Electronics and Electron Physics* 68, p. 245.

De Broglie, Louis. 1964. *The Current Interpretation of Wave Mechanics, A Critical Study.* New York: Elsevier.

Dennett, Daniel. 1991. *Consciousness Explained.* New York: Little Brown.

D'Espagnat, Bernard. 1989. *Reality and the Physicist: Knowledge, Duration and the Quantum World.* Cambridge: Cambridge University Press.

Deutsch, D. 1985. "Quantum Theory, the Church-Turing Principle and the Universal Quantum Computer." *Proc. Royal Soc. London* A400, pp. 97–117.

DeWitt, Bryce S., and Neill Graham, eds. *The Many-Worlds Interpretation of Quantum Mechanics,* Princeton: Princeton University Press.

Dicke, R. H. 1961. "Dirac's Cosmology and Mach's Principle." *Nature* 192, p. 440.

Dickson, Michael. 1993. *Stud. Hist. Phil. Sci.* 24, pp. 791–814.

Dirac, P. A. M. 1951. *Nature* 168, p. 906.

Dobyns, York H. 1992. *J. Scientific Exploration* 6, pp. 23–45.

———. 1993. *J. Scientific Exploration* 7, pp. 259–69.

Dowker, Fay, and Adrian Kent. 1994. "On the Consistent Histories Approach to Quantum Mechanics." Electronically disseminated paper DAMTP/94-48, NI 94006, gr-qc/9412067.

Drees, Willem B. 1990. *Beyond the Big Bang: Quantum Cosmologies and God.* La Salle, Ill.: Open Court.

Druckman, Daniel, and John A. Swets, eds. 1987. *Enhancing Human Performance: Issues, Theories and Techniques.* Washington, D.C.: National Academy Press.

Dunbar, D. N. F., R. E. Pixley, W. A. Wenzel, and W. Whaling. 1953. *Physical Review* 92, p. 649.

Dunne, Brenda J., and Robert G. Jahn. 1992. *J. Scientific Exploration* 6, pp. 311–32.

Eberhard, P. 1978. *Nuovo Cimento* 46B, p. 392.

Eberhard, Phillippe H., and Ronald R. Ross. 1989. *Found. Phys. Lett.* 2, p. 127.

Eccles, J. 1986. "Do Mental Events Cause Neural Events Analogously to the Probability Fields of Quantum Mechanics?" *Proc. Royal Soc. London* B227, pp. 411–28.

———. 1990. "A Unitary Hypothesis of Mind-Brain Interaction in the Cerebral Cortex." *Proc. Royal Soc. London* B240, pp. 433–51.

Eddington, Sir Arthur. 1958. *The Nature of the Physical World.* Ann Arbor, Mich.: University of Michigan Press.

Edelman, Gerald M. 1992. *Bright Air, Brilliant Fire (On the Matter of Mind).* London: Allen Lane, Penguin.

Einstein, A., B. Podolsky, and N. Rosen. 1935. "Can the Quantum Mechanical Description of Physical Reality Be Considered Complete?" *Physical Review* 47, p. 777.

Elvee, Richard Q., ed. 1982. *Mind in Nature.* San Francisco: Harper and Row.

Everett III, Hugh. 1957. *Rev. Mod. Phys.* 29, p. 454.

Ferguson, Marilyn. 1980. *The Aquarian Conspiracy: Personal and Social Transformation in the 1980s.* Los Angeles: Tarcher.

Feynman, R. P. 1948. "Spacetime Approach to Non-relativistic Quantum Mechanics." *Rev. Mod. Phys.* 20, pp. 367–87.

———. 1949a. "The Theory of Positrons." *Phys. Rev.* 76, pp. 749–59.

———. 1949b. "Spacetime Approach to Quantum Electrodynamics." *Phys. Rev.* 76, pp. 769–89.

———. 1965a. *The Character of Physical Law.* London: Cox & Wyman.

———. 1965b. "The Development of the Space-Time View of Quantum Electrodynamics." *Nobel Lectures Physics 1963–1970.* New York: Elsevier, 1992.

———. 1986a. *Surely You're Joking, Mr. Feynman.* London: Unwin. First published by W. W. Norton, 1985.

———. 1986b. "Quantum Mechanical Computers." *Foundations of Physics* 16, pp. 507–31.

Feynman, R. P., and A. R. Hibbs. 1965. *Quantum Mechanics and Path Integrals.* New York: McGraw-Hill.

Franck, Philipp. 1947. *Einstein—His Life and Times.* New York: Knopf.

French, A. P., and P. J. Kennedy, eds. 1985. *Niels Bohr: A Centenary Volume.* Cambridge, Mass.: Harvard University Press.

Gardner, Martin. 1967. *The Magic Numbers of Dr. Matrix.* New York: Simon and Schuster.

———. 1981. *Science: Good, Bad and Bogus.* New York: Avon.

Gell-Mann, Murray, and James P. Hartle. 1991. "Time Symmetry and Asymmetry in Quantum Mechanics and Quantum Cosmology." In the *Proceedings of the 1st International A. D. Sakarov Conference on Physics, Moscow, May*

27–31, 1991, and in the *Proceedings of the Nato Workshop on the Physical Origin of Time Asymmetry, Mazagon, Spain, September 30-October 4, 1991,* J. Haliwell, J. Perez-Mercader, and W. Zurek, eds. Cambridge: Cambridge University Press, 1992.

Gell-Mann, Murray. 1994. *The Quark and the Jaguar: Adventures in the Simple and the Complex.* New York: W. H. Freeman.

Ghirardi, G. C., A. Rimini, and T. Weber. 1986. *Phys. Rev.* D 34, p. 470.

Gisin, N. 1990. *Phys. Lett.* A 143, p. 1.

Gödel, Kurt. 1931. *Monatshefte für Mathematik und Physik* 38, pp.173–98.

Goldstein, Herbert. 1980. *Classical Mechanics,* 2nd ed. Addison-Wesley.

Goswami, Amit. 1993. *The Self-Aware Universe: How Consciousness Creates the Material World.* New York: G.P. Putnam's Sons.

Gould, Stephen J. 1989. *Wonderful Life: The Burgess Shale and the Nature of History.* New York: W. W. Norton.

———. 1994. "The Evolution of Life on the Earth." *Scientific American,* October, pp. 85–91.

Gleik, James. 1987. *Chaos: Making a New Science.* New York: Penguin.

Greenberger, D. M., M. A. Horne, A. Shimony, and A. Zeilinger. 1990. *Am. J. Phys.* 58, p. 1131.

Gribbon, John. 1980. *In Search of Schrödinger's Cat.* New York: Bantam; London: Corgi.

Gribbon, John, and Martin Rees. 1989. *Cosmic Coincidences: Dark Matter, Mankind, and Anthropic Cosmology.* New York: Bantam.

Griffiths, R. J. 1984. *J. Stat. Phys.* 36, p. 219.

———. 1986. "Correlations in Separated Quantum Systems: A Consistent History Analysis of the EPR Problem." *Am. J. Phys.* 55(1), pp. 11–17.

Grünbaum, Adolph. 1989. "The Pseudo-Problem of Creation in Physical Cosmology." *Philosophy of Science* 56, pp. 373–94.

Guth, A. 1981. "Inflationary Universe: A Possible Solution to the Horizon and Flatness Problems." *Phys. Rev.* D 23, pp. 347–56.

Guth, A., and P. Steinhardt. 1984. "The Inflationary Universe." *Scientific American* 215(5), pp. 116–28.

Haldane, J. B. S. 1934. *Philosophy of Science,* pp. 78–98.

Hall, Oliver W. "Further Implications of Anomalous Observations for Scientific Psychology." *American Psychologist* 41(1), pp. 1170–72.

Hameroff, S. R. 1994. "Quantum Coherence in Microtubules: A Neural Basis for Emergent Consciousness?" *Journal of Consciousness Studies* 1(1), pp. 91–118.

Hameroff, S. R., and R. Penrose. 1996. "Orchestrated Reduction of Quantum Coherence in Brain Microtubules: A Model for Consciousness." In *Toward a Science of Consciousness—Contributions from the 1994 Tucson Conference,* S. R. Hameroff, A. Kaszniak, and A. C. Scott, eds. Cambridge, Mass.: MIT Press.

Hansel, C. E. M. 1989. *The Search for Psychic Power: ESP and Parapsychology Revisited.* Amherst, N.Y.: Prometheus Books.

Harrison, Edward R. 1970. *Physics Today* 30, December.

Hartle, J. B. 1968. "Quantum Mechanics of Individual Systems." *Am. J. Phys.* 36(8), p. 704.

Haugeland, John. 1985. *Artificial Intelligence: The Very Idea.* Cambridge, Mass.: MIT Press.

Hawking, S. W. 1985. *Physical Review* D32, p. 2989.

———. 1988. *A Brief History of Time: From the Big Bang to Black Holes.* New York: Bantam.

Hegerfeldt, G. C. 1985. *Phys. Rev. Lett.* 54, p. 2395.

Heisenberg, W. 1927. *Zeit. Phys.* 43, p. 172.

———. 1958. *Physics and Philosophy.* New York: Harper and Row.

———. 1958a. *Daedalus* (Stockholm) 87, p. 95.

Herbert, N. 1982. *Found. Phys.* 12, p. 1171.

———. 1993. *Elemental Mind: Human Consciousness and the New Physics.* New York: Dutton.

Hiley, B. J., and F. David Peat, eds. 1987. *Quantum Implications: Essays in Honor of David Bohm.* London: Routledge & Kegan Paul.

Holland, John H. 1992. "Genetic Algorithms." *Scientific American* , July, p. 66.

Home, D., and M. A. B. Whitaker. 1992. "Ensemble Interpretations of Quantum Mechanics. A Modern Perspective." *Physics Reports* 210, pp. 213–317.

Hoyle, Fred. 1965. *Galaxies, Nuclei, and Quasars.* London: Heinemann.

Hume, D. 1757. *Dialogues concerning Natural Religion,* part 2, N. Kemp Smith, ed. 1977. Indiana: Bobbs Merrill.

Humphrey, Nicholas. 1992. *The History of Mind.* New York: Simon and Schuster.

Hyman, R. 1994. "Anomaly of Artifact?" *Psychological Bulletin* 115(1), pp. 19–124.

Jahn, Robert G., ed. 1981. *The Role of Consciousness in the Physical World.* Boulder, Colo.: Westview Press.

Jahn, Robert G., and Brenda J. Dunne. 1986. "On the Quantum Mechanics of Consciousness, with Application to Anomalous Phenomena." *Foun. Phys.* 16, pp. 721–72.

———. 1987. *Margins of Reality: The Role of Consciousness in the Physical World.* New York: Harcourt Brace Jovanovich.

Jahn, R., Y. Dobyns, and B. Dunne. 1991. *J. Sci. Expl.* 5, p. 205.

———. 1992. *J. Sci. Expl.* 6, p. 311.

Jaki, S. L. 1978. *The Road of Science and the Ways to God.* Chicago: University of Chicago Press.

Jammer, Max. 1974. *The Philosophy of Quantum Mechanics: The Interpretations of Quantum Mechanics in Historical Perspective.* New York, John Wiley.

Jarrett, J. 1984. *Noûs* 18, pp. 569–89.

Jastrow, Robert. 1980. *God and the Astronomers.* New York: Warner Books.
Jefferys, William H. 1992a. *J. Scientific Exploration* 6, pp. 47–57.
Jefferys, William H., and James O. Berger. 1992b. "Ockham's Razor and Bayesian Analysis." *American Scientist* 80, pp. 64–72.
Joos, E., and H. D. Zeh. 1985. *Zeit. Phys.* B59, p. 223.
Johnson, George. 1986. *The Machinery of the Mind.* New York: Times Books.
Johnson, Paul. 1987. *A History of the Jews.* New York: Harper and Row, p. 290.
Josephson, Brian D. 1991. "Biological Utilisation of Quantum Nonlocality." *Foun. Phys.* 21, pp. 197–207.

Kafatos, Menas, ed. 1989. *Bell's Theorem, Quantum Theory and Conceptions of the Universe.* Dordrecht: Kluwer Academic.
Kafatos, Menas, and Robert Nadeau. 1990. *The Conscious Universe: Part and Whole in Modern Physical Theory.* New York: Springer-Verlag.
Kauffman, Louis H., and H. Pierre Noyes. 1994. "Discrete Physics and the Derivation of Electromagnetism from the Formalism of Quantum Mechanics." SLAC-PUB–6697. Submitted to *Proceedings of the Royal Society A.*
Kauffman, Stuart A. 1991. "Antichaos and Adaptation." *Scientific American,* August, p. 78.
———. 1993. *The Origins of Order: Self-Organization and Selection in Evolution.* Oxford: Oxford University Press.
Kazanas, D. 1980. *Astrophysical J.* 241, pp. L59–63.
Kochen, S., and E. P. Specker. 1967. *J. Math. Mech.* 17, pp. 59–87.
Kolb, Edward W., and Michael S. Turner. 1990. *The Early Universe.* Reading, Mass.: Addison-Wesley.
Kosko, Bart. 1994. *Fuzzy Thinking: The New Science of Fuzzy Logic.* London: Harper Collins.
Krippner, Stanley. 1994. "Humanistic Psychology and Chaos Theory: The Third Revolution and the Third Force." *Journal of Humanistic Psychology* 34(3), pp. 48–61.
Kuhn, Thomas. 1970. *The Structure of Scientific Revolutions.* Chicago: University of Chicago Press.

Lamont, Corliss. 1990. *The Illusion of Immortality*, 5th ed. New York: Continuum.
Langton, Christopher G., ed. 1989. *Artificial Life. The Proceedings of an Interdisciplinary Workshop on the Synthesis and Simulation of Living Systems held September 1987 in Los Alamos, New Mexico,* vol. 6. Redwood City, Calif.: Addison-Wesley.
Langton, Christopher G., Charles Taylor, J. Doyne Farmer, and Steen Rasmussen, eds. 1992. *Artificial Life II: Proceedings of the Workshop on Artificial Life held February 1990 in Santa Fe, New Mexico.* Redwood City, Calif.: Addison-Wesley.
Lanza, Robert. 1992. "The Wise Science." *The Humanist,* November/December.

Laszlo, Ervin. 1987. *Evolution: The Grand Synthesis.* Boston: Shambhala.

Laudan, Larry. 1990. *Science and Relativism: Some Key Controversies in the Philosophy of Science.* Chicago: University of Chicago Press.

Lederman, Leon, and David Schramm. 1989. *From Quarks to the Cosmos.* New York: W. H. Freeman.

Lederman, Leon, with Dick Teresi. 1993. *The God Particle: If the Universe Is the Answer, What Is the Question?* New York: Houghton Mifflin.

Lerner, Eric. J. 1991. *The Big Bang Never Happened.* New York: Times Books.

Leslie, J. 1982. "Anthropic Principle, World Ensemble, Design," *American Philosophical Quarterly* 19, pp. 141–51.

Levy, Steven. 1992. *Artificial Life: The Quest for a New Creation.* New York: Pantheon Books.

Linde, A. D. 1982. *Physics Letters* 108B, p. 389.

———. 1987. "Particle Physics and Inflationary Cosmology." *Physics Today* 40, pp. 61–68.

———. 1990. *Particle Physics and Inflationary Cosmology.* New York: Academic Press.

———. 1994. "The Self-Reproducing Inflationary Universe." *Scientific American,* November, pp. 48–55.

Lodge, Sir Oliver. 1914. *Continuity. The Presidential Address to the British Association for the Advancement of Science, 1913.* New York: Putnam.

———. 1920. *Beyond Physics.* London: Alana and Unwin.

———. 1929. *Why I Believe in Personal Immortality.* New York: Doubleday.

Mandelung, E. 1927. "Quantentheorie in Hydrodynamischer Form," *Zeit. Phys.* 43, pp. 354–57.

Marlow, A. R., ed. 1978. *The Mathematical Foundations of Quantum Mechanics.* New York: Academic Press.

Marshall, T. W., E. Santos, and F. Selleri. 1983. *Physics Letters* 98A, p. 5.

McMullin, Ernan. 1993. "Indifference Principle and Anthropic Principle in Cosmology." *Stud. Hist. Phil. Sci.* 24(3), pp. 359–89.

Mermin, N. David. 1985. "Is the Moon There When Nobody Looks? Reality and the Quantum Theory." *Physics Today* 38, p. 38.

———. 1990. *Am. J. Phys.* 58, 731.

Minsky, Marvin. 1985. *Society of Mind.* New York: Simon and Schuster.

Moreland, J. P., and Kai Nielson. 1993. *Does God Exist? The Debate between Theists and Atheists.* Amherst, N.Y.: Prometheus Books.

Noyes, H. Pierre. 1994. "Decoherence, Determinism and Chaos Revisited." SLAC-PUB–6718. Submitted to *Law and Prediction in (Natural) Science in the Light of New Knowledge from Chaos Research.* Salzberg, Austria, July 7–10.

Omnès, R. J. 1988. *J. Stat. Phys* 53, pp. 893, 933, 957.
———. 1992. *Rev. Mod. Phys* 64, p. 339.
———. 1994. *The Interpretation of Quantum Mechanics.* Princeton: Princeton University Press.
O'Raifeartaigh, L., N. Straumann, and A. Wipf. 1991. *Comments Nucl. Phys.* 20, p. 15.
Oteri, Laura, ed. 1975. *Quantum Physics and Parapsychology, Proceedings of an International Conference Held in Geneva.* New York: Parapsychological Foundation.

Page, D. 1985. *Physical Review* D32, p. 2496.
Pagels, Heinz R. 1982. *The Cosmic Code: Quantum Physics as the Language of Nature.* New York: Bantam Books.
———. 1985. *Physics Today,* January, p. 97.
———. 1989. *Dreams of Reason: The Computer and the Rise of the Sciences of Complexity.* New York: Bantam.
Pais, A. 1979. *Reviews of Modern Physics* 51, p. 863.
———. 1982. *Subtle Is the Lord: The Science and the Life of Albert Einstein.* Oxford: Oxford University Press.
Palmer, John. 1986. "Have We Established Psi?" *American Society for Psychical Research* 81, pp. 111–23.
Palmer, John A., Charles Honorton, and Jessica Utts. 1989. *Journal of the American Society for Psychical Research* 83, p. 31.
Park, James L. 1968. "Nature of Quantum States." *Am. J. Phys.* 36, p. 21.
Penrose, Roger. 1979. In *General Relativity: An Einstein Century Survey,* S. Hawking and W. Israel, eds. Cambridge: Cambridge University Press.
———. 1989. *The Emperor's New Mind: Concerning Computers, Minds, and the Laws of Physics.* Oxford: Oxford University Press.
———. 1994. *Shadows of the Mind: A Search for the Missing Science of Consciousness.* Oxford: Oxford University Press.
Pius XII. 1972. "Modern Science and the Existence of God." *The Catholic Mind* 49, March, pp. 182–92.
Polanyi, M. 1961. *The Logic of Personal Knowledge.* Glencoe, Ill.: Free Press.
Polchinski, Joseph. 1991. *Phys. Rev.* 66, p. 397.
Popper, Karl R. 1982. *Quantum Theory and the Schism in Physics.* Totowa, N.J.: Rowan and Littlefield.
———. 1987. "Science: Conjectures and Refutations," in *Scientific Knowledge,* Janet A. Kournay, ed. Belmont, Calif.: Wadsworth.
Popper, K. R., and Eccles, J. C. 1977. *The Self and Its Brain.* Berlin: Springer.
Poundstone, William. 1985. *The Recursive Universe.* New York: Morrow.
Press, W. H., and A. P. Lightman. 1983. *Phil. Trans. R. Soc. Lond.* A 310, p. 323.
Prigogine, Ilya, and Isabella Stengers. 1984. *Order Out of Chaos.* New York: Bantam.

Przibam, K., ed. 1967. *Letters on Wave Mechanics.* New York: Philosophical Library.

Puharich, Andrija, ed. 1979. *The Iceland Papers.* Amherst, Wis.: Essentia Research Associates.

Rae, Alstair. 1986. *Quantum Physics: Illusion or Reality?* Cambridge: Cambridge University Press.

Randi, James. 1975. *The Magic of Uri Geller.* New York: Ballentine Books.

———. 1982. *The Truth About Uri Geller.* Amherst, N.Y.: Prometheus Books.

Redhead, Michael. 1987. *Incompleteness, Nonlocality, and Realism.* Oxford: Clarendon Press.

Rhine, J. B., and J. G. Pratt. 1954. "A Review of the Pearce-Pratt Distance Series of ESP Tests." *Journal of Parapsychology* 18, pp. 165–77.

Riordan, Michael, and David Schramm. 1991. *Shadows of Creation: The Dark Matter and Structure of the Universe.* New York: W. H. Freeman.

Russell, Robert John. 1985. "The Physics of David Bohm and Its Relevance to Philosophy and Theology." *Zygon: Journal of Religion and Science* 20, pp. 135–58.

Sarfatti, J. 1977. "The Case for Superluminal Information Transfer." *MIT Technological Review* 79, no. 5, p. 3.

Schilpp, Paul Arthur, ed. 1949. *Albert Einstein: Philosopher-Scientist.* Evanston, Ill.: The Library of Living Philosophers.

Schmidt, Helmut. 1969. "Quantum Processes Predicted?" *New Scientist,* 16 October, pp. 114–15.

———. 1992. *J. Scientific Exploration* 7, pp. 125–32.

———. 1993. *J. Parapsychol.* 57, p. 351.

Schnabel, Jim. 1994. *Round in Circles: Physicists, Poltergeists, Pranksters and the Secret History of the Cropwatchers.* London: Hamish Hamilton.

Searle, J. R. 1980. "Minds, Brains, and Programs." In *The Behavioral and Brain Sciences,* vol. 3. Cambridge: Cambridge University Press.

———. 1992. *The Rediscovery of the Mind.* Cambridge, Mass.: MIT Press.

Selleri, Franco, ed. 1988. *Quantum Mechanics Versus Local Realism: The Einstein-Podolsky-Rosen Paradox.* New York: Plenum Press.

Sharpe, Kevin J. 1993. "Holomovement Metaphysics and Theology." *Zygon: Journal of Religion and Science* 28, pp. 47–60.

Shiu, F. H. 1982. *The Physical Universe.* Mill Valley, Calif.: University Science Books.

Siegelmann, Hava T. 1995. "Computation Beyond the Turing Limit." *Science* 268, pp. 545–48.

Smolin, Lee. 1992. "Did the Universe Evolve?" *Classical and Quantum Gravity* 9, pp. 173–91.

Sommers, Paul. 1994. "The Role of the Future on Quantum Theory." High Energy Astrophysics Institute, University of Utah, gr-qc.xxx.lanl.gov electronic bulletin board paper number 9404022 (unpublished).

Squires, Euan. 1990. *Conscious Mind in the Physical World.* New York: Adam Hilger.

Stapp, H. P. 1971. *Physical Review* D3, p. 1303.

———. 1972. "The Copenhagen Interpretation." *Am. J. Phys.* 40, p. 1098.

———. 1985. *Am. J. Phys.* 54, p. 306.

———. 1990. *British Journal for the Philosophy of Science* 41, pp. 59–72.

———. 1993. *Mind, Matter, and Quantum Mechanics.* New York: Springer-Verlag.

———. 1994. *Phys. Rev.* A 50, p. 18.

Stenger, Victor J. 1988. *Not By Design: The Origin of the Universe.* Amherst, N.Y.: Prometheus.

———. 1988/89. "God and Stephen Hawking?" *Free Inquiry* 9(1), pp. 59–61.

———. 1990. *Physics and Psychics: The Search for a World Beyond the Senses.* Amherst, N.Y.: Prometheus.

———. 1990a. "The Universe: The Ultimate Free Lunch." *Eur. J. Phys.* 11, p. 236.

———. 1990b. "The Spooks of Quantum Mechanics." *Skeptical Inquirer* 15, Fall, pp. 51–61.

———. 1992. "Is the Big Bang a Bust?" *Skeptical Inquirer* 16, Summer, pp. 412–15.

———. 1992a. "Can One Worship a Hypothesis?" *Free Inquiry* 12(3), pp. 55–57.

———. 1992/93. "The Face of Chaos." *Free Inquiry* 13(1), pp. 13–14.

———. 1993. "The Myth of Quantum Consciousness." *The Humanist,* May/June 1993, pp. 13–15.

———. 1995. Review of *The Physics of Immortality* by Frank. J. Tipler. *Free Inquiry*, 15, Spring, pp. 54–55.

Stückelberg, E. C. G. 1941. "Remarques à propos de la création de paires de particules en théorie de la relativité." *Helv. Phys. Acta* 14, pp. 588–94.

———. 1942. "La méchanique du point matériel en théorie de la relativité." *Helv. Phys. Acta* 15, pp. 23–37.

Sture, Allén, ed. 1986. *Proceedings of the Nobel Symposium 65: Possible Worlds in Arts and Sciences.* Stockholm, August 11–15.

Talbot, Michael. 1991. *The Holographic Universe*. New York: Harper Collins.

Teilhard de Cardin, Pierre. 1975. *The Phenomenon of Man.* New York: Harper & Row.

Temple, G. 1934. *An Introduction to Quantum Theory*. New York: D. Van Nostrand.

Tipler, Frank J. 1994. *The Physics of Immortality: Modern Cosmology and the Resurrection of the Dead.* New York: Doubleday.

Trickett, David G. 1982. Review of *Wholeness and the Implicate Order* by David Bohm. *Process Studies* 12, pp. 50–54.

Trusted, Jennifer. 1991. *Physics and Metaphysics: Theories of Space and Time.* London: Routledge.

Tryon, E. P. 1973. "Is the Universe a Quantum Fluctuation?" *Nature* 246, pp. 396–97.

Tufillaro, Nicholas B., Tyler Abbott, and Jeremiah Reilly. 1992. *An Experimental Approach to Nonlinear Dynamics and Chaos.* Redwood City, Calif.: Addison-Wesley.

Turing, A. M. 1937. "On Computable Numbers, with an Application to the *Entscheidungsproblem.*" *Proc. Lond. Math. Soc.* (ser. 2), 42, pp. 230–65; corrected in 43, pp. 544–46.

Unruh, W. G. 1994. "Is the Wavefunction Real?" In V. de Sabbata and H. Tso-Hsiu, eds., *Cosmology and Particle Physics.* The Netherlands: Kluwer Academic Publishers, pp. 257–69.

Updike, John. 1986. *Roger's Version.* New York: Fawcett Crest.

Van der Merwe, Alwyn, and Franco Selleri, eds. 1988. *Microphysical Reality and Quantum Formalism,* vol. 1. Kluwer Academic.

Von Neumann, John. 1955. *The Mathematical Foundations of Quantum Mechanics.* Princeton: Princeton University Press. The original edition in German was published in 1932 by J. Springer, Berlin.

Von Neumann, John, and Oskar Morgenstern. 1953. *Theory of Games and Economic Behavior.* Princeton: Princeton University Press.

Waldrop, M. Mitchell. 1992. *Complexity: The Emerging Science at the Edge of Order and Chaos.* New York: Simon and Schuster.

Weinberg, S. 1989. *Ann. Phys.* (N.Y.) 194, p. 336.

———. 1992. *Dreams of a Final Theory: The Search for the Fundamental Laws of Nature.* New York: Pantheon Books.

Weyl, H. 1919. *Ann. Physik* 59, p. 101.

Wheeler, J. A., H. Wojciech, and W. H. Zurek, eds. 1983. *Quantum Theory and Measurement,* Princeton, N.J.: Princeton University Press.

Wilber, Ken, ed. 1984. *Quantum Questions: Mystical Writings of the World's Great Physicists.* Boulder, Colo.: Shambhala.

Will, Clifford M. 1986. *Was Einstein Right? Putting General Relativity to the Test.* New York: Basic Books.

Witt-Miller, Harriet. 1994. "Searching for New Centers: The Legacy of Copernicus." *Griffith Observer,* April, p. 2.

Wolf, Fred Alan. 1984. *Star Wave Ψ*Ψ: Mind, Consciousness, and Quantum Physics.* New York: Macmillan.

———. 1991. *Parallel Universes: The Search for Other Worlds.* London: Paladin.

Zadeh, L. A. 1965. "Fuzzy Sets." *Information and Control* 8, pp. 338–53.

Zeh, H. 1971. *Found. Phys.* 1, p. 69.

Zohar, Danah. 1990. *The Quantum Self. Human Nature and Consciousness Defined by the New Physics.* New York: Morrow.

Zukav, Gary. 1979. *The Dancing Wu Li Masters: An Overview of the New Physics.* New York: Morrow.

Zurek, W. H. 1981. *Phys. Rev.* D24, p. 1516.

———. 1982. *Phys. Rev.* D26, p. 1862.

———. 1991. "Decoherence and the Transition from Quantum to Classical." *Physics Today* 36, pp. 36–44.

———, ed. 1990. *Complexity, Entropy, and the Physics of Information: SFI Studies in the Sciences of Complexity,* vol. 7. New York: Addison-Wesley.

Index

Abelard, 278
Abrams, Nancy, 214
Action at a distance, 38, 39, 62–63, 69, 134, 140, 148, 149, 171, 183
Adaptive behavior, 272
Advanced waves, 145, 146, 150, 151, 155
Aether, 33, 34, 36, 37, 40, 42, 45, 46, 109, 145, 150, 165, 189, 209, 254
Aether waves, 45, 46
African religions, 242
Afrocentric, 24
Ageless Body, Timeless Mind, 25
Aharonov, Yakir, 167, 168, 169, 170, 190
Aharonov-Bohm effect, 167, 168, 169, 170
Albert, David, 96, 99, 125, 287
Algorithmic, 263, 275, 281, 285
Alternate histories, 176, 180, 181, 182, 198, 201, 202, 271, 272, 283
Alternative medicine, 12
American Association for the Advancement of Science, 97
Analog shift map, 282
Anandan, Jeeva, 16, 170
Anthropic coincidences, 11, 19, 230, 231, 233, 235, 239
Anthropic principle, 233, 237, 240
Antielectrons, 56, 145
Antichaos, 259–60, 264
Antimatter, 56
Antineutrons, 56
Antiparticles, 56, 146, 149, 151, 206
Antiprotons, 56
Aquinas, St. Thomas, 278
Argument from design, 218, 223, 230
Aristotle, 7, 34, 45, 129
Armstrong, Karen, 278
Arrow of time, 143, 144, 154, 187–89, 206, 211, 265
Artificial life, 259, 262, 264, 280, 285
Artificial mind, 262
Artificial intelligence, 262, 263, 274, 279
Aspect, Alain, 118, 119, 120
Aspect experiment, 119, 120
Atomic theory, 104, 110, 201, 211, 236

Attractor, 256
Augustine, St., 278
Axiomatic foundation of quantum mechanics, 106

Ballentine, Leslie, 15, 86, 88
Barrow, John, 233, 240
Basin of attraction, 257
Bayesian statistics, 31
BBC, 127
Beatles, 268
Behavioral and Brain Sciences, 279
Being-in-itself, 40
Bell, John, 39, 40, 85, 100, 109, 110, 111, 112, 113, 114, 116, 117, 118, 119, 120, 121, 122, 123, 124, 125, 127, 135, 140, 152, 165, 182, 195, 270
Bell's inequality, 112, 114, 116, 117, 118, 119, 120, 121, 122, 123, 127, 135, 140, 270
Bell's theorem, 39, 40, 110, 111, 112, 114, 117, 120, 121, 152, 165, 270
Berger, Stuart, 25
Berkeley, Bishop, 17, 166
Berkeley, California, 26, 83
Bible, 214, 279
Bifurcation, 256, 257, 258, 259, 264
Big bang, 11, 19, 187, 214, 217, 219, 220, 226, 228, 229, 230, 231, 232, 233, 240
Big crunch, 240
Binding energy, 244, 245, 246
Birkbeck College, 83
Birmingham University, 110
Black bodies, 48
Black hole, 216, 227, 228, 229, 237, 239
Bohm, David, 27, 28, 39, 67, 83, 84, 85, 95, 106, 107, 108, 109, 110, 111, 112, 120, 121, 123, 124, 127, 128, 129, 130, 135, 136, 141, 145, 151, 152, 157, 159, 166, 167, 168, 169, 170, 173, 174, 182, 195, 196, 197, 199, 228, 277
Bohm-EPR experiment, 83, 84, 85, 92, 112, 120, 151, 152, 174, 182, 197
Bohr, Niels, 21, 37, 44, 48, 51, 55, 58, 59, 60, 63, 66, 67, 68, 69, 70, 71, 72, 73, 74, 77, 78, 79, 81, 93, 94, 95, 101, 103, 105, 106, 180, 192, 206, 211, 244, 245, 246
Bohr atom, 51, 211
Boltzmann, Ludwig, 104, 143, 265
Boolean networks, 260, 261
Bootstrap theory, 26
Born, Max, 54, 56, 57, 60, 67, 68, 86, 100, 109, 140
Brahmin, 23, 129
Brain, 35, 40, 45, 49, 50, 58, 64, 96, 97, 127, 129, 133, 150, 253, 262, 263, 273, 274, 275, 279, 280, 282, 283, 284, 285, 286, 288, 289
BrainWorks, 261
Broken symmetry, 265
Brownian motion, 103
Buddhism, 242
Butterfly effect, 241, 257

Calculus of variations, 277
Cambridge University, 251
Canonically conjugate variables, 60, 61
Capra, Fritjov, 23, 24, 25, 26, 27, 128, 290, 291, 292
Carbon, 231, 232, 233
Carley, Adam, 288
Carnap, Rudolph, 7
Catholic church, 36
Catholicism, 278
Causal anomalies, 160
Causality, 142, 152, 228
Causal precedence, 142, 143, 144, 145, 146, 150, 156, 184, 196, 228
Cause and effect, 141, 142, 144, 146, 150, 184, 196, 206, 228, 251, 271
Celestial mechanics, 153

Cellular automata, 259
CERN, 110
Chaos, 11, 19, 159, 217, 223, 224, 225, 227, 229, 230, 241, 250, 251, 252, 254, 255, 256, 257, 258, 259, 260, 262, 263, 264, 265, 266, 272, 280, 282, 286, 288
Chaplin, Charlie, 268
Chopra, Deepak, 25
Christianity, 18, 19, 36, 129, 278, 279, 291
Church, Alonzo, 276, 278
Ch'i, 34
Circle map, 258
Circular polarization, 89, 90, 91, 92, 117
Clarke, Arthur C., 28
Classical electrodynamics, 167, 168, 169
Classical mechanics, 54, 58, 61, 82, 101, 107, 122, 154, 169, 177, 189, 200, 201, 202, 204, 207, 252, 259, 265, 266, 289
Classical wave theory, 68, 71, 156, 200
Clauser, John F., 117
Clausius, Rudolph, 265
Clinton, Bill, 241
Closed system, 143, 154, 218, 228, 265
Cloud chamber, 53
COBE (Cosmic Background Explorer) satellite, 217
Coherence, 186, 206
Coherence length, 187
Colatitude, 216
Cold fusion, 30, 269
Cold War, 110
Combinatorial explosion, 262
Communism, 268
Commutative law, 61
Complementarity, 58, 59, 60, 64, 70, 78, 125
Completeness, 12, 84, 85, 112, 120, 121, 197
Complex numbers, 56, 171, 178, 209
Complexity, 11, 19, 231, 232, 233, 239, 251, 252, 259, 263, 265, 266, 272
Compton, Arthur H, 67
Computable, 274, 275, 282, 283, 285
Computer science, 11, 251
Computer simulation, 140, 241, 259, 260, 261, 272
Comte, Auguste, 63
Confirmation waves, 155, 156, 157, 158, 206
Connectivity, 37, 260
Conscious Universe, The, 40
Consciousness, 10, 11, 16, 19, 22, 23, 24, 25, 33, 62, 63, 64, 72, 95, 96, 97, 121, 124, 129, 130, 135, 160, 171, 172, 173, 174, 175, 176, 184, 193, 195, 197, 213, 226, 269, 270, 271, 273, 275, 276, 283, 286, 287, 288, 291, 292
Conservation of angular momentum, 80
Conservation of energy, 218
Conservation of matter, 148
Conservation of momentum, 148
Consistent histories, 176, 179, 180, 181, 183, 184, 185, 186, 199
Contextuality, 111, 135, 154, 183, 204, 205
Continuity, 34, 37, 48, 107, 201, 209, 210, 211, 289
Continuum, 208, 209, 210, 211, 289
Copenhagen interpretation, 21, 40, 60, 61, 63, 64, 94, 97, 103, 104, 121, 124, 130, 157, 172, 190, 192, 193, 201, 287
Copernican principle, 234
Copernicus, 17, 214, 233
Correlation function, 114, 115, 116, 118
Cosmic Background Explorer. *See* COBE
Cosmic consciousness, 24, 25, 213, 291
Cosmic microwave background, 187, 217, 239, 253

Cosmic mind, 19, 20, 25, 31, 32, 40, 41, 165, 175, 240, 291
Cosmological constant, 221, 222
Cosmology, 11, 13, 19, 154, 175, 197, 213, 214, 216, 217, 219, 221, 224, 228, 230, 231, 233, 235, 236, 239, 250, 279, 290
Cosmovision, 250, 251, 252
Cosmythology, 213
Costa de Beauregard, Olivier, 150, 152
Counterfactual definiteness, 77, 102, 111
Craig, William Lane, 232, 234
Cramer, John, 15, 60, 63, 155, 156, 157, 158, 206
Creatio ex nihilo, 214
Creation, 54, 214, 217, 218, 223, 224, 230
Crick, Francis, 268, 288
Crookes, Sir William, 35
Crop circles, 251
Curvature of spacetime, 9, 221, 222
Curvature energy, 226, 229

Dalton, John, 50
Dancing Wu Li Masters, The, 24, 135
Dark matter, 219, 220
Darwin, Charles, 17, 50, 214, 215, 238, 263
Darwinian selection, 272, 285
Davies, Paul, 216, 277
De Baere, W., 120
De Broglie, Louis, 39, 49, 51, 52, 53, 67, 69, 73, 93, 104, 105, 106, 107, 108, 109, 148, 159, 197, 278
De Broglie wavelength, 49, 148, 149
Decoherence, 184, 186, 187, 188, 189, 190, 199, 200, 202,, 203, 206, 210, 265, 272, 275, 283, 284, 287
Decoherent histories, 176, 185, 187, 189, 193, 198, 199, 203, 207
Deep Thought, 275
Deism, 20
Dennett, Daniel, 288
Descartes, René, 24
Detection efficiency, 119
Determinism, 26, 120, 121, 123, 152, 158, 182, 197, 198, 199, 200, 207, 286
Deterministic chaos, 252, 255, 256
Deuterium, 231, 232
DeWitt, Bryce, 175
Dicke, R. H., 231
Dickson, Michael, 121
Differential equations, 54–55, 258, 289, 290
Differential operators, 61
Diffraction, 48, 49, 52, 68, 69, 70, 149, 200
Dirac, Paul, 56, 67, 145, 156
Dirac equation, 156
Discreteness, 201, 208, 209, 210, 289
Dispersion, 52, 53
Dissipation, 258
DNA, 241
Dobyns, York, 16, 31
Double slit, 70, 95, 108, 155, 157, 158, 167, 179, 180, 181, 183, 184, 185, 202, 204, 205, 206
Drees, Willem, 217, 290
Driving parameters, 259, 260
Dualism, 213
Duke University, 30

Early universe, 223, 227, 228, 229, 232, 233, 236, 239
Eastern Orthodox Christians, 278
Eastern mysticism, 20, 23, 24, 128
Eastern philosophy, 26, 282, 292
Eccles, Sir John, 283
Eddington, Sir Arthur, 276, 278
Edelman Gerald M., 276, 288
Edge of chaos, 260, 262, 286
Ehrenfest, Paul, 67
Einstein, Albert, 8, 9, 10, 11, 23, 27, 36, 37, 38, 39, 46, 47, 48, 49, 50,

56, 57, 58, 59, 62, 66, 67, 68, 70, 71, 72, 73, 74, 76, 77, 78, 79, 83, 84, 93, 95, 101, 102, 105, 106, 109, 118, 130, 131, 132, 134, 141, 142, 144, 150, 165, 169, 171, 174, 194, 195, 196, 198, 199, 209, 215, 218, 219, 221, 222, 226, 268, 278
Einstein causality, 142
Electromagnetic waves, 35, 37, 45, 56, 68, 89, 139, 145, 149
Electromagnetism, 37, 45, 46, 51, 201, 231, 289
Electroweak force, 231
Ellis, George, 242
Emergence, 11
Emergent properties, 142, 206, 252, 261
Emulation, 241
Enfolded order, 128
Enlightenment, 20
Ensemble, 75, 86, 87, 88, 89, 91, 94, 101, 115, 117, 148, 165, 166, 167, 172, 177, 179, 200, 224
Ensemble interpretation, 86, 91
Entanglement, 115
Entropy, 26, 143, 144, 154, 188, 206, 224, 225, 226, 227, 228, 229, 241, 265
Entropy of a black hole, 237
Entropy of the universe, 227, 228, 229
Entscheidungsproblem ("decision problem"), 276
Epicurus, 63
EPR experiment, 78, 83, 88, 91, 93, 110, 112, 113, 120, 122, 135, 136, 137, 139, 140, 151, 152, 157, 173, 174, 182, 183, 197, 206, 284
EPR paradox, 74, 77, 80, 95, 111, 120, 122, 150, 158, 169
Equation of continuity, 107
Equation of motion, 54, 107, 175, 221, 222
Extrasensory perception (ESP), 24, 25, 27, 29, 30, 32, 33, 107, 125, 252, 289
Euclid, 209, 264, 281
Everett, Hugh, 172, 173, 175, 176, 177, 180, 181, 184, 185, 198, 201, 202, 287
Evolution, 18, 35, 57, 96, 124, 153, 184, 198, 214, 215, 233, 237, 239, 252, 260, 272, 287, 291
Expectation value, 115
Expert systems, 282
Explicate order, 128
Extraterrestrial civilizations, 272

Face of God, 217
False vacuum, 236
Faraday, Michael, 50
Faster-than-light, 159
Feigenbaum, Mitchell, 256, 258
Ferguson, Marilyn, 24, 290, 291
Fermat's principle, 177
Ferromagnetism, 233
Feynman, Richard, 70, 71, 95, 145, 146, 148, 151, 156, 176, 177, 178, 179, 180, 181, 201, 206, 207, 271
Feynman diagrams, 145, 146, 148, 180
Feynman path integrals, 177, 178, 181, 201, 207
Firmament, 225, 226
First law of thermodynamics, 218, 219, 222, 223, 224
Fitness landscape, 286
Fitzgerald, George, 46
Fractal, 257
Frame of reference, 8, 65, 132, 142, 151, 197
Francis of Assisi, 278
Franck, Phillipp, 67
Free lunch, 218
Functionalism, 274
Fuzzy logic, 282

Gaia, 19

Galilean relativity, 46
Galileo, 8, 46, 214, 269
Gamma rays, 187
Gandhi, 268
Ganzfeld experiment, 29, 31
Gardner, Martin, 28
Gedankenexperiment, 68, 69
Gell-Mann, Murray, 154, 176, 180, 183
Geller, Uri, 28, 135
General relativity, 9, 10, 177, 219, 220, 222, 223, 241
Genetic algorithms, 259
Genome, 259
Geodesic equation, 177
Geometrical acceptance, 119
Gerlach, W., 80
God, 11, 18, 19, 20, 57, 66, 67, 101, 129, 200, 213, 277, 278, 291
Gödel's theorem, 274
Goldstein, Rabbi Herbert, 215
Goswami, Amit, 32
Gould, Stephen J., 272
Grand Unified Theory (GUT), 25
Granularity, 189, 201, 203, 204, 207, 208, 209, 210, 253
Gravitational field, 220
Gravitational forces, 34, 254
Gravitational lensing, 96
Gravitational red shift, 72
Gravity, 9, 34, 44, 50, 134, 199, 215, 252, 253, 264, 275
Greenberger, D. M., 117
Griffiths, Robert, 176, 179, 180, 183

Haldane, J. B., 48
Hameroff, Stuart, 283, 284
Hamilton-Jacobi equation of motion, 107
Hamilton's principle, 177
Hamlet, 285
Handwriting of God, 214
Hartle, James, 87, 88, 90, 154, 176, 180, 183, 282
Harvey, Paul, 239, 242
Hawaii, University of, 232
Hawking, Stephen, 154, 176, 215, 216
Heard, John, 23, 24
Heat death, 225
Hebrew, 241
Heidegger, Martin, 40
Heisenberg, Werner, 21, 53, 54, 55, 56, 58, 59, 60, 63, 64, 67, 68, 95, 103, 130, 192, 278
Helium, 232, 233
Herbert, Nick, 135, 283, 284, 286, 287
Hibbs, A. R., 178, 179
Hidden variables, 21, 39, 40, 54, 67, 83, 93, 102, 103, 104, 105, 106, 109, 110, 111, 112, 117, 119, 121, 122, 123, 124, 127, 131, 135, 148, 155, 159, 182, 193, 195, 196, 197, 199, 270, 271
Higgs boson, 240
High-energy physics, 110
High-temperature superconductors, 284
Hilbert, David, 276
Hindus, Hinduism, 242, 278
Hitler, Adolph, 241, 268
Holism, 39, 40, 41, 129, 130, 133, 134, 135, 171, 173, 200
Holistic medicine, 133
Holistic paradigm, 251, 261, 263, 264
Hologram, 128, 129
Holomovement, 128, 129, 130, 196
Homeostatic mechanism, 284
Homo sapiens, 240
Horne, Michael A., 117
Horseshoe map, 258
Howard, Don, 120
Hume, David, 8, 63
Hydrogen, 48, 143, 211, 231, 232, 233
Hydrogen atom, 48, 55, 220, 230, 245
Hydrogen binding energy, 244, 245, 246

Idealism, 166

Imaginary numbers, 216
Immortality, 11, 18, 36, 239, 240, 242
Implicate order, 83, 110, 123, 128, 129, 196
Impulse, 147, 148
Inclusive *or,* 281
Incompatible variables, 61
Indeterminacy, 103
Indeterminism, 57, 197, 198, 210, 286
Inflation, 221, 239
Inflationary universe, 222
Information, 15, 29, 37, 38, 39, 75, 88, 90, 93, 107, 108, 109, 128, 135, 137, 138, 139, 170, 173, 183, 188, 198, 200, 203, 224, 227, 228, 235, 259, 262, 285
Information theory, 251
Intelligence, 251, 262, 263, 272, 274, 275, 279, 280
Intelligent design, 213, 229, 230, 235
Interference, 68, 70, 71, 95, 103, 105, 108, 128, 156, 157, 158, 168, 169, 178, 179, 185, 186, 200, 202, 204, 205, 206, 210
Internal energy, 219, 220
Invariant observables, 171
Irrational numbers, 208, 209, 210
Islam, 241

Jahn, Robert, 29, 160
James, William, 62, 286
Jammer, Max, 59, 64, 67
Jarrett, Jon, 120, 121
Jastrow, Robert, 214
Jeans, Sir James, 278
Jefferys, William H., 16, 31
Jenson, Paul, 64
Jesus, 19, 36, 291
Jews, 241, 278
Jordan, Pascual, 54, 64
Josephson, Brian, 50, 277
Judeo-Christian, 213, 215, 241
Judgment Day, 36, 241

Kabbalah, 214
Kafatos, Menas, 40
Kali, 24
Kauffmann, Stuart, 259, 260, 286
Kennedy, John F., 144
Klein-Gordon equation, 156
Kochen, S., 111
Kochen-Specker theorem, 111
Koestler, Arthur, 28, 33
Kosko, Bart, 282
Krippner, Stanley, 250
Kuhn, Thomas, 194

Lagrangian, 177
Langton, Christopher, 260
Lanza, Robert, 23
Las Vegas, 166
Laudan, Larry, 192
Law of large numbers, 143
Lawrence Berkeley Laboratory, 63
Le Verrier, Joseph, 194
Lederman, Leon, 25
Leibnitz, Gottfried Wilhelm, 289
Lenin, 268
Light cone, 118, 131, 132, 133, 135, 146
Lilliquist, Michael, 16, 283
Limit cycle, 256, 260
Linde, André, 236
Linear theory, 159
Linearity, 160, 256
Lithium, 232, 233
Local variables, 39
Locality, 120
Locke, John, 8
Lodge, Sir Oliver, 35, 36, 37, 48
Logical positivism, 8
Logic Theorist, 262
Logistic map, 255, 256, 257, 258, 259
Lorentz, Hendrik, 44, 46, 67
Lorenz, Edward, 257
Los Alamos National Laboratory, 260

Mach, Ernst, 37, 63
Macintosh, 161
MacLaine, Shirley, 26, 291
Magnetic field, 20, 44, 51, 80, 105, 136, 139, 167, 168, 169, 170, 199
Magnetic flux, 168
Magnetic moment, 20, 79, 194, 195
Magnetic solenoid, 167
Magnetism, 34, 35, 44, 45, 50, 80, 134, 195, 253
Maharishi Mahesh Yogi, 25
Main sequence star, 231, 244, 245, 246
Malthusian exponential growth, 256
Malus, Etienne Louis, 139
Malus's law, 139, 140
Mandelbrot set, 251, 257
Mandelung, E., 106, 107
Many minds, 287
Many universes, 236
Many worlds, 172, 173, 175, 176, 180, 181, 182, 184, 193, 197, 198, 199, 235, 270
Matrix mechanics, 54, 55
Maxwell, James Clerk, 35, 44
Maxwell's equations, 37, 45, 46, 48, 49, 54, 145, 146, 150, 169, 289
McCullough, Daryl, 279
Me Decade, 290
Mermin, N. David, 136, 137, 161
Metabolism, 262
Metaphysics, 7, 8, 9, 11, 12, 22, 64, 83, 85, 104, 129, 175, 194, 199, 250, 255, 266
Michelson, Albert, 36, 46
Microtubules, 283, 284, 285
Microwave background, 187, 203, 217, 229, 239, 253, 265
Mind , 8, 10, 11, 12, 17, 18, 19, 20, 23, 24, 25, 27, 28, 29, 30, 31, 32, 34, 36, 37, 40, 49, 50, 58, 63, 64, 66, 72, 93, 96, 97, 103, 104, 110, 124, 125, 130, 135, 158, 160, 165, 166, 172, 175, 180, 184, 194, 199, 211, 213, 216, 240, 242, 251, 260, 262, 263, 266, 269, 270, 272, 273, 274, 276, 277, 278, 279, 280, 283, 286, 287, 288, 291
Mind-brain duality, 49
Mind of God, 11, 19, 215, 216
Mindwalk, 23
Minsky, Marvin, 250, 262
Mirror symmetry, 265
Momentum conservation, 84, 91, 92, 146, 148, 151
MonkeyGod, 236, 242, 245
Monroe, Marilyn, 268
Morley, Edward, 36, 46
Morphenes, 157
Moses, 241
Moslems, 278
Mystical experience, 27
Mysticism, 11, 12, 20, 23, 24, 25, 28, 83, 110, 124, 128, 130, 135, 173, 175, 193, 199, 213, 261, 274, 278, 289, 292
Mystics, 10, 96, 97, 129, 157, 165, 184, 261, 274, 276, 278, 284
Mythology, 20, 24, 111, 213, 215, 292

Nadeau, Robert, 40
Nand, 279
National Academy of Sciences, 29
National Research Council, 29
Native American religions, 242
Natural law, 153, 225, 226
Natural selection, 234, 237, 239, 259, 260, 263
Nature, 242
Negative energy, 56, 145, 146
Negative entropy, 26, 188, 224, 228
Negentropy, 224
Neo-Platonic, 85, 277
Nervous system, 261, 285
Network, 259, 260, 262, 263, 280, 282, 286
Neural junctions, 283

Neural network, 45, 262, 263, 281, 282, 283, 286, 288
Neurobiology, 279
Neurons, 253, 262, 263, 283, 285, 286
Neuroscience, 279, 287
Neutrino, 203, 233
Neutron, 230, 232, 233, 245
New Age, 18, 19, 24, 34, 40, 97, 110, 127, 144, 194, 214, 251, 264, 291, 292
New Age Christianity, 19, 291
Newall, Allen, 262
Newton, Patricia, 26
Newton, Sir Isaac, 8, 24, 26, 34, 50, 85, 134, 175, 264, 169, 289
Newtonian mechanics, 54, 58, 103, 109, 187, 283
Newtonian physics, 85, 134, 207, 218, 252, 264
Newtonian world machine, 54, 207
Newton's laws, 44, 253
Newton's second law, 175
Newton's theory of gravity, 44, 134, 252
Nonlinear quantum mechanics, 159, 160
Nonlinear system, 256, 258, 259, 264, 265, 272
Nonlinearity, 159, 160, 252, 256, 258, 259, 261
Nonlocality, 33, 38, 39, 40, 60, 80, 110, 121, 122, 123, 130, 134, 135, 140, 144, 145, 146, 148, 149, 152, 157, 160, 166, 169, 175, 182, 195, 196, 197, 199, 203, 226, 250, 252, 270, 271, 286, 289
Nuclear energy, 211, 215
Nuclear power, 20, 211
Nuclear radiation, 211
Nucleon, 245
Nucleon binding energy, 244, 245, 246
Nucleon radius, 244, 245, 246
Nucleosynthesis, 237
Numerology, 231

Objective reduction (OR), 274, 275
Objective reality, 10, 23, 50, 73, 74, 76, 166, 169, 171, 193, 199, 202, 278, 283
Occam's razor, 121, 199, 235, 236
Occult, 8, 251
Offer waves, 155, 156, 157
Omega Point, 241, 242
Omnès, Roland, 180
Ontological interpretation, 10
Open system, 265
Orchestrated objective reduction, 283
Oswald, Lee Harvey, 144
O'Connor, Cardinal, 215
O'Raifeartaigh, L., 169

Pais, Abraham, 73, 79
Paleontology, 214
Paradigm shift, 10, 251
Parallel computer networks, 259
Parallel processor, 282
Parallel universes, 144, 173, 175, 176, 181, 270, 271
Paranormal, 33, 97
Parapsychology, 28, 29, 33, 97
Park, James L., 86, 87, 88
Parsimony, 10, 223, 236
Path integral quantum mechanics, 178
Pauli, Wolfgang, 57, 59, 67, 68, 106, 110, 115, 197, 278
Pendulum, 159, 258
Penrose, Roger, 11, 154, 229, 234, 263, 273, 274, 275, 276, 277, 279, 280, 281, 283, 284, 285, 287
Penrose mysticism, 11, 274, 276
Perceptron, 262
Period doubling, 256, 257
Periodic Table, 55, 211
Perturbations, 254
Phase space, 234
Photodetector, 92, 95, 96, 118, 119

Photoelectric effect, 47
Photoelectricity, 252
Physical reality, 11, 37, 40, 74, 75, 77, 78, 79, 80, 84–85, 112, 116, 118, 122, 169, 195, 199
Physical Review, 33, 74, 83, 106, 160
Physics and Psychics, 28, 30, 274, 278
Physics of Immortality, 242
Pilot wave, 66, 105, 106, 107, 108
Placebo effect, 26, 133
Planck, Max, 37, 47, 67, 198, 199, 202, 278
Planck length, 209, 219, 220, 227, 228, 290
Planck mass, 246
Planck scale, 210, 219, 234
Planck's constant, 48, 53, 284
Planck time, 208, 209, 220, 221, 222, 223, 227, 228, 229, 234, 290
Plato, 7, 9, 20
Platonic realm, 275, 277, 290
Platonic reality, 279, 290
Platonic truth, 289
Platonic world, 7, 273
Podolsky, Boris, 37, 74, 79, 83, 84
Polar angle, 216
Polarimeter, 23, 118
Polarization, 22, 23, 88, 89, 90, 91, 92, 93, 117
Polarized positron emitter, 151
Polarizer, 89, 90, 91, 118, 119, 135, 204
Pontifical Academy, 214
Pope Pius XII, 97
Popper, Karl, 97
Population explosion, 255, 268
Positivism, 10, 60, 63, 180, 193, 195
Positrons, 56, 145, 146, 151, 152
Post-Everett quantum mechanics, 176, 177, 184, 185, 199, 201, 202
Pragmatism, 130, 193
Prana, 34
Precession of the perihelion of Mercury, 194
Presley, Elvis, 215, 268
Prigogine, Ilya, 264, 265
Primack, Joel, 214
Princeton Engineering Anomalies Research Center (PEAR), 29, 30, 31, 160
Principle of least action, 177, 178, 207
Probability amplitudes, 178, 179, 181
Probability current, 107
Probability density, 57, 100, 107
Probability postulate, 58, 140
Projection postulate, 57, 61
Propensity, 107
Protective measurement, 170
Protestantism, 278
Pseudoscience, 97, 150, 193, 242
Psi, 27, 28, 29, 31, 32, 33, 107
Psyche, 279
Psychic energy, 27, 34
Psychic force, 32, 34, 288–89
Psychic phenomena, 28, 31, 125, 289
Psychic plane, 157, 158
Psychokinesis, 125, 289
Psychological impact, 241
Psychons, 157, 158
Psychosomatic illness, 133
Pythagoras, 20

Quadratic map, 255
Quantization of angular momentum, 80
Quantum aether, 109, 145
Quantum animism, 286
Quantum coherence, 283
Quantum computation, 279
Quantum consciousness, 121, 184, 270, 286, 291, 292
Quantum cosmology, 154, 175, 197
Quantum field theory, 102, 146, 148, 159, 195, 203, 209
Quantum fluctuations, 29, 58, 145, 149, 226, 230, 236, 283, 285, 286, 288
Quantum gravity, 200, 275
Quantum healing, 12, 25

Quantum holism, 41, 129, 291
Quantum jump, 146, 148, 289
Quantum of action, 48, 53, 81
Quantum of angular momentum, 80
Quantum potential, 37, 39, 40, 106, 107, 108, 109, 124, 127, 129, 130, 145, 166, 199, 203
Quantum state, 87, 88, 94, 101, 265, 266, 275, 282
Quantum theory of radiation, 167
Quantum-to-classical transition, 177, 180, 265
Quark, 26, 240, 252, 254
Quasistable states, 259
Quintessence, 34, 37

R-process, 274
Radiation pressure, 221
Radioactive dating, 211
Rae, Allister, 175
Randi, James, 28
Random event generator (REG), 29, 30, 31, 160
Random number generator (RNG), 29, 33, 141, 160, 174
Rationality, 207, 235
Rational numbers, 208, 209, 210
Raub, L. David, 175
Realism, 193
Reality, 7, 9, 10, 11, 19, 20, 22, 23, 24, 25, 28, 33, 37, 38, 40, 48, 50, 59, 60, 63, 72, 73, 74, 75, 76, 77, 78, 79, 80, 85, 95, 97, 101, 102, 104, 110, 112, 116, 118, 122, 124, 128, 130, 134, 140, 165, 166, 169, 171, 175, 182, 189, 190, 193, 194, 195, 199, 200, 202, 210, 211, 214, 215, 217, 253, 269, 273, 277, 278, 279, 289, 290, 291
Real variables, 39
Redhead, Michael, 102, 112, 120
Reductionism, 24, 26, 200, 261
Reductionist materialism, 133
Reference frame, 58, 79, 132, 136, 142, 144, 146, 152, 196
Relative State Formulation of Quantum Mechanics, 172
Relativism, 193, 194, 195
Relativistic quantum theory, 145
Relativity, 8, 9, 10, 36, 39, 46, 47, 56, 58, 72, 102, 118, 131, 132, 141, 144, 145, 169, 177, 193, 194, 195, 196, 215, 216, 219, 220, 221, 222, 223, 226, 241, 253
Renormalization, 209
Rest energy, 46, 219, 220, 221
Rest mass, 46, 79, 166
Resurrection, 36, 239
Retarded waves, 155
Reviews of Modern Physics, 111
Rhine, Joseph Banks, 30
Robinson, Jackie, 268
Robots, 240, 241, 242
Roger's Version, 213
Rosen, Nathan, 37, 74, 79, 84
Rosenblatt, Frank, 262
Rosenfeld, Léon, 63, 65, 130
Rutherford, Ernest, 47, 48

Sagan, Carl, 216
Sarfatti, Jack, 27, 28, 135, 159
Schmidt, Helmut, 29, 30, 160
Schrödinger, Erwin, 21, 51, 52, 54, 55, 56, 67, 68, 74, 93, 94, 95, 278
Schrödinger's cat, 93, 94, 95, 157, 171, 173, 210
Schrödinger's equation, 55, 56, 57, 62, 106, 107, 108, 109, 123, 154, 156, 172, 208, 289
Schrödinger wave, 170
Science of wholeness, 252
Sciences of complexity, 11, 251, 252, 259, 263, 265, 266
Searle, J. R., 275
Second law of thermodynamics, 143, 206, 218, 223, 224, 225, 227, 229, 265

Self-esteem, 291
Self-organization, 18, 259, 286, 291
Semiclassical domain, 183
Semiconductor Quantum Interference Device (SQUID), 169, 200
Separability, 120, 121, 122, 123, 197
Sharpe, Kevin, 129
Shaw, Cliff, 262
Shimony, Abner, 117
Siegelmann, Hava T., 282
Significance level, 31
Simon, Herbert, 262
Simultaneity, 132
Simultaneous reality, 76, 85, 112
Singlet, 84, 91, 112, 113, 115, 118, 120, 136, 137, 139, 151, 152
Singularity, 219, 220
Smolin, Lee, 237, 239
Smoot, George, 215
Soft computing, 282
Solipsism, 77, 194
Solvay conference, 67, 68, 72, 105
Sommers, Paul, 154
Soul-body duality, 49
Spacelike, 132, 139, 145, 146, 148, 149, 150
Spacetime, 9, 63, 109, 128, 131, 132, 133, 145, 146, 148, 149, 150, 179, 199, 207, 210, 215, 216, 221, 222, 236, 253, 275, 289
Spacetime curvature, 221, 222
Specker, E. P., 111
Spectral lines, 211
Speed of light, 10, 35, 38, 39, 45, 46, 49, 56, 93, 95, 118, 130, 131, 133, 135, 141, 142, 175, 195, 208, 221, 271
Spherical coordinates, 216
Spin, 55, 56, 79, 80–83, 85, 86, 90, 92, 94, 102, 112, 113, 114, 115, 116, 117, 118, 129, 136, 137, 138, 139, 141, 151, 152, 161, 166, 169, 170, 173, 182, 183
Spin meter, 112, 113, 114, 115, 116, 136, 137, 138, 139, 140, 157, 161, 173, 175
Spinoza's God, 215
Spiritual experiences, 275
Spiritualism, 36
Squires, Euan, 287
Standard Model, 26, 102, 230
Standing waves, 51
Stanford Linear Accelerator, 111
Stapp, Henry P., 33, 63, 64, 97, 121, 130, 160, 283, 286, 287
State vector, 60, 61, 90, 94, 115, 169
Statistical interpretation, 60
Statistical mechanics, 87, 104, 105, 110, 123, 251
Statistical postulate, 56
Stellar evolution, 235
Stellar lifetimes, 231, 235, 238, 244, 245, 246
Stern, Otto, 68, 80
Stern-Gerlach apparatus, 81
Stern-Gerlach magnet, 81, 84
Stochastic hidden variables, 104
Straumann, N., 169
Strong interaction constant, 244
Strong locality, 120
Strong nuclear interaction, 236
Stückelberg, E. C. G., 145
Subluminal, 133, 141, 142, 210
Subnuclear forces, 102
Sub-quantum theory, 40, 77, 130, 195, 196
Sub-quantum variables, 103
Super computer, 209
Superconducting Quantum Interference Devices (SQUIDs), 200
Superconductor, 20, 284
Superluminal, 39, 100, 118, 121, 127, 132, 133, 134, 135, 139, 140, 142, 150, 196
Superluminal communication, 91, 139, 140, 157, 159, 160, 174

Superluminal connections, 39, 122, 133, 140, 195, 271
Superluminal motion, 39, 70, 140, 142, 144, 146, 149, 150–51, 171, 196, 197, 270, 271
Superluminal signals, 38, 39, 70, 120, 149, 176, 182, 195
Supernatural, 28, 214, 217, 221, 223, 225, 230, 235, 236, 239, 240, 242, 274, 288
Superposition, 51
Superstrings, 25
Swordy, Simon, 215
Synapse, 283, 284
Synaptic gap, 283, 285
Synaptic transmitter, 284
Systems theory, 24

Tachyon, 141
Taoism, 242
Tao of Physics, The, 23, 26, 128
Tardyons, 141
Teilhard de Chardin, Pierre, 241
Telekinesis, 97
Telepathy, 35
Thales, 51, 153
Theory of Everything (TOE), 234
Theory of the double solution, 105
Thermal motion, 284
Thermodynamics, 104, 105, 143, 206, 215, 218, 219, 222, 223, 224, 225, 229, 230, 258, 264, 265
Thought experiment, 68, 70, 72, 73, 118
Three-body problem, 255
Time asymmetry, 153, 154
Time irreversibility, 264
Timelike, 132
Time reflection symmetry, 154
Time's arrow, 141, 144, 188, 265
Time-symmetric quantum mechanics, 154, 159
Time symmetry, 152, 153, 155, 158, 204, 205, 206
Tipler, Frank, 233, 239, 240, 241, 242
Top quark, 240
Transactional interpretation, 155, 156, 157, 206, 271
Transcendental Meditation (TM), 25
Transcendent reality, 217
Transition to chaos, 257, 264
Transmissivity, 119
Transmitter molecules, 283, 284
Trickett, David, 129
Triplet, 152
True Basic, 161, 242
Turbulence, 258
Turing, Alan, 276, 277, 281
Turing limit, 282
Turing machine, 276, 281, 282

UFOs, 272
Ullman, Liv, 24
Ultraviolet, 47
Un-American Activities Committee, 83
Uncertainty principle, 53, 54, 56, 59, 60, 69, 70, 72, 76, 77, 101, 103, 145, 147, 148, 159, 178, 189, 209, 220, 284
Unification, 102, 231
Unified field theory, 101
Updike, John, 213

Vacuum aether, 145, 150
Vacuum fluctuation, 146, 147, 148, 149
Vaidman, Lev, 168
Vesicles, 282, 284, 285
Vigier, J. P., 129, 145, 146, 148, 149
Virgin Mary, 36
Virgo cluster, 240
Virus, 253
Von Neumann, John, 56, 57, 61, 106, 109, 110, 111

Wallace, Alfred Russel, 35
Waterston, Sam, 23
Wave function, 33, 37, 40, 55, 56, 57,

58, 61, 62, 63, 67, 68, 69, 70, 75, 76, 85, 86, 87, 88, 89, 91, 94, 96, 100, 101, 105, 106, 107, 108, 124, 130, 139, 155, 156, 159, 160, 165, 166, 167, 168, 169, 170, 171, 172, 173, 175, 178, 181, 186, 193, 196, 198, 200, 205, 208, 270, 287
Wave function collapse, 61, 62, 63, 68, 70, 94, 157, 160, 170, 171, 172, 174, 176, 197, 199, 270, 274, 275, 283, 287
Wave function of a photon, 167, 168
Wave function of the universe, 172, 173, 198, 199, 200
Wave function reduction, 61, 274
Wave mechanics, 54, 55
Wave packet, 51, 52, 53, 59, 60, 61, 62, 68, 105, 148, 160
Wave-particle duality, 38, 49, 50, 51, 57, 60
Wave-particle unity, 47, 50, 67
Wavicle, 49
Weak nuclear force, 51, 231
Weinberg, Steven, 159, 160, 192
Western science, 250, 251, 252, 269
Weyl, Hermann, 231
Wheeler, John Archibald, 95, 96, 97, 118, 145, 156, 157
Wholeness, 26, 33, 50, 60, 61, 128, 133, 250, 252, 263, 269, 286
Wigner, Eugene P., 96, 97
Wilber, Ken, 278
Wipf, A.,169
Wireless telegraphy, 35
Witt-Miller, Harriet, 250, 251, 252
Wolf, Fred Alan, 135, 157

X-rays, 211
xor (exclusive or), 281

You-niverse, 250
Young, Thomas, 70, 95

Zeilinger, Anton, 117
Zero point energy, 220, 221
Zigzagging, 145, 146, 149, 150
Zilsel, Edgar, 64
Zohar, Danah, 37, 97, 128, 290, 291
Zukav, Gary, 24, 25, 110, 128, 135, 290